RIEMANNIAN GEOMETRY
FIBER BUNDLES
KALUZA-KLEIN THEORIES
AND ALL THAT....

World Scientific Lecture Notes in Physics

ISSN: 1793-1436

---

*Published titles**

Vol. 67: Quantum Scaling in Many-Body Systems
M A Continentino

Vol. 69: Deparametrization and Path Integral Quantization of Cosmological Models
C Simeone

Vol. 70: Noise Sustained Patterns: Fluctuations and Nonlinearities
Markus Loecher

Vol. 71: The QCD Vacuum, Hadrons and Superdense Matter (2nd ed.)
Edward V Shuryak

Vol. 72: Massive Neutrinos in Physics and Astrophysics (3rd ed.)
R Mohapatra and P B Pal

Vol. 73: The Elementary Process of Bremsstrahlung
W Nakel and E Haug

Vol. 74: Lattice Gauge Theories: An Introduction (3rd ed.)
H J Rothe

Vol. 75: Field Theory: A Path Integral Approach (2nd ed.)
A Das

Vol. 76: Effective Field Approach to Phase Transitions and Some Applications to Ferroelectrics (2nd ed.)
J A Gonzalo

Vol. 77: Principles of Phase Structures in Particle Physics
H Meyer-Ortmanns and T Reisz

Vol. 78: Foundations of Quantum Chromodynamics: An Introduction to Perturbation Methods in Gauge Theories (3rd ed.)
T Muta

Vol. 79: Geometry and Phase Transitions in Colloids and Polymers
W Kung

Vol. 80: Introduction to Supersymmetry (2nd ed.)
H J W Müller-Kirsten and A Wiedemann

Vol. 81: Classical and Quantum Dynamics of Constrained Hamiltonian Systems
H J Rothe and K D Rothe

Vol. 82: Lattice Gauge Theories: An Introduction (4th ed.)
H J Rothe

Vol. 83: Field Theory: A Path Integral Approach (3rd ed.)
Ashok Das

Vol. 84: Foundations of Quantum Field Theory
Klaus Dieter Rothe

*For the complete list of published titles, please visit
http://www.worldscientific.com/series/wslnp

World Scientific Lecture Notes in Physics Vol. 16

RIEMANNIAN GEOMETRY FIBER BUNDLES KALUZA-KLEIN THEORIES AND ALL THAT....

ROBERT COQUEREAUX

Centre de Physique Theorique
Section II, CNRS-Luminy, Case 907
F-13288 Marseille
France

ARKADIUSZ JADCZYK

Centrum Fizyki Teoretycznej
Universytet Wroclawski
ul. Cybulskiego 36, 50-205 Wroclaw
Polska

World Scientific
Singapore • New Jersey • Hong Kong

Published by

World Scientific Publishing Co. Pte. Ltd.
5 Toh Tuck Link, Singapore 596224
USA office: 27 Warren Street, Suite 401-402, Hackensack, NJ 07601
UK office: 57 Shelton Street, Covent Garden, London WC2H 9HE

British Library Cataloguing-in-Publication Data
A catalogue record for this book is available from the British Library.

World Scientific Lecture Notes in Physics — Vol. 16
RIEMANNIAN GEOMETRY, FIBER BUNDLES, KALUZA-KLEIN THEORIES AND ALL THAT...

ISBN-13 978-9971-5-0426-7
ISBN-10 9971-5-0426-X
ISBN-13 978-9971-5-0427-4 (pbk)
ISBN-10 9971-5-0427-8 (pbk)

Pour Anne-Marie.

Dla Anny.

Acknowledgments

We are grateful to several institutes for gaving us the opportunity to meet and to work on this book, in particular the CERN Theory division, in Geneva, the Centre de Physique Théorique in Marseille, IHES in Bures sur Yvette, the University in Hamburg and the Centrum Fizyki Teoretycnej in Wroclaw.

We would also like to thank Professors R. Haag and D. Kastler for their interest and encouragement.

Most of the writing (and typing) of this book took place between 9 P.M. and 2 A.M. We are grateful to Anne-Marie, Valérie, Eric, Anna and Agneczka for putting up with it, and for their love.

CONTENTS

RIEMANNIAN GEOMETRY
FIBER BUNDLES
KALUZA-KLEIN THEORIES
AND ALL THAT....

I

GENERAL REMARKS AND PREREQUISITES

1-1 Introduction

Physical motivations

It is nowadays believed that it is possible and useful to describe physics in an "extended" space-time. Events that we see and that we measure are usually described by 3+1 numbers labelling the position and the time. Forces acting on objects and influencing their trajectories have also been described in the past in term of various tensor fields defined on a four dimensional manifold modelising the "space of events". It happens that, in many cases,the theory takes a simpler form if we assume that what we observe is just a shadow (projection) of something that takes place in space-time which has more than 4 dimensions. As an example, it has been recognised long ago that coupled gravitational and electromagnetic fields respectively described by a (4 dimensional) hyperbolic metric and a Maxwell field (a U(1) connection), could be also described by a U(1) invariant metric on a five-dimensional space. It is very possible (and it is the belief of the authors) that a correct formalization of the physics of "our universe" should involve an infinite dimensional manifold, and that for reasons which are still unknown, what we see classically looks four -dimensional. The fact that we do not see the extra dimensions (those of the so called "internal space") can be described, if not explained,by the fact that the metric of our multidimensional universe singles out some directions along which it is invariant or at least equivariant (in some sense).

Many papers have been published recently in the physical literature presenting many different constructions,sometimes under the same headings (Kaluza-Klein theories, Dimensional reduction, Symmetries of gravitational and gauge fields , etc.); very often the generality of the described situation was not studied. One of the aims of the present book is to present the geometrical and analytical aspects of "dimensional reduction" and to discuss with more generality several situations which have been considered in the past.

Content of the book

What the book is really about is Riemannian geometry of those spaces on which a group action is given (with a view on applications to physical theories -"unified theories"-). This study involves in particular the geometry of group manifolds, homogeneous spaces, principal bundles, non principal bundles (with group action),..., but also, in order to study the different kinds of "fields" defined on those spaces, it requires an appropriate generalization of (non abelian) harmonic analysis.

Each chapter of the book begins with a summary section which stresses the main ideas in plain terms ; the reader willing to make his knowledge more precise should then read the rest of the chapter where a more detailed discussion (using a more precise mathematical language) is given. The summary introductions do not usually require any knowledge of fiber bundles; however,we use freely the corresponding terminology and results in the core of each chapter. Indeed, although the summary section usually describes everything in a "local" way (e.g. using coordinates), we always want to render our considerations global.

The remaining sections of this first chapter recall some standard definitions of differential (Riemannian) geometry and has also the purpose of setting our conventions. Most of the results discussed here will be used freely in all chapters of the book; however, we should mention that the reader who wants to recast Riemannian geometry in the general framework of the theory of connections should jump directly to Ch.6, where these notions are developed from scratch.

Anybody willing to construct physical models generalizing the "old" Kaluza-Klein ideas should be first acquainted with some basic facts about the Riemannian structure(s) of Lie groups (Ch.2) and homogeneous spaces (Ch.3). The study of G-invariant metrics on groups and homogeneous spaces is also compulsory if one wants to analyse the situation when the space is a (generally only local) product of some manifold **M** times a group G or a homogeneous space G/H. G-invariant metrics on principal bundles and non-principal bundles carrying a G action are discussed respectively in Ch.4 and Ch.5. A general study of the Riemannian geometry of "matter fields", i.e., vector valued functions (or forms) defined on a manifold (in particular the covariant derivative acting on tensors, spinors, p-forms valued in some vector space...) is made in Ch.6 (this chapter could be read independently of the rest of the book). It is well known that, when a real (or complex) valued function is defined on a group or on a homogeneous space, it is possible to "expand" it (think of the usual spherical harmonics); however, when the underlying space is only a (local) product of some manifold **M** by G or G/H, the formalism has to be generalised and this is done in Ch.7. The particular case where such matter fields are usual tensors or spinors is studied in Ch.8 (G-spin-structures are naturally obtained there as a result of a process of "dimensional reduction"). The techniques described in particular in chapters 5, 7 and 8 provide us with a "general-purpose-tool" that we may use in several situations; as an example of such a use, we study in Ch.9 the dimensional reduction of Einstein-Yang-Mills systems i.e. analyse the geometry of a manifold on which both metric and connection are given, along with the action of a symmetry group. We will study this case by showing how it can be reduced to the situation studied in Ch.5. Finally, in Ch.10, we consider a more general situation where there is no global action of a finite dimensional Lie group G but where we can nevertheless define "interesting metrics" which are invariant under a "bundle of groups" (infinite dimensional groups of automorphisms of bundles are defined and studied in section 4.11).

Each main section of the book ends with a paragraph entitled "Pointers to the literature"; indeed, references are usually not given within each chapter but collected at the end.

New results

Before ending this introduction, we should maybe mention what is "original" in this book and what can be found elsewhere. It is clear that the whole discussion of homogeneous metrics on Lie groups and coset spaces can be found in a scattered way inside many papers of the mathematical literature. However the general discussion of metrics leading to "dimensionnal reduction", and in particular the general study of metrics on bundles with homogeneous fibers is probably new: although known "in principle", many explicit constructions and calculations carried out here do not seem to have been discussed elsewhere in the mathematical or in the physical literature (but by the authors themselves).

Also, let us mention a few other mathematical (or physical) constructions hardly to be found elsewhere: generalization of Frobenius and Peter-Weyl theorems (in Ch.7), intrinsic definition of the Lichnerowicz operator (in Ch.6), non-standard discussion of Einstein-Cartan theory with spinors (in Ch.6), link between G-spin structures and dimensional reduction (in Ch.8), generalization of the Wang theorem on G-invariant connections (in Ch.9), definition and study of "local" action of groups (bundle of groups, in Ch.4.11 and Ch.10).

How to read the book?

Method 1: from the beginning to the end.

Method 2: read only the summary sections.

The following diagram illustrates the interdependency of the chapters:

⟶ denotes a compulsory logical link

→ denotes an optional logical link

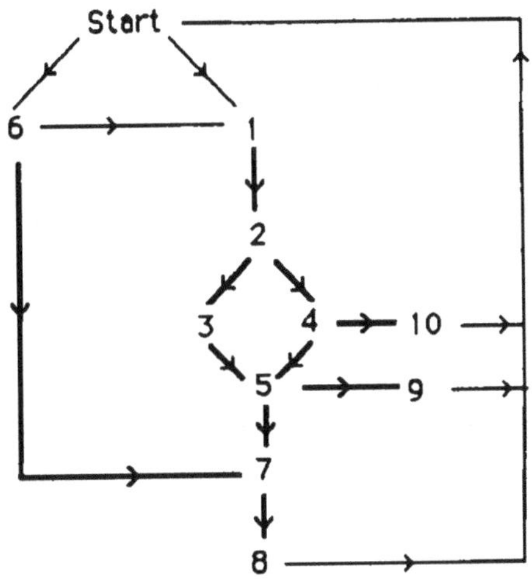

1.2 Differentiable manifolds

For this,we refer to the standard literature. Notice that a topological space is not necessarily a differentiable manifold (it has to be smooth !). Also, a given topological manifold may be endowed with none or several differentiable structures. For example, the number of inequivalent differentiable structures for spheres is 1 for S^p (p<6), 28 for S^7, 8 for S^9, 2 for S^{10}, 992 for S^{11}, while R^4, being truly exceptional, is believed to have even uncountably many different differentiable structures. The concept of differentiable structure should not be confused with that of metric structure (the later beeing defined after the former). Unless otherwise specified, by "manifold" we will always mean here "differentiable manifold with a given smooth structure". Many concrete manifolds we shall deal with will be homogeneous spaces; unless otherwise specified, by a homogeneous space we will always mean here homogeneous space with the smooth structure induced by the quotient space definition. In most examples of low dimensionality, the smooth structure is unique, anyway. Notice in particular that when we discuss "non standard" metrics on some spheres -as in sect.3.4-3.6-, we will assume that the sphere has a fixed differentiable structure.

Since there exists several conventions in the definitions of exterior deivative and exterior products, we give here those that we will follow.

f being a function on the manifold M, (a zero-form), we write its differential as $df = \partial_\mu f\, dx^\mu$ in the coordinate basis $\{dx^\mu\}$. Let now ω be a k-form, we write it as $\omega = 1/k!\ \omega_{i1 \ldots ik}\ dx^{i1} \wedge \ldots \wedge dx^{ik}$, and its differential $d\omega$ (a k+1-form) is given by

$$d\omega = 1/k!\ d\omega_{i1 \ldots ik}\ dx^{i1} \wedge \ldots \wedge dx^{ik}.$$

Observe that we know what $d\omega_{i1.....ik}$ is since $\omega_{i1.....ik}$ is a function. The exterior differential **d** is the unique operator such that, for all forms ω_1, ω_2,

$$d(\omega_1+\omega_2) = d\omega_1 + d\omega_2 ,$$

$$d(\omega_1\wedge\omega_2) = d\omega_1 \wedge \omega_2 + (-1)^k \omega_1 \wedge d\omega_2 \qquad (\omega_1 \text{ being a k-form})$$

$$d^2 = 0$$

Moreover, if $\xi_1, \xi_2,...,\xi_{k+1}$ are vector fields on **M**, we have

$$d\omega(\xi_1, \xi_2,...,\xi_{k+1}) = \Sigma_{i=1..k+1} (-1)^{i+1} \xi_i(\omega(\xi_1,...,\hat{\xi}_i,...\xi_{k+1})$$
$$+ \Sigma_{1\le i\le j\le n} (-1)^{i+j} \omega([\xi_i,\xi_j],\xi_1, ...,\hat{\xi}_i,...,\hat{\xi}_j,...,\xi_{k+1})$$

The group of permutations Σ_k acts as follows on k-upples of vectors

$$\sigma\in\Sigma_k : \quad \sigma(v_1,v_2,...,v_k)=(v_{\sigma(1)},v_{\sigma(2)},...,v_{\sigma(k)})$$

Let T be a covariant tensor of rank k; then we get a fully antisymmetrised tensor Alt T via the equation

$$\text{Alt } T = 1/k! \ \Sigma_{\sigma\in\Sigma k} \ \epsilon_{\sigma}. T.\sigma$$

This equation also defines the operator **Alt**. Notice that thanks to the presence of 1/k!, we have $\text{Alt}((\text{Alt } \omega\otimes\varphi)\otimes\beta) = \text{Alt}(\omega\otimes\varphi\otimes\beta)$

$$= \text{Alt}(\omega\otimes\text{Alt}(\varphi\otimes\beta)),$$

ω,φ and β being covariant tensors.

Let us now take ω a k-form (completely antisymmetric tensor of rank k) and α a l-form; then we define their exterior product as follows

$$\omega\wedge\alpha = (k+l)!/k!\,l! \quad \text{Alt}(\omega\otimes\alpha)$$

In particular, if ω and α are 1-forms, we get $\omega\wedge\alpha = \omega\otimes\alpha-\alpha\otimes\omega$. Notice that if ω is a k-form, then $\text{Alt}(\omega)=\omega$ and that $\wedge$ has the following properties (call Ω^k the space of k-forms)

i) $\wedge$ is distributive over + from the left and from the right

ii) $a(\omega\wedge\alpha) = a\omega\wedge\alpha = \omega\wedge a\alpha$ with $a\in\mathbb{R}$.

iii) $\omega\wedge\alpha = (-1)^{kl} \alpha\wedge\omega$ where $\omega \in \Omega^k$, $\alpha \in \Omega^l$, (in particular, if ω is odd $\omega\wedge\omega = 0$).

If φ denotes a smooth map from a manifold **M** to a manifold **N**, we will write φ_* (sometimes $d\varphi$) for the tangent map (locally it can be written as the Jacobian matrix of partial derivatives) and φ^* for the cotangent map. Notice that vectors can be pushed forward in the direction of the map but forms are pulled back (if v^μ are the components of a vector of **M**, then $\partial_\mu\varphi^i.\,v^\mu$ are those of its push-forward in **N**, whereas if α_i are the components of a 1-form in **N**, then $\partial_\mu\varphi^i.\,\alpha_i$ are those of its pull back in **M**). Finally, if ω and τ are forms in N, we have $\varphi^*(\omega\wedge\tau)=\varphi^*(\omega)\wedge\varphi^*(\tau)$.

1.3 Riemannian manifolds

We now give some notations and state without much discussion some well known formulae of Riemannian geometry. This serves mainly the purpose of setting our conventions (for a more precise definition of the concepts involved, the reader may consult chapt. 6).

1.3.1 Metrics, connections and curvatures

Metrics

The metric $g(y)$ at the point y describes a scalar product $g(\ ,\ )$ in the tangent space at y and can be represented as a matrix $(g_{\mu\nu}(y))$ or as

$$g(y)=g_{\mu\nu}(y)\ \ dy^\mu \otimes dy^\nu$$

in a coordinate basis $\{dy^\mu\}$. It can also be represented as a matrix g_{ij} or as

$$g(y)=g_{ij}(y)\ \ \omega^i(y) \otimes \omega^j(y)$$

in a moving frame of forms $\{\omega^i(y)\}$.

Calling $(g^{\mu\nu}(y))$ the inverse of the matrix $(g_{\mu\nu}(y))$, we may consider the following object which defines a scalar product in the cotangent space at y

$$g^{-1}(y)=g^{\mu\nu}(y)\ \ \partial/\partial y^{\mu} \otimes \partial/\partial y^{\nu}$$

in a coordinate basis $\{\partial/\partial y^{\mu}\}$. By a slight abuse of the word "inverse", we will call it "the inverse metric" $g^{-1}(y)$ at the point y. It can also be represented as as matrix $(g^{ij}(y))$ or as

$$g^{-1}(y)=g^{ij}(y)\ \ e_i(y) \otimes e_j(y)$$

in a moving frame of vectors $\{e_i(y)\}$.

Connections

For an arbitrary linear connection , we write the covariant derivative of e_j in the direction e_i as

$$\nabla_i e_j = \Gamma_i{}^k{}_j\, e_k$$

We will always assume that the linear connection is compatible with the metric and define the torsion as

$$T_i{}^k{}_j = \Gamma_i{}^k{}_j - \Gamma_j{}^k{}_i - f_i{}^k{}_j$$

The structure functions of the moving frame[1] being

$$[e_i, e_j] = f_i{}^k{}_j\ \ e_k$$

[1] When the manifold is a Lie group G and when the moving frame (e_i) is made out of left-invariant vector fields (hence a basis in the Lie algebra of G), we will write $[e_i,e_j] = C_{ij}{}^k\ \ e_k$; therefore $C_{ij}{}^k = f_i{}^k{}_j$. The structure functions are actually structure constants in this case. It is however better to use the notation $f_i{}^k{}_j = (f_i)^k{}_j$ in order to refer to matrix elements of the adjoint representation of Lie(G): e_i " (ad e_i) with (ad e_i)(e_j) := $[e_i,e_j]$.

$T_i{}^k{}_j$ and $f_i{}^k{}_j$ are obviously antisymmetric on indices i and j. We also define

$$T_{ijk} = g_{jl}\, T_i{}^l{}_k \qquad \text{and}$$
$$f_{ijk} = g_{jl}\, f_i{}^l{}_k$$

The Christoffel symbols Γ_{ijk} are then defined by

$$\Gamma_{ijk} = g_{jl}\, \Gamma_i{}^l{}_k$$

They can be computed from the following formula

$$\begin{aligned} 2\,\Gamma_{ikj} = \; & (\partial_i g_{jk} + \partial_j g_{ik} - \partial_k g_{ij}) \\ & + (T_{ikj} - T_{jik} + T_{kji}) \\ & + (f_{ikj} - f_{jik} + f_{kji}), \end{aligned}$$

where

$$\partial_i g_{jk} := e_i(g_{jk}) = e_i{}^\mu \partial_\mu g_{jk}\,.$$

In an orthonormal moving frame made of vectors fields such that g_{ij} is constant,the term in the first parenthesis vanishes, and if the linear connection is <u>the</u> Levi-Civita connection, the term in the second parenthesis vanishes as well since this connection is torsionless. From this time on we shall always deal with the Levi-Civita connection.

Curvatures

The Riemann curvature tensor is given by

$$R_{ij}{}^k{}_l = \partial_i\, \Gamma_j{}^k{}_l - \partial_j\, \Gamma_i{}^k{}_l + (\,\Gamma_i{}^k{}_m \Gamma_j{}^m{}_k - \Gamma_j{}^k{}_m\, \Gamma_i{}^m{}_l) - \Gamma_m{}^k{}_l\, f_i{}^m{}_j$$

The Ricci curvature tensor is defined by

$$R_{ij} = R_{ki}{}^k{}_j$$

The scalar curvature is

$R=R^i{}_i$

Lie derivatives

If $X=X^i e_i$ is a vector field, then the Lie derivative of a tensor, say $t_{ij}{}^k$ is given by

$$L_X t_{ij}{}^k = X(t_{ij}{}^k) - \partial_l X^k \, t_{ij}{}^l + \partial_i X^l \, t_{lj}{}^k + \partial_j X^l \, t_{il}{}^k ,$$

where $X(t_{ij}{}^k)=X^l \partial_l t_{ij}{}^k$, and ∂_i are derivatives along e_i. In particular for the metric tensor

$$L_X(g_{ij}) = X^r \partial_r g_{ij} + \partial_i X^l \, g_{lj} + \partial_j X^l \, g_{il}$$

In an orthonormal frame the term $X^r \partial_r g_{ij}$ vanishes. If $L_X(g_{ij})=0$ then X (if non-zero) is called a Killing vector for g. The set of all Killing vector fields for a given metric is a (finite dimensional) Lie algebra. Killing vectors generate one-parameter local groups of isometries. A generic metric has no isometries and no Killing vectors at all.

Riemannian spaces with transitive group of isometries

Homogeneous spaces which we will study will be equipped with metrics having transitive local groups of isometries. On such a space one can have a local basis of Killing vectors. If e_i is such a basis then, in this basis, the curvature tensor is given by

$$4\, R_{ijlk} = -\{(g([[e_i,e_j],e_k],e_l) - g([[e_i,e_j],e_l],e_k)) -$$

$$(g([[e_k,e_l],e_i],e_j) - g([[e_k,e_l],e_j],e_i))\}$$

$$+(f_i{}^m{}_k \, f_{jml} + 2\, f_i{}^m{}_j \, f_{kml} - f_j{}^m{}_k \, f_{iml}) \quad +$$

$$(S_{ik}{}^m S_{jlm} - S_{jk}{}^m S_{ilm}),$$

where

$$S_{ijm} = -(f_{mij}+f_{mji}).$$

These formulae will be useful for computing curvatures of invariant metrics on groups and homogeneous spaces.

1.3.2 Particular spaces

Spaces of constant curvature :

These are the spaces of constant sectional curvature. Given any two non-parallel vectors **v**,**w** at y one defines sectional curvature $K(y;\mathbf{v},\mathbf{w})$ at y with respect to **v**,**w** by

$$K(y;\mathbf{v},\mathbf{w}) = R_{ijlk}v^i w^j v^k w^l/((g_{ik}g_{jl}-g_{il}g_{jk})v^i w^j v^k w^l).$$

One then proves that if, for each y, $K(y;\mathbf{v},\mathbf{w})$ does not depend on the choice of **v** and **w**, then it is also constant with respect to y. This is equivalent to

$$R_{ijlk}=c(g_{ik}g_{jl}-g_{il}g_{jk}),$$

where c is a constant (In this case $R_{ij} = c(d-1)g_{ij}$ and $R = cd(d-1)$, d being the dimension of the space).

Einstein spaces :

In the mathematical literature they are also known as spaces of constant Ricci curvature, the Ricci curvature $K(y;\mathbf{v})$ being defined as

$\Sigma_m K(y;v,e_m)$, where the sum is taken over some orthonormal basis. The condition that $K(y;v)$ is constant is equivalent to

$$R_{ij}=k\, g_{ij}$$

k is called the Einstein constant.It is often convenient to introduce the Einstein tensor

$$E_{ij}=R_{ij} - g_{ij}\, R/2$$

Notice that

$$R_{ij}= E_{ij} - Tr(E_{ij}).\, g_{ij} \,/\, (n-2).$$

Therefore Einstein tensor and Ricci tensor determine each other.If (M,g) is an Einstein space with Einstein constant k, then

$$E_{ij}+\Lambda\, g_{ij} = 0$$

where Λ, called cosmological constant in physics is

$$-\Lambda = k - R/2$$

R being the scalar curvature.

Spaces of constant scalar curvature :

As it is obvious from their name,these are the spaces for which the scalar curvature R is constant over the manifold. Observe that constant curvature implies constant Ricci curvature, which in turn implies constant scalar curvature. We have therefore the following sequence of conditions (from the more to the less stringent)

$R_{ijkl} = c\,(g_{ik}g_{jl}-g_{jk}g_{il})$,

$R_{ij} = k\, g_{ij}$,

R = constant.

It is clear from the above that Einstein spaces are "between" spaces of constant curvature and spaces of constant scalar curvature.This observation explains partly why spaces of constant curvature are completely classified,why Einstein spaces are partly classified and why spaces of constant scalar curvature are not classified at all.

The reader has probably in mind several examples of homogeneous spaces endowed with homogeneous metrics (for example, any Lie group with a left invariant metric): these are examples of spaces with constant scalar curvature.

Einstein metrics as saddle points .

We will need at times the following result: Consider the space of Riemannian metrics on a compact manifold **S** with fixed volume form *dv*. Then the critical points of the functionnal

$$A[g] = \int R[g]\, dv$$

are precisely the Einstein metrics on S. In physical terms this means that we allow only for fluctuations of the conformal structure and not of the length scale. If we vary the same functional without keeping the volume fixed, we get the action principle of general relativity: the saddle points are the metrics satisfying the (in our case vacuum) Einstein equations $E_{ij}=0$ or, equivalently, $R_{ij}=0$; the space is Ricci flat.

II

RIEMANNIAN GEOMETRY OF LIE GROUPS

2.0 Summary

Modulo discrete factors, any compact Lie group can be written as a product of simple Lie groups and U(1) factors. By using left or right multiplication, any element **g** of a group G defines a left and a right action; these two actions may of course be identical if **g** belongs to the center of G. The Jacobians (matrix of partial derivatives) of these two actions at the identity are linear transformations mapping the tangent space at the identity of G (called the Lie algebra of G and denoted by **Lie(G)**) onto the tangent space at the point **g** of G. Choosing for example the right action and its associated linear map, we can associate to each element T_i of **Lie(G)** a vector $\mathbf{e}_i^R(\mathbf{g})$ at the point **g**; for fixed T_i these vectors constitute a vector field: $\mathbf{g} \to \mathbf{e}_i^R(\mathbf{g})$ called "right fundamental vector field". This vector field is also called "left invariant vector field". Of course there are corresponding statements if we permute the words "left" and "right".

When no metric is specified, one should think of a Lie group G as a smooth manifold (locally like $\mathbb{R}^n$) with well defined topology (holes, handles), and where multiplication of points (in a given order) is a smooth operation. However, its metric is not yet given: this is another extra structure (which, intuitively specifies the "shape") . There exist many metrics on a Lie group; some are maximally symmetric, most have no symmetries at all and others are in between: the isometry group of a metric on a simple compact G is always of the type H×K ,where H and K are subgroups of G acting respectively from the left and from the right (they may be equal to the identity or to G itself).

At one end we have therefore the metrics which are invariant under both left and right actions and for this reason are called bi-invariant metrics. Their isometry group, in the case of a semi-simple Lie group, is G×G (when G is not semi-simple, the U(1) factors should be counted only once; it may also happen in this case that the isometry group is an extension of G -think of the translation group). The Killing fields generating the above isometries are precisely the left and right fundamental vector fields. When G is simple, any such metric is proportional to the usual Killing metric; when G is semi-simple, one can scale the simple components of G with different scaling factors.

Then comes the family of those metrics which are invariant under the right (or the left) action and are called G-invariant metrics.Their isometry group is at least G and these isometries are generated by the corresponding fundamental fields. A G-invariant metric is fully characterized by the values it takes at the identity of the group; for this reason we could say that there is a "dimensional reduction above a point".

It is well known that, by choosing a subgroup **K** of G, we can write a coset decomposition of G i.e., we write G as the union of a family of (left-translated) subgroups **K** parametrized by G/**K**. Starting then from a bi-invariant metric on G, we can "squash" it by introducing a scaling parameter in the direction of **K** and get in such a way a one parameter family of G×**K**-invariant metrics.

If we add a left invariant metric (**h**) and a right invariant metric (**k**), we get a metric (**h** + **k**) which has usually no invariances at all, however it is "dimensionally reducible" in the sense that its value is known everywhere provided we know the scalar products ($\mathbf{h}_{ij}$) and ($\mathbf{k}_{ij}$) at the origin of the group.

The curvature tensors and curvature scalar for metrics which are at least G-invariant will be given in sect. 2.5 .

For a simple G the Killing metric is always an Einstein metric (a "standard" one), but an arbitrary G-invariant metric needs not be such; however, given a subgroup **K** of G, one can often find a value of the squashing parameter for which the corresponding metric on G is a non standard Einstein metric with isometry group G×**K**.

It is useful to know how to compute volumes of Lie groups. In physical applications, in particular, one has to integrate over the "internal space", which, often, is a group. To each metric on G we can associate a corresponding measure whose integral is the corresponding volume. The value of the volume is easy to get in the case of a G-invariant metric where it can be expressed in term of some standard volume. Classical Lie groups are transitive on spheres and spheres have a standard volume as subset of R^n; by computing first the lower dimensional cases and by using spheres as an intermediate step one can indeed define standard volumes of simply connected classical Lie groups. Such a volume is associated to a certain standard bi-invariant metric which is proportional to the Killing metric. The method however fails for exceptional Lie groups since they are not transitive on spheres in general. The standard volumes of non simply connected groups is easily gotten from the standard volumes of their covering group by dividing by the order of the covering.

The next section discusses the above notions in the particular case of the group SU(2); a more detailed analysis of the general case then follows.

We recall that, unless stated otherwise, we shall always assume that our groups are compact Lie groups. However, with appropriate care, our formulae are valid also in a non compact case.

2.1 The case of SU(2)

SU(2) is topologically a three-sphere S^3 (proof: parametrize the point p of SU(2) by 2×2 matrices with complex coefficients α, β

$$p = \begin{pmatrix} \alpha , & \beta \\ -\beta^*, & \alpha^* \end{pmatrix}$$

then, we obtain

$$1 \;=\; p^* p \;=\; \det(p).\, I$$
$$\det(p) = Re^2(\alpha) + Im^2(\alpha) + Re^2(\beta) + Im^2(\beta) = 1$$

i.e., the equation of a 3-sphere).

It is well known that SU(2) is the 2-fold covering group of SO(3); the Lie algebra of SU(2) is generated by X_1, X_2, X_3 with the commutation properties

$$[X_1, X_2] = -X_3 \quad \text{,etc.,by cyclic permutation.}$$

One take for example $X_i = i/2\ \sigma_i$, σ_i being the Pauli matrices.

We can parametrize SU(2) by Euler angles ψ, θ, φ, writing

$$p = R_3(\psi)\, R_1(\theta)\, R_3(\varphi) \quad ,0 < \psi \leq 4\pi,\ 0 < \theta \leq \pi,\ 0 < \varphi \leq 2\pi,$$

where $R_i(x)$ is a rotation of angle x around the axis i. We consider in SU(2) the curves obtained by <u>right</u> translation

$$D_i(t) = p\, R_i(t) \quad ,\ i \in \{1,2,3\}.$$

We call then $e_i^R(p)$ the fundamental vector fields generating these right translations (they are also called left invariant); explicitly

$$e_i^R(p) = d/dt\ (p.\exp(tX_i))_{t=0}$$

$$\begin{pmatrix} e_1^R \\ e_2^R \\ e_3^R \end{pmatrix} = N \begin{pmatrix} \partial/\partial\theta \\ \partial/\partial\psi \\ \partial/\partial\varphi \end{pmatrix} \quad , \quad N = \begin{pmatrix} \cos\varphi & \sin\varphi/\sin\theta & -\sin\varphi\, \mathrm{ctg}\theta \\ \sin\varphi & -\cos\varphi/\sin\theta & \cos\varphi\, \mathrm{ctg}\theta \\ 0 & 0 & 1 \end{pmatrix}$$

The commutation relations read

$$(2.1) \qquad [e_1^R, e_2^R] = -e_3^R \quad , \text{etc.}$$

The dual basis $\{\omega^{iR}\}$ satisfies

$$\langle \omega^{iR}, e_j^R \rangle = \delta^i_j \quad , \text{and}$$

$$[\omega^{1R},\ \omega^{2R},\ \omega^{3R}] = [\, d\theta,\ d\psi,\ d\varphi\,]\, N^{T-1}$$

We also could consider in SU(2) curves obtained by left translations as well and their corresponding fundamental vector fields e_i^L and dual basis ω^{iL}. We will not do it explicitly: they can be gotten from the previous calculations by replacing everywhere φ by ψ. One can then check that

(2.1')
$$[e_1^L, e_2^L] = +e_3^L \qquad \text{, etc.}$$

and that

$$[e_i^R, e_j^L] = 0 \ .$$

A bi-invariant metric on SU(2) can be written

(2.2)
$$\begin{aligned} g &= \rho^2 \ (\omega_1^R \otimes \omega_1^R + \omega_2^R \otimes \omega_2^R + \omega_3^R \otimes \omega_3^R) \\ &= \rho^2 \ (\omega_1^L \otimes \omega_1^L + \omega_2^L \otimes \omega_2^L + \omega_3^L \otimes \omega_3^L) \\ &= \rho^2 \ (d\theta \otimes d\theta + d\varphi \otimes d\varphi + d\psi \otimes d\psi + \cos\theta \ (d\psi \otimes d\varphi + d\varphi \otimes d\psi)) \end{aligned}$$

The inverse metric can be written

$$\begin{aligned} g^{-1} &= \rho^{(-2)} (e_1^R \otimes e_1^R + e_2^R \otimes e_2^R + e_3^R \otimes e_3^R) \\ &= \rho^{(-2)} \ (e_1^L \otimes e_1^L + e_2^L \otimes e_2^L + e_3^L \otimes e_3^L) \\ &= \rho^{(-2)} (\partial/\partial\theta \otimes \partial/\partial\theta + \partial/\partial\varphi \otimes \partial/\partial\varphi + \partial/\partial\psi \otimes \partial/\partial\psi + \\ &\qquad 1/\sin\theta^2 \ (\partial/\partial\psi + \cos\theta \ \partial/\partial\varphi)^2) \end{aligned}$$

Left and right actions of SU(2) on the vector fields e_i^R and e_i^L just rotate them, therefore it is obvious that such a metric is indeed G×G invariant. This can also be seen at the infinitesimal level by noticing that the Lie derivative of g with respect to both left and right fundamental fields vanishes (it is technically simpler to see it on the inverse metric[1] since taking into account that the Lie derivative

---

[1]Remember that, by "inverse metric", we mean the bilinear form on the cotangent space whose matrix representative is $(g^{ij}) = (g_{ij})^{-1}$.

satisfies the usual Leibnitz rule with respect to tensor products, and, for a vector field v, we get $L_u v = [u,v]$)

$$L_{e_i^R}(g^{-1}) = L_{e_i^L}(g^{-1}) = 0$$

These fields are therefore the Killing fields of the metric (they generate isometries).

Notice that the components of the Killing metric in the frame e_i^R are

$$\begin{aligned} g_{ij} &= - C_{ik}{}^l\, C_{jl}{}^k \\ &= 2\,\delta_{ij} \end{aligned}$$

The Killing metric is therefore a member of the previous family, for $\rho^2 = 2$.

Left invariant metrics are of the form

$$g^{-1} = g^{ij}\, e_i^R \otimes e_j^R$$

where (g^{ij}) is a symmetric matrix of constant coefficients (the only dependence on the point p of the group comes in the coordinates θ, φ, ψ hidden in the vector fields e_i^R). From their very form, it is clear that such metrics are indeed G-invariant. At the infinitesimal level, we can remark that the vector fields e_i^L are Killing vector fields for this metric since left and right invariant vector fields commute.

Right invariant metrics are of the form

$$g^{-1} = g^{ij}\, e_i^L \otimes e_j^L$$

where (g^{ij}) is a symmetric matrix of constant coefficients. We have the same properties as above by interchanging the words "right" with "left" and e_i^L with e_i^R.

In both cases, it is worth noticing that we know the metric everywhere as soon as we know the scalar product (g_{ij}) defined in the Lie algebra. This would not be the case for an arbitary metric which can be written

$$g^{-1}=g^{ij}(p)\, e_i^L \otimes e_j^L$$

The matrix (g_{ij}) is here a function of $p \in G$, i.e. a periodic function of the Euler angles ψ,θ,φ.

One can always diagonalise a symmetric matrix by an orthogonal transformation. Therefore, given two scalar products, it is always possible to find a basis which is simultaneously orthonormal for the first scalar product and orthogonal for the second one; thus in particular, if k is a bi-invariant metric on G (for example the Killing metric), and if g is an arbitrary left-invariant metric, it is always possible to choose a basis E_i^R of left invariant fields for which

$$k^{-1} = E_1^R \otimes E_1^R + E_2^R \otimes E_2^R + E_3^R \otimes E_3^R$$

and

$$g^{-1}=\lambda_1^{(-2)} E_1^R \otimes E_1^R + \lambda_2^{(-2)} E_2^R \otimes E_2^R + \lambda_3^{(-2)} E_3^R \otimes E_3^R$$

Now, for SU(2) the automorphism group of Lie(SU(2)) coincides with the orthogonal group in 3 dimensions (the Lie algebras of SU(2) and SO(3) are isomorphic). Therefore, in this particular case, E_i^R will have the same structure constants as e_i^R. We can therefore even call $E_i^R = e_i^R$.

Using these properties, it is easy to exhibit several examples of metrics on SU(2) with isometry groups subgroups of SU(2)×SU(2).

The following are metrics for which the isometry group is SU(2)×U(1)

$$g^{-1}= e_1^R \otimes e_1^R + e_2^R \otimes e_2^R + \lambda_3^{(-2)}\, e_3^R \otimes e_3^R$$

Indeed, it is clear that e_i^L and e_3^R are Killing fields for the metric.

The following is a metric for which the isometry group is $U(1)\times U(1)$

$$g^{-1} = e_1^R \otimes e_1^R + e_2^R \otimes e_2^R + e_3^L \otimes e_3^L$$

Here, we see that e_3^R and e_3^L are Killing fields for this metric.

The reader should notice that it is usually cumbersome and unnecessary to express explicitly the left and right fundamental fields in term of local coordinates on the group; here, for example, we introduced Euler angles only for pedagogical reasons. To convince the reader, let us consider the following left invariant metric

$$g = \lambda_1^2 \ \omega^{1R} \otimes \omega^{1R} + \lambda_2^2 \ \omega^{2R} \otimes \omega^{2R} + \lambda_3^2 \ \omega^{3R} \otimes \omega^{3R}$$

and let us express it in the coordinate basis associated to Euler angles; we get

$$g = [\lambda_1^2 \cos^2\varphi + \lambda_2^2 \sin^2\varphi] \, d\theta \otimes d\theta + [[\lambda_1^2 \sin^2\varphi + \lambda_2^2 \cos^2\varphi] \sin^2\theta + \lambda_3^2 \cos^2\theta] \, d\psi \otimes d\psi + \lambda_3^2 \ d\varphi \otimes d\varphi + [\sin\varphi \cos\varphi \sin\theta \ (\lambda_1^2 - \lambda_2^2)] (d\theta \otimes d\psi - d\psi \otimes d\theta) + \lambda_2^2 \cos\theta \, (d\psi \otimes d\varphi + d\varphi \otimes d\psi)$$

The G-invariance property is now far from being obvious!

As we saw above, a left invariant metric can be easily expressed in terms of left invariant vector fields but, at any point p of G, right invariant vector fields also constitute a basis; one could of course express a left invariant metric in term of right invariant fields, but the matrix (g_{ij}) would then depend upon the point p.

The following metric

$$g = h_{ij} \ \omega^{iL} \otimes \omega^{jL} + k_{ij} \ \omega^{iR} \otimes \omega^{jR}$$

where $\mathbf{h}_{ij}$ and $\mathbf{k}_{ij}$ are arbitrary (positive definite) symmetric matrices, could be written only in terms of the ω^{iL}. This metric has no particular invariances in general; however, it is not arbirary since it is well defined as soon as it is known at the origin of the group. These kind of metrics do not seem to have a name in the mathematical literature; however, they play an important role in Kaluza-Klein theories. We will call them "dimensionally reducible metrics" (of course, G-invariant metrics are a particular important subcase of this family).

In the case of G-invariant metrics, the Riemann, Ricci and scalar curvatures can be easily computed by using the general results recalled in sect. 1.3.1. For a metric

$$4\,g = \lambda_1^2\ \omega^{1R} \otimes \omega^{1R} + \lambda_2^2\ \omega^{2R} \otimes \omega^{2R} + \lambda_3^2\ \omega^{3R} \otimes \omega^{3R}$$

we find

$$R^1{}_1 = [-2\sigma_1{}^2 + 4\ \sigma_2 + 4(\lambda_1{}^2)^2 + 4\lambda_2{}^2\lambda_3{}^2] \,/\, \sigma_3$$
$$R^2{}_2 = [-2\sigma_1{}^2 + 4\ \sigma_2 + 4(\lambda_2{}^2)^2 + 4\lambda_1{}^2\lambda_3{}^2] \,/\, \sigma_3$$
$$R^3{}_3 = [-2\sigma_1{}^2 + 4\ \sigma_2 + 4(\lambda_3{}^2)^2 + 4\lambda_1{}^2\lambda_2{}^2] \,/\, \sigma_3$$

and

$$R \ = R^1{}_1 + R^2{}_2 + R^3{}_3 = 2\,[\,4\,\sigma_2 - \sigma_1{}^2\,]/\sigma_3$$

where

$$\sigma_1 = \lambda_1{}^2 + \lambda_2{}^2 + \lambda_3{}^2$$
$$\sigma_2 = \lambda_1{}^2\,\lambda_2{}^2 + \lambda_2{}^2\,\lambda_3{}^2 + \lambda_3{}^2\,\lambda_1{}^2$$
$$\sigma_3 = \lambda_1{}^2\,\lambda_2{}^2\,\lambda_3{}^2$$

These metrics describe anisotropic 3-spheres.
Notice that when $\lambda_1=\lambda_2=1$, $\lambda_3=\lambda$, the metric is $SU(2)\times U(1)$ invariant and that the scalar curvature is

$$R = 2(4-\lambda^2)\,,$$

quantity which may well be negative although the space is compact.

All bi-invariant metrics on SU(2) -they are all proportional to the Killing metric- are Einstein metrics. It can be shown that they are no other Einstein metrics on SU(2). [92].

The volume form associated to the above metric is

$$dv=(1/2^3).\ \lambda_1.\lambda_2.\lambda_3\ .\omega^{1R}\wedge\omega^{2R}\wedge\omega^{3R}$$
$$=(1/8).\lambda_1.\lambda_2.\lambda_3.\ \sin\theta\ d\theta\ d\psi\ d\varphi$$

The corresponding volume is

$$V=1/8.\ \lambda_1\lambda_2\lambda_3 \int_0^{4\pi} d\psi \int_0^{2\pi} d\varphi \int_0^{\pi} \sin\theta\ d\theta$$
$$=2\ \pi^2\ .\lambda_1.\lambda_2.\lambda_3$$

Notice that we have choosen a scale factor 4 in front of the above metric in order to find the volume agreeing with the standard volume on S^3 for $\lambda_1=\lambda_2=\lambda_3=1$.

2.2 Left and right fundamental fields.

The Lie algebra Lie(G) of a Lie group G coincides as a vector space, with the space tangent to G at the origin $e \in G$. To each element **X** of **Lie(G)** there correspond two vector fields $X^R(a)$ and $X^L(a)$, $a \in G$, the right and left fundamental vector fields generated by **X**. They are defined as

$$(2.3)\qquad X^R(a) = d/dt\ (a.e^{tX})|_{t=0}$$
$$X^L(a) = d/dt\ (e^{tX}.a)|_{t=0}$$

It follows from this definition that

$$X^R(e) = X^L(e) = X$$

and the usual definition of the Lie bracket in Lie(G) gives

$$[X^R, Y^R] = [X,Y]^R\ ,\ [X^L, Y^L] = -[X,Y]^L,\ [X^R, Y^L] = 0$$

It is convenient to make the following notational abuse of (2.3)

$$X^R(a) = aX$$
$$X^L(a) = X\,a$$

This explains why the right fundamental fields $X^R(a)$ are often called left invariant fields, indeed

$$X^R(ba) = bX^R(a)$$

and similarly the left fundamental vector fields are right invariant

$$X^L(ab) = X^L(a)b$$

Notice that

$$X^L(a) = a^{-1}X^R(a)a = Ad(a^{-1})[X^R(a)]$$

where

$$Ad(a)[s] = a\ s\ a^{-1}$$

If T_i is a basis in Lie(G) then

(2.4) $$[T_i\,, T_j] = C_{ij}{}^k\, T_k \quad = f_i{}^k{}_j\, T_k$$

where $C_{ij}{}^{k}$ are the structure constants of G [2]. The matrix $Ad(a)^{i}{}_{j}$ of the adjoint representation of G on Lie(G) is given by

$$(2.5) \qquad a\, T_j\, a^{-1} = Ad(a)^{k}{}_{j}\, T_k \,.$$

Taking for **a** a one-parameter subgroup of G, $\{\exp(tT_i)\}$, generated by a basis vector T_i, and differentiating (2.5) with respect to t at t=0 we get

$$d/dt\, Ad(\exp(tT_i))^{k}{}_{j}|_{t=0} = C_{ij}{}^{k} \,.$$

Another way of writing the above equation[2] is

$$ad(T_i)\, T_j = C_{ij}{}^{k}\, T_k = f_i{}^{k}{}_{j}\, T_k,$$

where **ad** is the adjoint representation of Lie(G) on itself defined as

$$ad(X)Y = [\, X \,,\, Y \,]$$

In other words the structure constants can be considered as (the matrix element of) infinitesimal generators of the adjoint representation.

We have recalled the general definition of left and right invariant vector fields, we now recall the corresponding properties for invariant forms. The space of left invariant vector fields is a vector space identified with the Lie algebra; the space of left invariant forms is defined as its dual, its elements still carry an R upper index to remember that they are related to the fundamental right action. The usual (left invariant) Maurer-Cartan form ω^R is defined as the map which, to the left invariant field X^R associates its value at the origin,

---

2 Cf. footnote 1 of chapter 1.

namely $X \in Lie(G)$. Of course, one can also consider the (right) Maurer-Cartan form ω^L. It is usual to write these two forms as follows

$$\omega^R = g^{-1}\, dg$$
$$\omega^L = dg\, g^{-1}$$

One has then the so called Maurer-Cartan structure equations which are dual to (2.1),(2.1')

$$d\omega^{kR} = -1/2\; C_{ij}{}^{k}\, \omega^{iR} \wedge \omega^{jR}$$
$$d\omega^{kL} = +1/2\; C_{ij}{}^{k}\, \omega^{iL} \wedge \omega^{jL} ,$$

where ω^i are defined by $\omega = \omega^i T_i$

2.3 Principal fibration of a group with respect to a subgroup.

Let G/K denote the set of left classes of G along a subgroup K, i.e.

$$G/K = \{ aK : a \in G \}$$

The group G is the union of these K classes; this is the well known coset decomposition of G. We will always assume that K is a closed subgroup of G in order for G/K to be a separable topological space. The group G is now a (right) principal bundle with structure group (also fiber) K and base G/K. The projection map is

$$a \in G \rightarrow aK \in G/K$$

and the right action of the elements $k \in K$ on the bundle space G is

$$a \in G \rightarrow a.k \in G$$

Of course, there exists a left principal fibration of G over the space K\G of right classes along K

$$K\backslash G = \{Ka : a \in G\}$$

Notice that in most traditional textbooks, principal fiber bundles are defined via a right action of the structure group.

The group G acts on the principal bundle $G \to G/K$ (resp. $G \to K\backslash G$) from the left (resp. from the right) by bundle automorphisms i.e. it maps fibers onto fibers commuting with the action of the structure group because left and right actions commute.

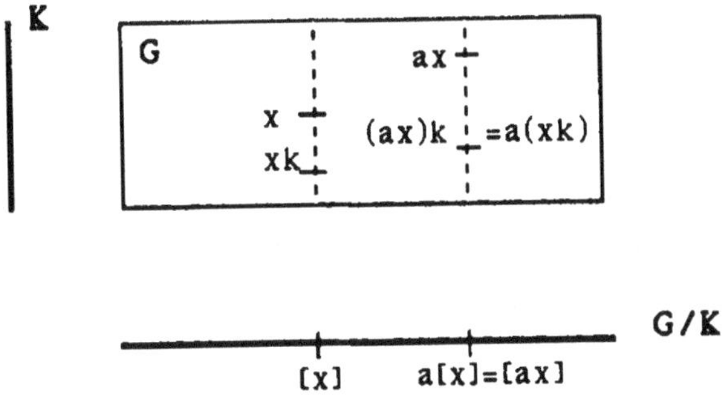

The group G acts on the principal bundle $G \mapsto G/K$ by bundle automorphisms. The element $\mathbf{a} \in G$ transforms the coset $[\mathbf{x}] = \mathbf{x}K$ onto the coset $\mathbf{a}[\mathbf{x}] = [\mathbf{ax}] = \mathbf{ax}K$; this action of the group G on the principal bundle (also G) commutes with the action of the structure group **K** of the bundle. Often it is convenient to use the notation $[\mathbf{a}]$ for a coset $K\mathbf{a}$ (or $\mathbf{a}K$) of $\mathbf{a} \in G$. We shall do that in cases where it is clear from the context as to which subgroup **K** is meant and whether left or right cosets are considered.Let us end this section with a few examples:

- $SU(2)$ is a $U(1)$ principal bundle over the sphere S^2.
- $SU(3)$ is an $SU(2)$ principal bundle over S^5.
- $U(2, \mathbb{H})$ is an $SU(2)$ principal bundle over S ($\mathbb{H}$ is the quaternion field).
- $G \times G$ is a G^d bundle over $G \times G/G^d \cong G$ where:

$$G^d = \{ (\mathbf{a},\mathbf{a}) : \mathbf{a} \in G \} .$$

This principal fibration is trivial; indeed, in each coset $[(\mathbf{a},\mathbf{b})]$ there is a unique representative of the form $(\mathbf{c},\mathbf{e})$. The map $[(\mathbf{a},\mathbf{b})] \to (\mathbf{ab}^{-1},\mathbf{e})$ defines thus a global section of the bundle $G \times G \to G \times G/G^d$

- **Spin**(9) is a **Spin**(7) bundle over S^{15}.
- We will see more examples in the section devoted to homogeneous spaces.

2.4 Bi-invariant metrics.

A Riemannian metric on a Lie group G is called bi-invariant if it is invariant under both left and right translations. There is a one to one correspondance between such metrics and Ad G invariant scalar products on Lie(G). Let $\mathbf{g}$ be a bi-invariant metric on G, let $\{T_i\}$ be a basis in the Lie algebra, and let $(\mathbf{g}_{ij})$ be the matrix of scalar products of the basis vectors: $\mathbf{g}_{ij} = (T_i, T_j)$.

Let $\{e_i^L\}$, (resp. $\{e_i^R\}$) be a basis of right, (resp. left), invariant vector fields on G generated by (T_i), and let $\{\omega^{iL}\}$, (resp. $\{\omega^{iR}\}$), be the dual basis (of forms). Then the metric $\mathbf{g}$ and its inverse $\mathbf{g}^{-1}$ can be represented as

$$\mathbf{g} = \mathbf{g}_{ij}\, \omega^{iL} \otimes \omega^{iL} = \mathbf{g}_{ij}\, \omega^{iR} \otimes \omega^{iR}$$

$$\mathbf{g}^{-1} = \mathbf{g}^{ij}\, e_i^L \otimes e_j^L = \mathbf{g}^{ij}\, e_i^R \otimes e_j^R$$

The property that the scalar product on Lie(G) is Ad G invariant can be described by

$$\text{(2.6)} \qquad Ad(a)^i{}_j\, Ad(a)^k{}_l\, \mathbf{g}^{jl} = \mathbf{g}^{ik} ,$$

or, infinitesimally by

$$\text{(2.7)} \qquad C_{jk}{}^i\, \mathbf{g}_{il} + C_{jl}{}^i\, \mathbf{g}_{ik} = 0.$$

In case of a simple or semi-simple group the Killing metric is the bi-invariant metric characterized by the scalar product matrix $\mathring{\mathbf{g}}_{ij}$

$$\text{(2.8)} \qquad \mathring{\mathbf{g}}_{ij} = - C_{ik}{}^l\, C_{jl}{}^k .$$

If the group G is simple, all other bi-invariant metrics are just proportional to this one

$$g = \rho^2 \, \mathring{g} \quad , \rho \in R$$

If the group G is semi-simple, all other bi-invariant metrics are obtained by introducing several scaling coefficients (Cf. for example [O'Neill, ch.11, p. 304]). For example if $G=G_1 \times G_2$ is a product of two simple components, we can write

$$\mathring{g} = \mathring{g}_1 \oplus \mathring{g}_2$$

and

$$g = \rho_1{}^2 \, \mathring{g}_1 + \rho_2{}^2 \, \mathring{g}_2 \quad ; \quad \rho_1, \rho_2 \in R \, .$$

(The scaling coefficients need not be positive if G is not compact).

If g is a bi-invariant metric on G, then Ricci and scalar curvature are given by

(2.9) $$R[g]_{ij} = 1/4 \; \mathring{g}_{ij} \; ,$$

(2.10) $$R[g] \; = 1/4 \; g^{ij} \; \mathring{g}_{ij} \; ,$$

(for derivation of these formulae see Sec. 2.5.).

In particular G with the Killing metric is an Einstein space with Einstein constant 1/4 and scalar curvature (dim(G))/4.

For a simple compact Lie group G, the connected isometry group of a bi-invariant metric is always $G \times G$. This need not be true in a general case as the example with $G = R \oplus R$ clearly shows.

Notice that, if $g' = \rho^2 \, g$, then, $R' = R \, / \, \rho^2$; for example, in the case of $G = SU(2) = S^3$, we get $R[\, \mathring{g}] = 3/4$, but the "standard" metric introduced in sect. 2.1 (for which the volume is $2\pi^2$) differs from the Killing metric as follows

$$g' = 1/8 \; \mathring{g} \, , \qquad \text{therefore} \qquad R[g'] = 6.$$

2.5 Left and Right invariant metrics.

Here, we will only discuss right invariant metrics, mainly because this is the case which we will generalize when we study invariant metrics on manifolds. Of course everything would go through by replacing "left" by "right" and conversely. A right invariant metric on G is a metric which is invariant under the right translations of G.

Let us recall that we denote by $\{e_i^R\}$ a base of fundamental fields for the right action (they are left invariant) and by $\{e_i^L\}$ a base of fundamental fields for the left action (they are right invariant); also $\{\omega^{iR}\}$ and $\{\omega^{iL}\}$ denote the corresponding dual basis.

An arbitary right invariant metric can be written as

$$h = h_{ij}\, \omega^{iL} \otimes \omega^{jL}$$

and

$$h^{-1} = h^{ij}\, e_i^L \otimes e_j^L$$

Here, h_{ij} denotes a matrix of constant coefficients; notice that if $T_i = e_i^L(\text{Identity}) \in Lie(G)$, then the metric is also completely characterized by the scalar product $h_{ij} = \langle T_i\, , T_j \rangle$ in the Lie algebra. Notice that the fundamental fields e_i^R are Killing vector fields of this metric.

Of course, it is also possible to express the metric h at the point $a \in G$ in terms of right fundamental fields, using the relations of sect. 2.2 we find

$$h^{-1}(a) = h^{ij}(a)\;\; e_i^R(a) \otimes e_j^R(a)$$

but now, $(h^{ij}(a))$ is not a constant matrix; it depends upon a as follows

$$h^{ij}(a) = Ad(a^{-1})\,^i{}_k\;\; h^{kl}\, Ad(a)^j{}_l\, .$$

It is finally possible to write a right invariant metric in terms of local coordinates on the group but it is usually unnecessary and very cumbersome (cf. sect. 2.1 for such an example on SU(2)).

As we shall see later, the number of right invariant metrics on a Lie group is related to the number of scalar fields which can be introduced in the process of "dimensional reduction". As we saw before, G-invariant metrics are in one to one correpondance with scalar products in the Lie algebra; the dimension of the space of G-invariant metrics, for a Lie group of dimension n is therefore n(n+1)/2. In order to define such a scalar product, we can choose an arbitrary basis in the Lie algebra as an orthonormal basis of a metric, then any other basis obtained by the action of an element of GL(n) defines another scalar product, unless if this element of GL(n) belongs to O(n); the manifold of right invariant metrics on G is therefore isomorphic to the coset space GL(n)/SO(n), whose dimensionality is n(n+1)/2, as it should be.

The Riemann,Ricci and scalar curvatures of right invariant metrics on a group can be computed from the formulae of sect. 1.3.1

$$
\begin{aligned}
R_{ijlk} = 1/4 \,[\, & C_{ik,m}\, C_{jl,m} + 2\, C_{ij,m}\, C_{kl,m} - C_{jk,m}\, C_{il,m} \\
& - C_{ij,m}\, C_{mk,l} + C_{ij,m}\, C_{ml,k} - C_{kl,m}\, C_{mi,j} \\
& + C_{kl,m}\, C_{mj,i} + (C_{mi,k} + C_{mk,i})(C_{mj,l} + C_{ml,j}) \\
& - (C_{mj,k} + C_{mk,j})(C_{mi,l} + C_{ml,i}) \,] \qquad , \qquad (2.11)
\end{aligned}
$$

$$
\begin{aligned}
R_{jk} = & -1/2\, C_{mj,n}\, C_{nk,m} - 1/2\, C_{mj,n}\, C_{mk,n} + 1/4\, C_{mn,j}\, C_{mn,k} \\
& -1/2\, (C_{mj,k} + C_{mk,j})\, C_{mn}{}^{n} \qquad , \qquad (2.12)
\end{aligned}
$$

$$R = -1/4\, C_{mk,n} C_{mk,n} - 1/2\, C_{mk,n} C_{nk,m} - C_{mk,k} C_{mn,n} \qquad (2.13)$$

Here, $C_{ij}{}^k$ are the structure constants of the Lie algebra in an arbitrary frame and, $C_{ij,l} := h_{kl} C_{ij}{}^k$

Observe that the structure constants $C_{ij,k}$ are not, in general, antisymmetric with respect to the last two indices; this is because h is not assumed to be bi-invariant. Also, we use the convention that the summation over repeated indices on the same level is performed with h^{ij}, for example,

$$C_{ij,k} C_{ij,l} = h^{ii'} h^{jj'} h_{kk'} h_{ll'} C_{ij}{}^{k'} C_{i'j'}{}^{l'}$$

The last terms (containing the trace of $C_{ij}{}^k$) of the above formulae for curvatures vanish for compact groups and more generally for unimodular groups; in those cases we indeed get $C_{ij}{}^j = 0$, these terms will therefore be omitted in the following.

Suppose now h is a right invariant metric on G determined by a matrix h_{ij} as above. Let K be a (closed) subgroup of G (in particular we can take $K=G$), then a necessary and sufficient condition for h to be also left K-invariant is that h_{ij} is $Ad(K)$-invariant. Explicitly, we should have

$$Ad\,(a^{-1})^l{}_i\; h_{lm}\; Ad\,(a)^m{}_j = h_{ij} \qquad (2.14)$$

for all $a \in K$. Taking in particular a one-parameter subgroup $\exp(tT_\alpha)$, $T_\alpha \in Lie(K)$, and differentiating (2.14) with respect to t at t=0 we find the infinitesimal form of (2.14)

$$C_{\alpha i}{}^l h_{lj} + C_{\alpha j}{}^l h_{il} = 0 \; . \qquad (2.15)$$

2.6 Metrics with isometry group H×K where H,K are subgroups of G.

The reader will have no trouble in generalizing the example given in sect. 2.1 (a U(1)×U(1) invariant metric on SU(2)) -cf. also [50].

2.7 G×K invariant metrics related to a fibration of G.

Let **K** denote a (closed) subgroup of G (supposed to be compact, as usual). Then, we can write the following decomposition of Lie(G)

(2.16) $$\mathrm{Lie}(G) = \mathrm{Lie}(K) \oplus \mathcal{P}$$

where $\mathcal{P}$ is the orthogonal complement of Lie(K) for some (choosen) bi-invariant metric **g** on G (for example the Killing metric if G is semi-simple). We may now consider the following family of metrics

$$h_t = g(\text{proj. on } \mathcal{P}) \oplus t^2.g(\text{proj. on Lie}(K))$$

where t is a real parameter. For each value of t^2, we obtain a new metric which is no longer G×G invariant but G×K invariant, as it is almost obvious by looking at the expression of h^{-1} in terms of fundamental fields. Notice that one example given in sect. 2.1 (a family of SU(2)×U(1) invariant metrics on SU(2)) fits into this category. In the same way, we could construct:

- SU(3)×SO(3) invariant metrics on SU(3)
- SU(4)×S(U(2)×U(2)) invariant metrics on SU(4)
- SU(4)×U(2,ℍ) invariant metrics on SU(4)
- SU(4)×SO(4) invariant metrics on SU(4)
- SU(2)×SU(2)×SU(2) invariant metrics on SU(2)×SU(2)
- etc.

We will see in sect. 2.9 and 4.5.2 how to compute easily the scalar curvature of such metrics. The title of the present section is justified by the fact that we have given here an example of a standard way of obtaining new interesting metrics on a manifold: one starts with some given metric, chooses some fibering of the manifold (here, a K bundle over G/K) and start to distort the given metric in the direction corresponding to the fibering; we will meet several examples of this technique. The above example could also be generalized as follows: Let G be a simple Lie group, **g** the Killing metric of G, K a connected subgroup of G (not necessarily simple), then we write the following decompositions

$$\mathrm{Lie}(G) = \mathrm{Lie}(K) \oplus \mathcal{P}$$

and

$$\mathrm{Lie}(K) = \mathcal{K}_0 + \mathcal{K}_1 + \ldots + \mathcal{K}_s,$$

where $\mathcal{K}_0$ is the center of Lie(K) and $\mathcal{K}_i$,(i=1...s), are simple components of Lie(K). Then, if we modify the bi- invariant metric as above by introducing arbitrary real positive scaling parameters r_i, (i=0...s), for each component Ki, we obtain metrics which are not G×K invariant but G×Π(Ki) invariant.

2.8 Dimensionally reducible metrics (action of a bundle of groups).

The metrics on Lie groups discussed so far: bi-invariant metrics, left (or right) invariant metrics are "dimensionally reducible" in the sense that their expression at an arbitrary point **p** of G (an n dimensional manifold) can be obtained as soon as we know their expression at one particular point of G, for example at the origin of G (a point is a zero dimensional manifold). There is a dimensional reduction from n to 0. This can be visualized as follows. The fact that a metric **g** on G is (say) right G-invariant means that its Lie derivative with respect to all right fundamental vector fields X^R vanishes. For $X^R = e_i^R$ one gets thus the equation

$$0 = \underset{e_i^R}{L}{}_{e_i^R}(g_{kl}) = \partial_i g_{kl} - C_{ik}{}^m\, g_{ml} - C_{il}{}^m\, g_{km}$$

or

$$\partial_i g_{kj} = C_{ik}{}^m g_{ml} + C_{il}{}^m g_{km}.$$

The last equation is nothing but a "propagation law" for g_{ij} along G. Thus having given g_{ij} at one point (e.g. at the origin $e \in G$) the equation determines g_{ij} at any other point of G.

However, the previously discussed metrics are not the only ones to share this property; indeed, let

$$h_1{}^{-1} = h_1{}^{ij}\ e_i^R \otimes e_j^R$$

be a left invariant (inverse) metric and

$$h_2{}^{-1} = h_2{}^{ij}\ e_i^L \otimes e_j^L$$

be a right invariant (inverse) metric, then

$$h^{-1} = h_1{}^{-1} + h_2{}^{-1}$$
$$= (h_1{}^{ij} + a^{-1}\, h_2{}^{ij}\, a)\ e_i^R \otimes e_j^R$$

defines a metric on G which is usually neither left nor right invariant but which is obviously dimensionally reducible in the above sense.

We can calculate again the Lie derivative $L_i = L_{e_i^R}$ of h^{kl}. This time the result is

$$L_i\, h^{kl} = \partial_i\, h^{kl} + C_{im}{}^k\, h^{ml} + C_{im}{}^l\, h^{km} = h_1{}^{ml}\, C_{im}{}^k + h_1{}^{km}\, C_{il}{}^m$$

so that

$$\partial_i h^{kl} + h_2{}^{ml}\, C_{im}{}^k + h_2{}^{km}\, C_{im}{}^l = 0,$$

which again allows to propagate h from one point to the whole of G.

Let us end this section by giving a convenient geometrical interpretation of such metrics. We consider the group G×G and choose on this group an arbitrary G×G right invariant metric (it is not a priori bi-invariant on G×G, since such a bi-invariant metric would have G×G×G×G as isometry group). A right action of G×G on itself at the point (p_1, p_2) is defined as follows:

$$(p_1, p_2) \in G\times G\ ,\ (k_1, k_2) \in G\times G \ \longrightarrow\ (p_1 k_1, k_2{}^{-1} p_2) \in G\times G$$

The metric is in particular invariant under the subgroup **diag**(G×G) and therefore goes to the quotient $G\times G/\mathbf{diag}(G\times G) \cong G$ for the above action. The metric on G obtained in that way has usually no invariances left and is clearly a member of the family of "dimensionally reducible metrics" discussed above. We refer to Ch. 4.11 for a definition of a bundle of groups and will see in Ch. 10 in which sense the kind of metrics discussed here is invariant under a "bundle of groups".

We will also see later how to generalize this construction in the case of homogeneous spaces and specially in the case of fiber bundles with homogeneous fibers (here, the bundle was a bundle over a point).

Remark: *More general metrics*

We do not intend to discuss here more general metrics on groups, but the reader should remember that "most" metrics on groups, and more generally on manifolds have no isometries at all. The word "most" has here a very precise sense: it can be proven that the space of metrics with no more isometries than the identity itself is an open dense submanifold in the space of all metrics.[22]

2.9 Einstein metrics on groups.

As we already mentioned, the Killing metric on a simple Lie group is an Einstein metric. This is the "standard" Einstein metric.

Many non standard Einstein metrics on groups with isometry group $G\times K$ (K, a subgroup of G) can be obtained in the family of metrics discussed in sect. 2.7 Let us here suppose that G is simple. We first write G as a K principal bundle over G/K, and construct a one parameter family of $G\times K$ invariant metrics by rescaling the Killing metric in the direction of K. Explicit calculation of the scalar curvatures of G, K and G/K (for the induced metrics) shows that

$$R^G = R^{G/K} + R^K - 1/4 \,.\, C_{ab,\hat{c}}\, C_{ab,\hat{c}}$$

Here, we assume that the basis $\{T_i\}$ is adapted to the decomposition (2.16) and write: $\{T_i\} = \{T_a, T_{\hat{a}}\}$, $T_{\hat{a}} \in Lie(K)$, $T_a \in \mathcal{P}$

This formula resembles a "Kaluza-Klein reduction" where the "external space" is G/K, the "internal space" is K, and the field strengh is $C_{ab}{}^{\hat{c}}$; it is a particular case of a more general formula which will be obtained later (sect.4.5.2). For the Killing metric $\overset{\bullet}{g}$, the above formula reads

$$n/4 = s/2 + c\, k/4 - k(1-c)/4$$

where n=dim(G), k=dim(K), s= dim(G/K)=n-k and c, the index of K in G (cf. sect. 2.11) is defined by the following formulae

$$\overset{\bullet}{g}_{\hat{a}\hat{b}} = c\, g_{\hat{a}\hat{b}} \quad \text{, where}$$

$$\overset{\bullet}{g}_{\hat{a}\hat{b}} = - C_{\hat{a}\hat{c}}{}^{\hat{d}}\, C_{\hat{b}\hat{d}}{}^{\hat{c}}$$

is the Killing metric of K and where

$$g_{\hat{a}\hat{b}} = - C_{\hat{a}c}{}^{d}\, C_{\hat{b}d}{}^{c}$$

is the restriction to K of the Killing metric of G.

For the metrics $h = g(\text{proj.on } \mathcal{P}) + t^2 g(\text{proj.on } \mathcal{K})$, we get

$$R^G = s/2 + (c.k/4)1/t^2 - k(1-c)t^2/4$$

However, when t varies, the volume of G varies; in order to keep it fixed we just have to make a conformal rescaling and consider the family of metrics

$$\bar{h} = (1/t^2)^{k/n}\, h,$$

then $\det(\bar{h}) = (1/t^2)^k.(t^2)^k = \text{const.}$

The corresponding scalar curvature reads

$$R^G = (t^2)^{k/n} [\ s/2 + ck/4t^2 \ + k(c-1)t^2/4\ ]$$

In order to find Einstein metrics we now vary this expression with respect to t and look for saddle points (cf. sect. 1.3.2). Let us also suppose that (G,K) is a symmetric pair (i.e. $[\mathcal{P},\mathcal{P}] \subset Lie(K)$) to ease the calculation (in that case, c=1-s/2k).

$$dR^G/dt = -s/4\ .(2k+s)/(k+s)\ .\ t^{2k/n\ -3}.(t^2-1).\ (t^2-(2k-s)/(2k+s))$$

We find therefore two candidates for Einstein metrics corresponding to the values $t^2=1$ and $t^2= (2k-s)/(2k+s)$. The first value corresponds of course to the bi-invariant metric on G, the other to a non standard Einstein metric, a G×K invariant metric. For example, if G=SU(3) and K=SO(3), we get an SU(3)×SO(3) invariant Einstein metric on SU(3) for the value $t^2=1/11$. This method can be also generalized to the discussion of $G\times\Pi K_i$ invariant metrics introduced in sect. 2.7. [50]

Notice that varying the functional $\int R[g]\,dv$ in the space of <u>all</u> metrics is not the same as looking at its saddle points in the space of G-invariant metrics; however, one can check explicitly (for example by computing the Ricci tensor) that the above method gives indeed Einstein metrics on G. [50]

Besides, let us mention that very often, there exist also Einstein metrics on Lie groups which are <u>not</u> G-invariant and for which the above method obviously fails.

Notice that the previous construction also applies when G is not a simple group; for example, let G_1 be a simple group and take G = $G_1\times G_1$, we can then choose H = diag($G_1\times G_1$); G/H is diffeomorphic with G_1 but does not carry a (quotient) group structure since H is (in general) not normal in G [indeed, call N(H) the biggest subgroup of G in which H is normal - it is the normalizer of H in G, cf. Sect.3.3- then N(H)|H can easily seen to be isomorphic with the center of G_1; while normality of H in G would imply N(H)|H = G|H = G_1] . We can then write $G_1\times G_1$ as an H bundle over G/H and define a family of

$G_1 \times G_1 \times G_1$ -invariant metrics on $G_1 \times G_1$ by scaling the Killing metric in the H direction; very often we can find new Einstein metrics in this way (for example there exists an $SU(2) \times SU(2) \times SU(2)$ invariant metric on $S^3 \times S^3$ which is an Einstein metric and is not a product, this happens for the value $t^2 = 1/3$ of the scaling parameter, here $c = 1/2$).

2.10 On the classification of compact simple Lie groups

The purpose of this paragraph is just to remind the reader of the usual Cartan classification; this will be used in many examples of the book.

G	Dimension	Center	Complex extension
$A_m = SU(m+1, \mathbb{C})$	$m(m+2)$	$\mathbb{Z}_{m+1}$	$Sl(m+1, \mathbb{C})$
$B_m = \mathbf{Spin}(2m+1, \mathbb{R})$	$m(2m+1)$	$\mathbb{Z}_2$	$\mathbf{Spin}(2m+1, \mathbb{C})$
$C_m = U(m, \mathbb{H})$	$m(2m+1)$	$\mathbb{Z}_2$	$Sp(2m, \mathbb{C})$
$D_m = \mathbf{Spin}(2m, \mathbb{R})$	$m(2m-1$	$\mathbb{Z}_4$ if $m = 2l+1$	$\mathbf{Spin}(2m, \mathbb{C})$
		$\mathbb{Z}_2 \times \mathbb{Z}_2$ if $m = 2l$	
G_2	14	1	
F_4	52	1	
E_6	78	$\mathbb{Z}_3$	
E_7	133	$\mathbb{Z}_2$	
E_8	248	1	

The above groups G are simple,compact and simply connected real Lie groups. **Spin**(n) is the two-fold covering of SO(n). $U(n, \mathbb{H})$ denotes the set of $n \times n$ unitary matrices over the quaternion field $\mathbb{H}$; other favorite notations for these symplectic groups are

$U(n,\mathbb{H}) = USp(2n,\mathbb{C}) = Sp(2m,\mathbb{C}) \cap SU(2m,\mathbb{C})$

They are even sometimes denoted by Sp(n) or Sp(2n), depending upon the authors! We will stick to $U(n,\mathbb{H})$. Let us recall the following isomorphisms:

$SU(2)=Spin(3)=U(1,\mathbb{H})$

$Spin(5) = U(2,\mathbb{H})$

$Spin(6) = SU(4)$

$Spin(4) \quad = SU(2)\times SU(2)$ -this is the only non simple group in the above table-.

Decomposition of compact Lie groups.
We state the following theorem [183]:

Any compact connected Lie group G is isomorphic with a unique group of the form $(T_0 \times G_1 \times \ldots \times G_m) / \mathbb{K}$, where T_0 is the identity component of the centre of G and G_i are compact, simple, connected, simply-connected Lie groups, with $\mathbb{K}$ a finite subgroup of the center of the product.

2.11 Standard normalizations, indices etc.

-Conventional normalization of generators in Physics

Assume G compact and simple, and let $\{T_i\}$ be a basis of the Lie algebra of G such that $-\mathrm{Tr}(\mathbf{ad}\,(T_i))(\mathbf{ad}\,(T_j)) = i(G)\,\delta_{ij}$ where the coefficient 2i(G) is given by the following table

G	SU(n)	U(n,ℍ)	SO(n),n>4	SO(3)	G2	F4	E6	E7	E8
2i(G)	2n	2(n+1)	2(n-2)	4	8	18	24	36	60

This normalization is standard in Physics (the factor -1 is usually incorporated into generators $\sqrt{-1}\,T_j$). The above numbers have the following geometrical interpretation: let $\mathring{g}$ be the Killing metric on G, then

$$(17) \qquad g_{standard} = \mathring{g}/2i(G)$$

is another bi-invariant metric on G for which the length of the maximal root of Lie(G) is -2. This metric is sometimes called the *standard metric* on G (recall that a bi-invariant metric on G determines an AdG invariant bilinear symmetric form on Lie(G) and this form in turn determines a form on the dual vector space $(\mathrm{Lie}(G))^*$).

If G is semi-simple, $G=G_1\times G_2$, one sets $i(G)=i(G_1)+i(G_2)$.

- ***Indices of representations***

Assume G, compact and simple let ρ be a faithful (real or complex) representation of Lie(G) and $\{T_i\}$ a basis in Lie(G) normalized as above. Then, $\mathrm{Tr}(\mathrm{ad}(T_i)\,\mathrm{ad}(T_j))$ and $\mathrm{Tr}(\rho(T_i)\,\rho(T_j))$ are proportionnal and we define the index $i(\rho)=i(\rho,G)$ of the representation ρ as the number determined by

$$(18) \qquad - \mathrm{Tr}\, \rho(T_i)\rho(T_j) = i(\rho)\, \delta_{ij}$$

Remarks:

1) The number i(G) given above is the index of the adjoint representation of G

2) The quantity $2i(\rho)$ is sometimes called "index" in the mathematical literature [224].

3) If ρ is reducible, $\rho=\rho_1+\rho_2$ then, clearly, $i(\rho)=i(\rho_1)+i(\rho_2)$, and if ρ is trivial then $i(\rho) = 0$.

4) Tables for the indices $i(\rho)$ of irreducible representations exist [170]; these authors list the values of the quantity $I_2(\rho)/\mathrm{rank}(G)$ which is related to ours by the relation

$$2\, i(\rho) = I_2(\rho)/\mathrm{rank}(G).$$

For the fundamental (complex) representation ρ of SU(n) groups, we get in particular $i(\rho) = 1/2$.

- *Coupling constants in Yang-Mills theories*

In Physics, one introduces a "coupling constant" in the Lagrangian density to measure the strengh of interactions mediated via Abelian or non Abelian Yang-Mills fields. The part of the Lagrangian density which describes this kind of interactions should be a gauge invariant scalar quantity, at most quadratic in the derivatives. The only candidate which is not a total divergence is the trace of $F_\wedge{}^*F$, F being the curvature of the connection, in some (usually irreducible) representation. For a simple Lie group, traces in different representations are proportional, as already discussed. For an arbitrary semi-simple Lie group G, one writes the Yang-Mills action as

$$I[A] = (1/2g^2) \int F^i{}_\wedge{}^* F^j \, Tr(\rho(T_i) \rho(T_j))$$

where the curvature F is written $F = F^i T_i$ and where the value of the 'coupling constant' g depends upon the choice of a representation ρ.

For example, (and for historical reasons), in the case of SU(n) groups, one usually chooses the fundamental representation ρ (of smallest dimension) and normalizes its generators according to

$$Tr(\rho(T_i) \rho(T_j)) = - i(\rho) \delta_{ij} \quad \text{with } i(\rho) = 1/2 .$$

In SU(2) and SU(3), one takes respectively $\rho(T_i) = i \sigma_i/2$ and $\rho(T_i) = i \lambda_i/2$ where σ and λ are the Pauli and Gell Mann matrices. In this case, the action associated with the Yang-Mills potential A reads

$$I[A] = - (1/4g^2) \int \delta_{ij} F^i{}_\wedge{}^* F^j$$

The number g plays the role of a coupling constant as it is clear after the rescaling:

$$A \text{ ----------------- } \longrightarrow A^{new} = A/g$$

$$F = dA + 1/2\,[A \wedge A] \text{ ---------} \longrightarrow F^{new} = dA^{new} + g\,1/2[A^{new} \wedge A^{new}]$$

and one gets

$$I[A^{new}] = 1/2 \int dx\, tr(\rho(F^{new})\, \rho({}^*F^{new})).$$

The choice of the representation ρ being made, and remembering that

$$i(\rho)\, Tr(\, Ad\, T_i\, Ad\, T_j\,) = i(G)\, Tr(\, \rho(T_i)\, \rho(T_j)\,)$$

we can write also the Yang - Mills action as

$$I[A] = -\,(1/2g^2)\;\; i(\rho)/i(G)\;\; h_{Killing}(F,F)$$

or as

$$I[A] = -\,(i(\rho)/g^2)\, h_{Standard}(F,F)$$

If h is a metric on the group G then $h' = \lambda^2 h$ with constant λ is a new metric with volume $V' = \lambda^{\dim G} \times V$, it is therefore clear that the geometrical meaning of the value of a "coupling constant" is a measure of the inverse "radius" on the group G in some conventional units: measuring the coupling constant amounts to measuring the volume of G. We will return to this discussion in chapter 4 after discussing the dimensional reduction of the scalar curvature on a principal bundle $P = M \times G$ endowed with a G-invariant metric. We will get there a more general Yang-Mills term of the kind

$$I = -\,(1/4)\int F^i(x) F^j(x)\, h_{ij}(x)\quad ,$$

$h(x)$, being a G-invariant metric on the copy G_x of G at the point x of M.

- ***Casimir operators***

Assume G compact and semi simple, let ρ and T_i be as in 2.11-18, then the operator $C_\rho = \Sigma\, \rho(T_i)\rho(T_i)$ is called (by physicists) the Casimir operator in the representation ρ. Not surprisingly, most mathematicians call $2 \times C_\rho$ by the same name; we will stick to the physicist terminology. When ρ is irreducible, C_ρ is proportional to the unit matrix: $C_\rho = c(\rho).\mathbb{1}$ and the eigenvalue $c(\rho)$ is related to the index by the obvious relation

$$c(\rho) = i(\rho) \times (\dim G \, / \, \dim \rho).$$

For example, with the above conventions, we get:
for the representation j of dimension $2j+1$ of $SU(2)$,

$$i(j) = 2j(2j+1)(2j+2) \, / \, 12 \quad , \quad c(j) = j(j+1).$$

for the fundamental representation ρ (of dimension n) of $SU(n)$,

$$i(\rho) = 1/2 \, , \quad \text{and} \quad c(\rho) = (n^2 - 1)/2n.$$

- ***Embedding coefficient and Dynkin index***

When G and H are both simple Lie groups with H a Lie subgroup of G then, the embedding coefficient $c(G,H)$ is the number such that

$$g_{Killing}(H) = c(G,H) \, g_{Killing}(G)|_{\text{resticted to } H}, \quad \text{i.e.}$$

$$C_{i\mathfrak{k}}{}^{\mathfrak{l}} \, C_{j\mathfrak{l}}{}^{\mathfrak{k}} = c(G,H) \, C_{ik}{}^{l} \, C_{jl}{}^{k}$$

The knowledge of this number is important when we apply dimensional reduction techniques, for example when we write G as an H bundle over G/H, as in sec.2.7,.2.9. For example , in the case of the Killing metric on G, we get, for the scalar curvature (see Chapter. 4)

$$R^G = n/4 = R^{G/H} + c \dim(H) \, /4 - \dim(H) \, (1-c) \, /4 \; , \text{ with } c = c(G,H)$$

This last formula allows us to relate the scalar curvatures R^G and $R^{G/H}$ if c is known. Notice that for a symmetric irreducible space G/H endowed with the normal Killing metric (cf. Chapter.3), we have $R^{G/H} = \dim(G/H)/2$ and therefore $c(G,H) = 1 - \dim(G/H)/(2\dim(H))$.

The Dynkin index $[G:H]$ is defined in an analogous way to be the number such that

$$(19) \qquad g_{Standard}(H) = 1/[G:H] \, . g_{Standard}(G).$$

From the definition of the standard metric, we get immediately

$$[G:H] \, . \, c(G,H) = i(H)/i(G)$$

In most cases, called standard embeddings, we have $[G:H]=1$ and we get $c(G,H)$ from the table given at the beginning of this section. Let us mention Dynkin indices for some non standard embeddings: $[SO(n):SO(3)] = 2$ if $n > 4$, $[SU(n):SO(n)] = [U(n,H),SU(n)] = 2$ except $[SU(3):SO(3)]=4$.

In general, one can compute $[G:H]$ by using the following well known theorem: let ρ be any representation of G, call $i(\rho,G)$ its index and let $\rho_{/H}$ be its restriction to H, call $i(\rho_{/H},H)$ its index then $1/[G:H] = i(\rho,G)/i(\rho_{/H},H)$. In particular, we may choose ρ as the adjoint of G and we get

$$1/[G:H] = i(G)/i(AdG_{/H},H) \, .$$

For example if $G=SU(3)$ and $H=SU(2)$ so that G/H is then topologically S^5 (observe that G/H is not a symmetric space), we have $i(G)=3$ and the branching rule:

$8 \longrightarrow 3+1+2+2$ then $i(8,H)=i(3,H)+i(1,H)+i(2,H)+i(2,H) = 2 + 0 + 1/2 + 1/2 = 3$.

Thus $[G:H]=3/3 = 1$ and it is therefore a standard embedding.

However if $G=SU(3),H=SO(3)$, G/H being here an irreducible symmetric space, we have the branching rule:

$8 \longrightarrow 3+5$ then $i(8,H)=i(3,H)+i(5,H) = 2 + 10 = 12$

thus $[G:H]=12/3=4$ and this is not a standard embedding.

- ***Relation between normalization of volumes,scalar curvature etc.***

Whenever a metric g is choosen on a compact space G, one can then define and compute the corresponding volume $Vol(g)$ and the scalar curvature R[g]. If we scale g by a constant factor λ^2, i.e. if we define $g' = \lambda^2 g$, we get $Vol(g') = \lambda^n Vol(g)$ and $R'[g'] = R[g]/\lambda^2$, where n=dim G. For example, the bi-invariant metric g of eq. 2.2 defines a Riemannian structure on $SU(2)=S^3$ for which the volume is $2\pi^2$ and the scalar curvature is 6. However the Killing metric on SU(2) is proportional to g but has a scalar curvature dim(G)/4=3/4, therefore $\mathring{g} := g_{Killing} = 8\, g$ and $Vol(\mathring{g}) = 8^{3/2}\, Vol(g)$. This simple remark allows us to compute normalization factors in many cases provided we remember that the Killing metric defined on any simple compact group by $\mathring{g}_{ij} = - C_{ik}{}^l C_{jl}{}^k$ and the normal Killing metric $\mathring{g}_{\alpha\beta} = -C_{\alpha k}{}^l C_{\beta l}{}^k$ defined on a symmetric homogeneous space G/H have scalar curvatures equal to dim(G)/4 and dim(G/H)/2 respectively (cf. chapter 3).

-Standard volumes on spheres and projective spaces

Let E,M be two Riemannian spaces, then if $\pi: E \to M$ is a Riemannian submersion [158], we have [12]

$$(20) \qquad Vol(E) = \int vol(\pi^{-1}(x))\, dv(x).$$

In particular we will consider cases where π is the projection map associated to a fibration of E and at the same time a Riemannian submersion. For example if G is a simple Lie group ,π: $G \to G/H$ is the projection map, G is endowed with a bi-invariant metric and G/H with the induced G-invariant metric, we get Vol(G)=Vol(G/H) Vol(H), where Vol(H) denotes the volume of the typical fiber. Another example: G' is the universal covering group of G, with $G' \to G$ being the covering map, then Vol(G') = p.Vol(G), p=card(π^{-1}(x), $x \in G$.

We have defined previously the "standard" metric on a semi-simple group G, $g_{Standard} = g_{Killing}/2i(G)$. We can define, in the same way the "standard" metric on a homogeneous space G/H, it is the normal metric obtained from g_{Standard} on G by passing to the quotient. Owing to the fact that

$$R^G = \dim(G)/4 = R^{G/H} + c\ \dim(H)/4 - \dim(H)\ (1-c)\ /\ 4$$

for the Killing metric on G, with c=c(G,H) and using the relation

$$R^G{}_{\text{Standard}} = 2i(G)\ R^G,$$

we find

$$\text{Vol}_{\text{Standard}}(G) = [G:H]^{(\dim H\ /2)}\ \text{Vol}_{\text{Standard}}(H)\ \text{Vol}_{\text{Standard}}(G/H)$$

We have already warned the reader that $\text{Vol}_{\text{Standard}}(G_1/H_1)$ can be different from $\text{Vol}_{\text{Standard}}(G_2/H_2)$, even if G_1/H_1 and G_2/H_2 are diffeomorphic to the same manifold S; in particular $\text{Vol}_{\text{Standard}}$ $(SO(2n+2)/SO(2n+1))$ is not equal to $\text{Vol}_{\text{Standard}}$ $(SU(n+1)/SU(n))$ although these two coset are both the sphere S^{2n+1}. Fortunately everybody agrees with the definition of the "standard volume" of the n sphere, defined through its standard equation in R^{n+1}. Besides spheres, everybody seems also to agree on the standard volumes of projective spaces, irrespective of any coset structures; they are gotten by using (20) together with the following Hopf fibrations:

$$Z_2 \longrightarrow S^{2n} \longrightarrow RP^{2n} \quad \Rightarrow \quad \text{Vol}(RP^{2n}) = (2\pi)^n/(2n-1)!$$
$$Z_2 \longrightarrow S^{2n+1} \longrightarrow RP^{2n+1} \quad \Rightarrow \quad \text{Vol}(RP^{2n+1}) = \pi^{n+1}/n!$$
$$U(1) \longrightarrow S^{2n+1} \longrightarrow CP^n \quad \Rightarrow \quad \text{Vol}(CP^n) = \pi^n/n!$$
$$S^3 \longrightarrow S^{4n+3} \longrightarrow HP^n \quad \Rightarrow \quad \text{Vol}(HP^n) = \pi^{2n}/(2n+1)!$$

Notice that $V(CP^1)=\pi$ although $CP^1=S^2$ and $\text{Vol}(S^2)=4\pi$ also

$\mathrm{Vol}(HP^1)=\pi^2/3$ although $HP^1=S^4$ and $\mathrm{Vol}(S^4)=2^3\,\pi^2/3$
For spheres, we have the usual results:

$$\mathrm{Vol}(S^{2n}) = 2\,(2\pi)^n/(2n-1)!!$$
$$\mathrm{Vol}(S^{2n+1}) = 2\,\pi^{n+1}/n!$$

For the standard metrics on the sphere S^n, the scalar curvature is equal to $n(n-1)$.

2.12 Pointers to the literature

Invariant metrics, Curvature of Lie groups 8, 9, 50, 148, 160
Bi-invariant metrics 148, 155
Einstein metrics on Lie groups 11, 12', 24, 50, 111, 128, 129, 154, 241
Volume of Lie groups 12
Space of Riemannian structures 23
Table of indices, representations, .. 8, 171
Fundamental fields, invariant fields .. 127, 183
Maurer-Cartan forms, structure equations.. 73, 183
Transformation groups 94, 177

III

RIEMANNIAN GEOMETRY OF HOMOGENEOUS SPACES

3.0 Summary

It is essentially the same to think of a manifold, say S, on which a certain group, say G, acts transitively or to think of some coset space G/H (the difference between the two concepts is analogous to that between an affine and a vector space) . Indeed G/H is a space on which G acts transitively (by left translations) : **a** : [b]⟶ [ab], where [b] = bH; on the other hand, given a transitive action of a group G on a space S we can we can fix a point $s_0 \in S$ (an "origin" of S) and then S (as a G-space) becomes isomorphic to G/H, with H being the stability group of s_0 (i.e. the subgroup of G leaving invariant). If instead of s_0 we choose some other s_0', then we will construct an isomorphism between G/H and G/H', with H' a subgroup conjugated to H. For this reason often, when we talk about "homogeneous space G/H" it must be understood that H needs not be "that particular H" - it can be any conjugated to a given one. The point $s_0 \in S$, which corresponds to coset [e] = **e**H = H of G/H is called the origin of G/H; on the other hand H is determined by the choice of the origin s_0 - it is its stability (or "isotropy", or "little") group.

When G is a Lie group and H is a closed subgroup of H, we can construct two coset spaces : one denoted G/H of left classes [**a**] = **a**H, on which G acts from the left **a** : [**b**] $\rightarrow$ [**ab**], and another denoted H\G of right classes [**ba**]!-[**b**] : **a**. These two spaces are isomorphic (or, better, anti-isomorphic) but they should be considered as different, unless when H is a normal subgroup of G (this happens when for every **a** in G and every **h** in H, $aha^{-1} \in H$); in this case G|H (we will use then a vertical bar) carries a group structure : [**a**] [**b**] : = [**ab**] defines an unambigous group multiplication law for the cosets. In general however multiplication of two elements of G/H is ambigous : we only have the above mentioned action of G on G/H (sometimes called a "non-linear representation" in the physical literature); we will be more precise about it later. Among all possible homogeneous spaces (i.e. all possible pairs H, G with $H \subset G$, H defined up to a conjugacy H $\rightarrow aHa^{-1}$) some are called "irreducible". They play for homogeneous spaces especially for the symmetric ones a similar role as simple Lie groups play for Lie groups; a Lie group is simple if the adjoint representation of G on its Lie algebra Lie(G) is (real) irreducible. Similarly (but observe the difference !) G/H is *irreducible* (or better isotropy irreducible) if the adjoint representation of H on **S**, **S** being defined by a (reductive) decomposition Lie(G) = Lie(H) + **S**, is irreducible. There are cases where the space G/H is symmetric (it may be irreducible or not) ; this happens when there is a discrete automorphism ρ of G such that H is the set of all elements of G which are left invariant by ρ; at the infinitesimal level such spaces are characterized by the condition [**S**, **S**] $\subset$ Lie(H). Symmetric spaces have many nice properties which are important in the theory of σ-models ; from the point of view of the techniques of dimensional reduction, whether G/H is symmetric or not will not really matter.

- Vector fields on G/H

Consider for example the coset space $H\backslash G = \{Ha : a \in G\}$ of right classes along H. The group G acts from the right on this space. If we fix a point $y \in H\backslash G$ then we have a map $G \to G/H$ given by ya ! a. The Jacobian (matrix of partial derivatives) of this map associates a vector $e_i(y)$ to every element T_i of the basis in the Lie algebra of G. This happens for every $y \in H\backslash G$, and so we get n vector fields (n = dim G) $e_i(y)$ on G/H. These are the fundamental vector fields associated to the (right) action of G on $H\backslash G$; they coincide with the projections of the fundamental fields $e_i^R(a)$ associated with the right action of G on itself, under the (derivative of the) projection $G \to H\backslash G$. In particular these fields e_i on $H\backslash G$ satisfy the commutation relations of Lie(G), however they are not linearly independent (they form an overcomplete family). We will see that it is possible to choose a basis made out of a subset of these fundamental fields in a neighborhood of the origin (or, if one prefers, any other point) of $H\backslash G$; the vector fields associated to the right action of H, considered as a subgroup of G, on $H\backslash G$ all vanish at the origin, and the remaining basis vectors are linearly independent and span the tangent space to $H\backslash G$ at [e]. Notice that on G there are also fundamental vector fields of the left action of G on itself (right invariant vector fields). These fields are not left G-invariant, in general they are also not (left) H-invariant - therefore they do not pass naturally to the quotient. The above discussion goes through by replacing $H\backslash G$ by G/H and by permuting the words "right" and "left". We prefer $H\backslash G$ over G/H because the fundamental fields of any right action of G have the commutation relations of the Lie algebra of G itself which, conventionally is defined in terms of left-invariant vector fields.

We could also describe the situation by noticing that G is a right principal H-bundle over G/H but also a left principal H bundle over $H\backslash G$:

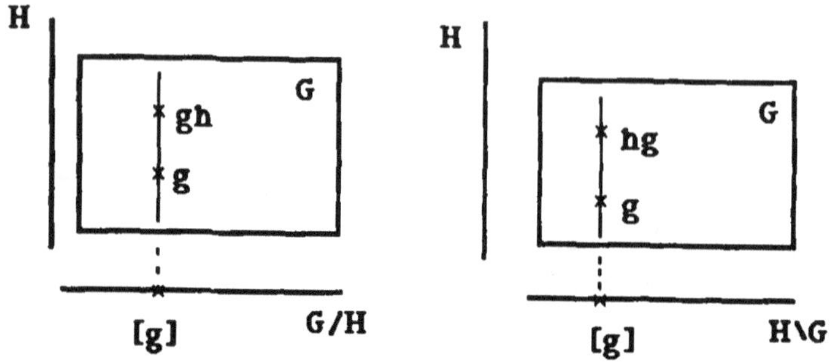

- The normalizer.

We have already mentioned that H is usually not an invariant (= normal) subgroup of G. It is certainly an invariant subgroup of itself. It is then interesting to ask for the biggest sugroup of G of which H is an invariant subgroup. Such a subgroup N exists, is unique, and is called the *normalizer* of H in G. N|H (we can use the vertical bar now) is then a group (in a sense, and intuitively, it is the "biggest group contained in H\G") which will play an extremely important role later on. N|H will appear as an effective gauge group emerging from the process of dimensional reduction. When the homogeneous space H\G is isotropy irreducible, we will see that N|H is a discrete group (it should be observed that N|H differs from the centralizer of H in G by possible U(1) factors).

- Metrics on H\G

There exists a multitude of metrics on a homogeneous space H\G, among them some are maximally symmetric but most of them have no symmetries at all. Of particular interest are those metrics which are G-invariant, and all these may be obtained as projections of (at least) H×G invariant metrics of G (the curvature tensors and curvature scalars of such metrics on H\G are given in Sect. 3.4). Some of these G-invariant metrics result by projecting the *bi-invariant* metrics of G (if G is not simple then it has many bi-invariant metrics). These metrics on H\G are called *normal*, they are particularly "round" - the isometry group of a normal metric contains (N|H)×G. The maximal isometry group of such a metric on H\G can be even bigger than that, this owing to the fact that two different pairs (G,H) and (G',H') may give the same (or isomorphic) homogeneous space (example : SU(3)/ SU(2) = SO(6)/ SO(5) = S^5). The formal "equation" G = H × (H\G) has its precise geometrical meaning: G can be fibrated over (H\G) (the base of the fibration) with H as a typical fibre (and as structure-group); G becomes then a principal bundle. One can write another "identity" : H\G = (N|H) × N\G and, in, fact, H\G itself can be fibrated over N\G with fibre N|H (now H\G, which was the base of the first fibration becomes itself a principal bundle). This second fibration is useful for constructing metrics on H\G which are not perfectly "round" - instead, they are "squashed" a bit. Starting from a normal metric on H\G (which is at least (N|H)×G invariant) one distorts it by introducing a scaling parameter in the direction of N|H; in this way one gets an interesting one-parameter family of (N|H)×G invariant metrics. There exists often a nonstandard value ($\neq 1$) of the squashing parameter for which the obtained metric is also an Einstein metric. At the end of this chapter we will give a discussion of (standard) volumes on homogeneous spaces, similar to the one carried out for groups in the previous section.

A homogeneous space is a manifold together with a transitive action of a Lie group G. Such a manifold can be always realized as a coset space G/H (if the group acts from the left) or H\G (if it acts from the right). A given manifold may have several homogeneous structures (for example SO(6)/ SO(5) and SU(5)/ SU(2) are both diffeomorphic to the sphere S^5). When we study a homogeneous space, we always understand that it is homogeneous for a given action at a given group G, i.e. we have made a specific choice of a pair (G,H).

- The subgroup H will always be a closed subgroup of G in order for G/H to be Hausdorff topological space.

- The action of G on G/H (or H\G) is transitive, therefore all isotropy subgroups are conjugate to H.

- We shall always assume that the action of G on G/H is *effective*, i.e. that the subgroup of G consisting of those elements of G "do nothing" to G/H is trivial or at most discrete. In practice that means that H does not contain a nondiscrete invariant subgroup of G.

- We will always assume that G/H is a reductive homogeneous space; let us recall that

[G/H is reductive] $\Leftrightarrow$ [There exists a (Cartan) decomposition Lie(G) = Lie(H) + $\mathbf{S}$ such that Ad(H) $\mathbf{S} \subset \mathbf{S}$]

$\Rightarrow$ [There exists a decomposition (as above) such that [Lie(H), $\mathbf{S}$] $\subset \mathbf{S}$]

The second property becomes sufficient if H is connected. It is often possible to find several decompositions of Lie(G) satisfying the above conditions; in the following we will always assume that one such decomposition has been chosen. In our applications G will usually be compact and we can therefore define $\mathbf{S}$ in these cases as the orthogonal complement to Lie(H) for some bi-invariant scalar product (i.e. for the Killing metric). In all cases we will have therefore a (real) representation of H (and of Lie(H)) on the vector space $\mathbf{S}$ - this representation will be called the isotropy representation.

- The space $\mathbf{S}$ can be in a natural way identified with the tangent space to G/H at the origin. Indeed, the map which sends each vector v tangent to G/H at [e] to its projection onto $\mathbf{S}$ is a linear isomorphism.

- A homogeneous space for which the connected component of the identity of H acts irreducibly on $\mathbf{S}$ is called an (isotropy) irreducible space. Such spaces may be symmetric or not (see below) and have been completely classified [234], [236].

-When H is the set of fixed elements of G by an involutive automorphism σ of G ($\sigma^2 = 1$) then, σ induces an involutive diffeomorphism on G/H (a "symmetry" with respect to the origin) and G/H is called a symmetric space. At the infinitesimal level, we have

$$\mathrm{Lie}(H) = \{ X \in \mathrm{Lie}(G) \text{ s.t. } X^\sigma = X \}$$

$$\mathbf{s} = \{ X \in \mathrm{Lie}(G) \text{ s.t. } X^\sigma = - X \}$$

Therefore $[\mathbf{s}, \mathbf{s}] \subset \mathrm{Lie}(H)$, a property which characterises those spaces. Symmetric spaces may be irreducible or not, those which are irreducible are completely classified [235]; the isometry classes of irreducible compact simply connected symmetric spaces are given by the following list (we omit the Lie group themselves): $SU(p+q)/(SU(p+q)\cap(U(p)\times U(q))$; $SU(n)/SO(n)$; $SU(2n)/Sp(n)$, $n>1$; $SO(p+q)/(SO(p)\times SO(q))$, $p+q>4$; $SO(2n)/U(n)$, $n>2$; $Sp(p+q)/(Sp(p)\times Sp(q))$, $p+q>2$; $Sp(n)/U(n)$, $n>2$; $G_2/SO(4)$; $F_4/Spin(9)$; $F_4/(Sp(3)\times Sp(1))$; $E_6/(SU(6)\times SU(2))$; $E_6/(SO(10)\times SO(2))$; E_6/F_4; $E_6/(Sp(4)/Z_2)$; $E_7/(SU(8)/Z_2)$; $E_7/(SO(12)\times SU(2))$; $E_7/(E_6\times SO(2))$; $E_8/SO(16)$; $E_8/(E_7\times SU(2))$.

Let us now give a few examples of isotropy irreducible non symmetric spaces : $G_2/SU(3)$ $(=S^6)$; $Spin(7)/G_2$ $(=S^7)$; $F_4/(SO(3)\times G_2)$; $E_8/((SU(3)\times E_6)/Z_3)$; .... .

It is useful to know what groups are transitive on spheres:

$$S^{2n} = SO(2n)/SO(2n-1)$$

$$S^{2n+1} = SO(2n+2)/SO(2n+1) = SU(n+1)/SU(n)$$

$$\begin{aligned} S^{4n+3} &= SO(4n+4)/SO(4n+3) = SU(2n+2)/SU(2n+1) \\ &= U(n+1,\mathbb{H})/U(n,\mathbb{H}) = U(n+1,\mathbb{H})\times SU(2)/U(n,\mathbb{H})\times SU(2) \\ &= U(n+1,\mathbb{H})\times U(1)/U(n,\mathbb{H})\times U(1) \end{aligned}$$

particular cases :

$S^6 = G_2/SU(3)$; $S^7 = Spin(7)/G_2$; $S^{15} = Spin(9)/Spin(7)$

3.1 The example of S^2

S^2, the two-sphere, will be defined as the coset space $U(1)\backslash SU(2)$. Our aim, in this section is only to illustrate the notions and the terminology introduced. Let us call $\{T_1, T_2, T_3\}$ the basis for **Lie(G)** with G = SU(2), call H = U(1) ={ $\exp tT_3$ }, the abelian subgroup generated by T_3, and call **S** the vector space spanned by T_1, T_2. We have **Lie(SU(2)) = Lie(U(1)) + S**. This is a reductive decomposition since we have

$$\exp(tT_3) \begin{pmatrix} T_1 \\ \\ T_2 \end{pmatrix} = \begin{pmatrix} \cos t & -\sin t \\ \\ \sin t & \cos t \end{pmatrix} \begin{pmatrix} T_1 \\ \\ T_2 \end{pmatrix}$$

This decomposition (Cartan decomposition) corresponds to the fact that we can decompose the 3 dimensional space $\mathbb{R}^3$ considered as a vector space centered at the north pole of S^2 as the sum of the 2-dimensional tangent space $\mathbf{S} \cong \mathbb{R}^2$ at the north pole and the one dimensional real line perpendicular to it; the commutation relation associated to this reductive decomposition corresponds to the fact that we have a real representaion of U(1) in the plane **S** (usual rotations in the plane) - Moreover this representation is clearly real irreducible. The relation $[T_1, T_2] = -T_3$ shows that $[\mathbf{S}, \mathbf{S}] \subset$ Lie(H) therefore this homogeneous space is also symmetric.

We have defined fundamental fields $\mathbf{e}_i^R$ generating right translations of SU(2) in sect. 2.2, these fields satisfy $\mathbf{e}_i^R(\mathbf{ab}) = \mathbf{a}\, \mathbf{e}_i^R(\mathbf{b})$ therefore $\mathbf{e}_i^R(\mathbf{ha}) = \mathbf{h}\mathbf{e}_i^R(\mathbf{a})$, $\mathbf{h} \in H$ and one can therefore define fundamental fields on S^2 by

$$\mathbf{e}_i([\mathbf{a}]) := H\, \mathbf{e}_i^R([\mathbf{a}]), \quad [\mathbf{a}] = H\mathbf{a} .$$

These fields obviously generate the right actions of SU(2) on S^2.

(Observe that the element (-1) of SU(2) does not affect the points of S^2 at all, and therefore one can effectively replace SU(2) by SO(3) = SU(2)/Z_2 in this example). At the north pole we have $e_3(\sigma = H) = 0$; e_1 and e_2 will stay linearly independent in an (open) upper hemisphere and calculations can be carried out there.

From the parametrization of points of SU(2) in term of Euler angles given in sect. 2.1 it is clear that points of H\G = S^2 can be parametrized by θ and φ, indeed

$$\bar{p} = \{R_3(\psi)\}_{\psi\in[0,4\pi]} \times R_3(\psi)R_1(\theta)R_3(\varphi) = Hp \in S^2$$

The form $\omega_3{}^L$ vanishes everywhere on S^2: $\omega_3{}^L = d\psi + \cos\theta\, d\varphi = 0$, but the left fundamental fields $e_1{}^L$ and $e_2{}^L$ of SU(2) do not project unambiguously on S^2 because they are not left invariant. Let us illustrate this in local coordinates. Using the relations given at the beginning of 2.1, we get

$$\omega_1{}^R = \cos\varphi\, d\theta - \sin\varphi \cos^2\theta\, d\varphi$$
$$\omega_2{}^R = \sin\varphi\, d\theta + \cos\varphi \cos^2\theta\, d\varphi$$
$$\omega_3{}^R = d\varphi + \cos\theta\, d\psi = d\varphi + \cos\theta\,(-\cos\theta\, d\varphi) = \sin^2\theta\, d\varphi$$

But we cannot find $\omega_1{}^L$ and $\omega_2{}^L$ unless we integrate the equation $d\psi + \cos\theta\, d\varphi = 0$ and chose a solution $\psi = \psi(\theta,\varphi)$ which will depend upon an integration constant: we call "σ" the choice of such a solution and ${}^\sigma\omega_1{}^L$, ${}^\sigma\omega_2{}^L$ the corresponding forms. then,

$${}^\sigma\omega_1{}^L = \cos(\psi(\theta,\varphi))\, d\theta + \sin(\psi(\theta,\varphi)) \sin\theta\, d\varphi$$
$${}^\sigma\omega_2{}^L = \sin(\psi(\theta,\varphi))\, d\theta - \cos(\psi(\theta,\varphi)) \sin\theta\, d\varphi$$
$$\omega_3{}^L = 0$$

In the neighborhood of the north pole, $\theta = 0$ and up to order θ^2, we have

$$\omega_1{}^R \simeq \cos\varphi\, d\theta - \sin\varphi\, d\varphi$$
$$\omega_2{}^R \simeq \sin\varphi\, d\theta + \cos\varphi\, d\varphi$$

$\omega_3{}^R \cong 0$

In order to write an expression for the standard metric on S^2 (the one inherited from the standard bi-invariant metric on SU(2) and going to the quotient), we can either use the canonical base $\omega_1{}^R$, $\omega_2{}^R$ in a neighborhood of the north pole or use the non canonical base $^\sigma\omega_1{}^L$, $^\sigma\omega_2{}^L$ around an arbitrary point. In both cases, we get

$$g = \rho 2 \; (\omega_1{}^{R,L} \otimes \omega_1{}^{R,L} + \omega_2{}^{R,L} \otimes \omega_2{}^{R,L} + \omega_3{}^{R,L} \otimes \omega_3{}^{R,L})$$
$$= \rho 2 \; (d\theta^2 + \sin 2\theta \; d\varphi^2) = \text{" } ds^2 \text{ "}$$

setting $x = \rho \sin\theta \sin\varphi$, $y = \rho \sin\theta \cos\varphi$, $z = \rho \cos\theta$, we recognize the standard metric of the usual sphere $(x^2+y^2+z^2= \rho^2)$ coming from the usual metric $(dx^2+dy^2+dz^2)$ on $\mathbb{R}^3$ via the embedding $S^2 \subset \mathbb{R}^3$.

3.2 The role of the normalizer N of H in G.

We recall that H is a closed subgroup of a compact Lie group G.

.) Definition . The normalizer N of H in G is defined as the biggest subgroup of G in which H is normal. Equivalently

$$N = \{ n \in G : nHn^{-1} \subset H \} = \{ n \in G : nHn^{-1} = H \}^1 .$$

Since H is normal in N the quotient N|H is a group.

.) The normalizer N of H in G should not be confused with the centralizer Z of H in G, which is defined as follows :

$$Z = \{ z \in G : zhz^{-1} = h, \; \forall h \in H \}.$$

It is clear that $C \subset Z \subset N$ (here C = C(G))is the center of G).

For example consider H = SU(2)×U(1) and G = SU(5) ; then, up to discrete factors, C(H) = U(1), Z = SU(3)×U(1), N = SU(3)×SU(2)×U(1)

[1]See Bredon Ch.0.1.9 for the last equality

and $K := N|H = SU(3)$. This example demonstrates at the same time the generic relation between centralizers and normalizers : modulo discrete factors we have

$$N|H \approx Z|C(H).$$

This relation, however, need not be true in a non compact case.

.) We will often use the letter N (resp. Z) to denote the normalizer (resp. centralizer) of H in G. This is an abuse of notation which can sometimes be confusing. "Normalizer" is always the normalizer of "something" (e.g. of H) in "something" (e.g. in G), and to avoid possible confusion one should denote it $N_G(H)$; similarly one should write $Z_G(H)$.

.) $K = N|H$ as the automorphism group of $H\backslash G$. The geometrical structure of a homogeneous $H\backslash G$ is determined by the action of G on it. Thus it is natural to define an *automorphism* of $H\backslash G$ as a map α: $H\backslash G \longrightarrow H\backslash G$ which *commutes* with this action of G

$$\alpha(\mathbf{za}) = \alpha(\mathbf{z})\mathbf{a} \ , \ \mathbf{z} \in H\backslash G, \ \mathbf{a} \in G \ .$$

The set of all automorphisms is a group under composition of maps. This group is isomorphic to $K = N|H$. Indeed, given $\mathbf{n} \in N(H)$ the map α_n given by $\alpha_n([\ \mathbf{a}\ ]) = [\ \mathbf{na}]$ is an automorphism of $H\backslash G$ which depends only on the class $[\ \mathbf{n}\ ]$ of $\mathbf{n}$ modulo H. Conversely, every automorphism α is of the form $\alpha = \alpha_n$ for some $\mathbf{n} \in N$; it is enough to take any $\mathbf{n} \in \alpha\ ([\mathbf{e}])$.

Of course a particular example of a homogenous space $H\backslash G$ is the group G itself. It corresponds to $H = \{\mathbf{e}\}$. But an automorphism of G as a homogeneous space is a different concept from that of an automorphism of G as a group. The first would require $\alpha(\mathbf{a})\mathbf{b} = \alpha(\mathbf{a})\mathbf{b}$ while the second $\alpha(\mathbf{ab}) = \alpha(\mathbf{a})\alpha(\mathbf{b})$. And it is clear that $H = \{\mathbf{e}\}$ implies $N(H) = G$ and thus $K = G$ while the group of all automorphisms of the group G may be different from G (for example $\mathrm{Aut}\ (U(1)) = Z_2$).

.) Let us summarize by saying that, on H\G we have an action of G from the right and an action of **K** = N|H from the left ; these two actions commute so that the group acting on H\G is **K** × G. The action of **K** × G need not be, in general, effective since the central elements of G act rom the left and from the right the same way. As we shall see later the group **K** will appear naturally as a gauge group from G-invariant dimensional reduction scheme. The group **K** × G (divided by the center of G) will appear in a more general scheme discussed in sect. 10.

.) The Lie algebra decomposition.

We recall that H is a closed subgroup of a compact Lie group G. The Lie algebras of H and G are denoted $\mathfrak{H}$ and $\mathfrak{G}$ respectively. For technical reasons we assume H is connected. Then Ad(H) invariance is the same ad($\mathfrak{H}$) invariance. If one meets a case with H consisting of several components, then one can replace H\G by its covering space to reduce the stability group to its connected component of the identity[2]

Usually, given a homogeneous space H\G one considers a *reductive* Lie algebra decomposition $\mathfrak{G} = \mathfrak{H} + \mathbf{S}$, i.e. one chooses a linear subspace $\mathbf{S}$ of $\mathfrak{G}$, complementary to $\mathfrak{H}$, and such that

(1) $\quad \mathrm{Ad}(H)\mathbf{S} \subset \mathbf{S}$

or (owing to the assumed connectedness of H) equivalently

(2) $\quad [\mathfrak{H},\mathbf{S}] \subset \mathbf{S}$

It is rarely realized that there is a finer decomposition which is naturally associated to a homogeneous space. This finer decomposition

[2]See e.g. Lichnerowicz (GGT) &3.1-25.3

comes from the normalizer N = N(H) of H in G. We denote by $\mathcal{N}$ its Lie algebra, and consider the following reductive decomposition (4) , (5).

(4) Let $\mathcal{G} = \mathcal{N} + \mathcal{L}$ with $Ad(N)\mathcal{L} \subset \mathcal{L}$

(5) Similarly, let $\mathcal{N} = \mathcal{H} + \mathcal{K}$, with $Ad(H)\mathcal{K} \subset \mathcal{K}$

For instance $\mathcal{L}$ can be taken as the orthogonal complement of $\mathcal{H}$ with respect to some Ad(G) invariant scalar product on G (e.g. the Killing form) - then automatically $Ad(H)\mathcal{L} \subset \mathcal{L}$, but reductive decompositions exist (in a non-compact, non-semisimple case) even when $\mathcal{G}$ does not admit such a scalar product.

In the particular case when $\mathcal{K}$ is defined to be the orthogonal complement of $\mathcal{H}$ in $\mathcal{N}$ with respect to an invariant scalar product, then $\mathcal{K}$, being an orthogonal complement of an invariant subspace (recall that H is an invariant subgroup of N), is itself invariant, i.e.

(6) $Ad(N)\mathcal{K} \subset \mathcal{K}$

At the infinitesimal level we then have

(7) $[\mathcal{N},\mathcal{K}] \subset \mathcal{K}$

and, in particular

(8) $[\mathcal{K},\mathcal{K}] \subset \mathcal{K}$

This last relation allows us to identify $\mathcal{K}$ with the Lie algebra of K := N|H in a straightforward way. We shall assume it to hold in all our discussion, this will simplify some arguments; in a general case one has first to identify $\mathcal{K}$, as a vector space, with the quotient $\mathcal{N}/\mathcal{H}$, and then equip it with the quotient Lie algebra structure.

We have then

(9) $$\mathcal{G} = \mathcal{H} + \mathcal{S}$$

where

(10) $$\mathcal{S} = \mathcal{K} + \mathcal{L}$$

which is automatically a reductive decomposition

$$Ad(H)\,\mathcal{S} \subset \mathcal{S}.$$

$\mathcal{S}$ can be identified with the space tangent to H\G at the origin [e]; an element $\xi \in \mathcal{S}$ being identified with the vector tangent to the path $t \to [e^{t\xi}]$ at $t=0$.

It follows now that (10) splits $\mathcal{S}$ into Ad(H)-singlets and the rest; we have

(11) $$\begin{aligned}\mathcal{K} &= \{\, \xi \in \mathcal{S} : [\xi,\tau] = 0 \ , \ \forall\ \tau \in \mathcal{H} \,\} \\ &= \{\xi \in \mathcal{S} :\ Ad(h)\,\xi = \xi \ , \ \forall\ h \in H \}.\end{aligned}$$

Indeed if $\xi \in \mathcal{K}$ and $\tau \in \mathcal{H}$ then $[\xi,\tau] \in \mathcal{K}$ because of (5), and $[\xi,\tau] \in \mathcal{H}$ since $\mathcal{H}$ is an ideal in $\mathcal{N}$. Therefore $[\xi,\tau] = 0$. Conversely, suppose $\xi \in \mathcal{S}$ with $[\xi,\tau] = 0$ for all $\tau \in \mathcal{H}$. Writing $\xi = \chi + \lambda$ with $\chi \in \mathcal{K}$ and $\lambda \in \mathcal{L}$ we find $[\lambda,\tau] = 0$. It follows that $\lambda \in \mathcal{N}$ (since $\mathcal{N} = \{ \zeta \in \mathcal{G} : [\zeta,\tau] \in \mathcal{H},\ \forall\ \tau \in \mathcal{H}\}$) and therefore, by (4), $\lambda=0$, thus proving that $\xi \in \mathcal{K}$. The second line of (11) follows owing to the assumed connectedness of H. The property (11) is important, it guarantees that *$\mathcal{K}$ and $\mathcal{L}$ are orthogonal to each other with respect to every* Ad(H) - *invariant scalar product on* $\mathcal{S}$ (here, of course, we make use of the compactness of H)

Corresponding to the decomposition

(12) $$\mathcal{G} = \overbrace{\mathcal{H} + \mathcal{K}}^{\mathcal{N}} + \mathcal{L}, \qquad \underbrace{\mathcal{K} + \mathcal{L}}_{\mathcal{S}}$$

discussed above we shall use the following index conventions : T_i will denote a basis in G adapted to the decomposition ; then we will write

$$\{T_i\} = \{T_{\hat{\alpha}} \in \mathcal{H},\ T_\alpha \in \mathcal{S}\}$$

(13)

$$\{T_\alpha\} = \{T_{\hat{a}} \in \mathcal{K},\ T_a \in \mathcal{L}\}$$

The property (11) then implies that the matrix $\Lambda = Ad(h)$, $h \in H$, of the adjoint representation of H on $\mathcal{S}$ has the block form

$$(14) \qquad (\Lambda^{\alpha}{}_{\beta}) = \begin{pmatrix} \delta^{\hat{a}}{}_{\hat{b}} & 0 \\ 0 & \Lambda^{a}{}_{b} \end{pmatrix}$$

.) A few examples :

A homogeneous space $S = H\backslash G$ is called isotropy irreducible (or sometimes, just irreducible) if the adjoint representation of H on $\mathcal{S}$ is real irreducible. It is clear from the property (11) that in such a case K is necessarily discrete. Indeed, by the irreducibility of $\mathcal{S}$, either $\mathcal{K}$ is trivial or else $\mathcal{L} = \{0\}$. In the latter case $\mathcal{S}=\mathcal{K}$ and, since H acts on $\mathcal{K}$ trivially, we have (at least locally) $G = H \times K$, what would contradict the assumed effectiveness of $H\backslash G$ for a non trivial K. Another class of examples is given by $H\backslash G$ a symmetric homogeneous space. In this case K must be necessarily Abelian; indeed $[\mathcal{S},\mathcal{S}] \subset \mathcal{H}$ implies $[\mathcal{K},\mathcal{K}] \subset \mathcal{H}$ and, by (8) $[\mathcal{K},\mathcal{K}] = 0$.

A few (explicit) examples:

Spheres

$G = SO(n+1)$, $H = SO(n)$, $S = S^n$, $K = Z_2$.

$G = SU(n+1)$, $H = SU(n)$, $S = S^{2n+1}$, $K = U(1)$.

$G = U(n+1, \mathbb{H}), H = U(n, \mathbb{H}), S = S^{4n+3}, K = SU(2)$.

Notice that $Spin(7)/G_2 = S^7$ and $G_2/SU(3) = S^6$ are isotropy irreducible (Lie $K = 0$), $Spin(9)/Spin(7) = S^{15}$ is not isotropy irreducible ($[15] \rightarrow [8]+[7]$) but nevertheless Lie $K = 0$. The above results are obvious when we represent these groups in terms of square matrices, they are also obvious by using well known branching rules for representations (take for example $S^5 = SU(2)\backslash SU(3)$, then $[8] \rightarrow [3]+[(2+\bar{2})] + [1]$ corresponding to Lie G = Lie $H + L +$ Lie K).

Projective spaces

$G = SO(n+1), H = SO(n)\times Z_2, S = RP^n, K = 1$, irreducible.

$G = SU(n+1), H = S(U(1)\times U(n)), S = CP^n, K = 1$, irreducible.

$G = U(n+1, \mathbb{H}), H = U(n, \mathbb{H})\times U(1), S = CP^{2n+1}, K = 1$, not irreducible.

$G = U(n+1, \mathbb{H}), H = U(n, \mathbb{H})\times U(1, \mathbb{H}), S = HP^n, K = 1$, irreducible.

$G = F_4$, $H = Spin(9), S = CaP^2, K = 1$, irreducible.

By using a table of isotropy-irreducible spaces, it is easy to build examples of non irreducible spaces and where K is not trivial : we choose G/N isotropy irreducible and N non simple, we then choose H such that $N = H.K$ and such that the representation of N on G/N is faithful and irreducible (N is then automatically the normalizer of H in G, modulo a discrete group); cf. also tables 1 and 2 in 5.6.3.

$G = SU(p.q)/Z_m, H = SU(p)/Z_p, S = G/H, K = SU(q)/Z_q$, m = l.c.m(p,q).

$G = E_6/Z_3$, $H = G_2$, $S = G/H$, $K = SU(3)/Z_3$.

$G = E_8$, $H = E_6$, $S = G/H$, $K = SU(3)/Z_3$.

3.3 Fundamental fields

We study the space H\G = of right classes along H. The group G acts from the right on H\G ; given y = [a]= **Ha** ∈ H\G we have the map b ∈ G →yb = [ab] ∈ H\G from G to H\G. The derivative (Jacobian) of this map at the identity of G is a linear map from $T_e(G)$ = Lie G onto the tangent space $T_y(H\backslash G)$. Therefore to every vector ξ ∈ Lie G we can associate a vector field $\tilde{\xi}$ on H\G : y ∈ H\G → $\tilde{\xi}(y)$ ∈ $T_y(H\backslash G)$. Explicitly, we have

$$(0) \qquad \tilde{\xi}(y) = d/dt\,(ye^{t\xi})|_{t=0}$$

These vector fields coincide with the projections of right fundamental vector fields (i.e. left-invariant vector fields) on G. Therefore we have

$$(1) \qquad [\,\tilde{\xi},\tilde{\tau}] = [\xi,\tau]^{\sim} \,.$$

Observe that the right fundamental vector fields on G project unambigously on H\G; this because being left-G-invariant they are, a fortiori, left-H-invariant.

Since the (right) action of H on H\G does not move the origin [**e**] of H\G (H is the stability group of this point), it is clear that the fundamental fields $\tilde{\xi}$ corresponding to ξ ∈ Lie(H) vanish at the origin y_0 = [**e**] :

$$\tilde{\xi}(y_0) = 0, \quad \xi \in \mathrm{Lie}(H).$$

Let (T_i) be a basis of the Lie algebra $\mathcal{G}$ of G ; the basis is assumed to be adapted to the decomposition $\mathcal{G} = \mathcal{H} + \mathcal{S}$, $\mathcal{S} = \mathcal{K} + \mathcal{L}$ (cf. 3.2) with the index conventions

$$T_i \in \mathcal{H}, \quad T_\alpha \in \mathcal{S}$$
$$T_{\hat{a}} \in \mathcal{K}, \quad T_a \in \mathcal{L}\,.$$

We denote by $\mathbf{e}_i$ the fundamental vector fields $\tilde{T}_i$ of the right action of G on H\G. We have in **Lie(G)** but also, owing to (3.3-1.)

(2) $\quad [e_i, e_j] = C_{ij}{}^k \, e_k$,

for the corresponding fundamental fields on H\G. The right action of the group on the fundamental vector fields is given by the adjoint representation:

(2a) $\quad e_i(y)a = e_j(ya) \, Ad(a)^j{}_i$

At the origin the vector fields e_i vanish, and e_α form a basis of the tangent space T_{y_0} H\G. Since linear independence of smooth vector fields is an open property (determinant different from zero), the same holds in a certain open neighborhood at the origin. The fields e_α will therefore be used as a basis of vector fields in a neighborhood of y_0. Of course if one moves too far from y_0 (where y_0 is characterized by having H as its stability subgroup), then e_α may cease to be linearly independent. On the other hand notice that e_i will, in general, cease to vanish for y arbitrarily close to the origin (except for special directions generated by the action of N(H)).

All our calculations of connections and curvatures of G/H, and later of bundles with G/H as fibers, will be done using the above moving frame, e_α. Therefore we need some of the properties of the structure functions of this frame, structure functions defined by the commutation relations

(3) $\quad [e_\alpha, e_\beta] \, (y) = f_\alpha{}^\gamma{}_\beta \, (y) \;\; e_\gamma$

Composing (3) with (2) at $y = y_0$ we find, first of all, the relation

(4) $\quad f_\alpha{}^\gamma{}_\beta \, (y_0) = C_{\alpha\beta}{}^\gamma$

(Observe that the same holds if y_0 is replaced by any other point of H\G with H as its stability group). We shall also need the values of the directional derivatives of $f_\alpha{}^\gamma{}_\beta$ in the directions of vector fields e_δ

calculated at y_0. To calculate them we take the commutator of both sides of (3) with e_δ with the result

$$[e_\delta,[e_\alpha,e_\beta]] = (\partial_\delta(f_\alpha{}^\gamma{}_\beta))\, e_\gamma + f_\alpha{}^\gamma{}_\beta\, [e_\delta,e_\gamma]$$

while on the other hand, from (2)

$$[e_\delta,[e_\alpha,e_\beta]] = C_{\alpha\beta}{}^i\, C_{\delta i}{}^j\, e_j\,.$$

Comparing now the values of both expressions at y_0, and taking into account 4, we get

$$\partial_\delta(f_\alpha{}^\gamma{}_\beta\,(y_0)) = C_{\alpha\beta}{}^i\, C_{\delta i}{}^\gamma - C_{\alpha\beta}{}^\chi\, C_{\delta\chi}{}^\gamma$$

or

(5) $$\partial_\delta(f_\alpha{}^\gamma{}_\beta\,(y_0)) = C_{\alpha\beta}{}^{\hat{i}}\, C_{\delta i}{}^\gamma\,.$$

<u>Remark</u> : We have seen that the fundamental fields e_i on $H\backslash G$ are obtained from the fundamental fields $e_i{}^R$ on G by projection; this projection is nothing but the derivative of the projection π:G⟶H\G sending $\mathbf{a} \in G$ into $[\mathbf{a}] \in H\backslash G$. It is not possible to project this way the fundamental fields $e_i{}^L$ (unless we replace H\G with G/H), since $e_i{}^L(\mathbf{ha})$ needs not be equal to $\mathbf{h}e_i{}^L(\mathbf{a})$. However, if we choose a "local embbeding" σ:H\G⟶G i.e. a local cross-section of the principal bundle π:G⟶H\G, then we define the fields ${}^\sigma e_\alpha(y)$ as projections of $e_\alpha{}^L(\sigma(y))$. These fields define also a moving frame in a neighborhood of y although they are not G-invariant (H\G would then be parallelizable, which is rarely the case) they transform under G-action by the adjoint representation of H. To see this property observe first that a section σ:H\G⟶G determines a mapping $(y,\mathbf{a}) \longrightarrow \chi(y,\mathbf{a})$ from H\G×G to H by the relation

(6) $\sigma(y)\mathbf{a} = \chi(y,\mathbf{a})\, \sigma(y\mathbf{a})$

with

(7) $\chi(y,a) \in H$

($\chi(y,a)$ satifies then the groupoid character relation $\chi(y,ab) = \chi(y,a)\,\chi(ya,b)$)

From this we find, for every $X \in$ Lie G, and $a \in G$

$$[e^{tX}\,\sigma(y)]\,a = [e^{tX'}\,\sigma(ya)]$$

where

$$X' = Ad(\chi(ya))^{-1}\;X$$

Specializing to $X = T_\alpha \in S$, we find

$$(8) \qquad {}^{\sigma}e_\alpha(y)a = (Ad(\chi(ya))^{-1})^{\beta}{}_{\alpha}\;{}^{\sigma}e_\beta(y).$$

3.4 <u>G-invariant metrics on H\G</u>

A metric on G will pass to the quotient H\G if it is left-invariant under H. The resulting metric on H\G is not that easy to describe as it requires a kind of a Kaluza-Klein mechanism (a "horizontal lift") to be discussed later. It is much easier to describe the resulting inverse metric. This is because the inverse metric is a *contravariant* tensor, and contravariant tensors are "pushed forward" in the direction of the map (π:G→H\G in our case) contrary to the covariant objects (like e.g. differential forms) which are "pulled back" against the direction of the map. Therefore, if g is a metric on G which is left-H-invariant, we first take its inverse g^{-1}, then project it by the projection π (more precisely, by the derivative map π_* acting on vectors and contravariant tensors), and finally take an inverse to get a (covariant) metric tensor $\hat{g}$ on H\G :

$$(1) \quad g \rightarrow \hat{g} := (\pi_* g^{-1})^{-1}$$

(There is one subtle thing beyond the above notation: it is that $\pi_* \mathbf{g}^{-1}$ gives an unambiguous *tensor field* only owing to the property that $\mathbf{g}^{-1}$ is H-invariant; c.f. the analogous problems of "φ-related vector fields", e.g. [127]). Observe that a metric on H\G obtained this way will have, in general, no invariance properties left. It may be also remarked that *every* metric on H\G can be obtained by the above method.

Among metrics on H\G there will be metrics obtained from *bi-invariant* metrics on G by passing to the quotient formula (1). These are called normal metrics on H\G. When G is simple there is only one normal metric up to scale; its isometry group will then at least be G×(N|H) (modulo central elements). To fix the scale we shall use the Killing metric of G. Notice that the notion of normal metric is relative to a pair (G,H), not to the coset manifold H\G; for instance the normal metric on SU(3)\SU(4) is not even proportional to the normal metric on SO(7)\SO(8) although both spaces are the seven sphere S^7. Assuming G simple and the normal metric on H\G obtained from the Killing metric on G, we will find later that the scalar curvature of this metric has a very simple expression when H\G is a symmetric space: $R = 1/2 \dim(H\backslash G)$.

We will be mainly interested in those metrics on H\G which are at least G-invariant. Such a metric is known as soon as its value at the origin (or any other point) is known (it is "dimensionally reducible" to a point). Now, while the origin itself does not move under the action of H, the curves passing though it will in general rotate when acted upon by the elements of H. Therefore H induces rotations of the tangent space $T_0(H\backslash G)$ at the origin. Writing down a reductive decomposition Lie G = Lie H + $\mathbf{s}$, and identifying $\mathbf{s}$ with $T_0(H\backslash G)$, we identify the above rotations of $T_0(H\backslash G)$ by H with the adjoint action of H on $\mathbf{s}$. If $\mathbf{g}$ is a G-invariant metric on H\G, then the scalar product that it defines on $T_0(H\backslash G)$ must be invariant under H-rotations. Therefore $\mathbf{g}$ defines an **Ad(H)**-invariant scalar product on $\mathbf{s}$, and using transitivity of the G-action, we can propagate it all over the space H\G to a unique G-invariant metric which extends ? [127]. So there is a natural one-to-one correspondence between G-invariant Riemannian metrics on the coset space H\G and Ad(H) - invariant scalar products on the vector space $\mathbf{S}$. Let us see now an example which illustrate the above property of G-invariant metrics. We choose a basis e_i^L of left-fundamental vector fields for G, orthonormal with respect to a biinvariant metric g of G. Then

$$h^{-1} := \underbrace{\delta^{ij}\, e_i^L \otimes e_j^L}_{\mathcal{H}} + \underbrace{h^{\hat{a}\hat{b}}\, e_i^L \otimes e_j^L}_{\mathcal{K}} + \underbrace{\delta^{ab}\, e_i^L \otimes e_i^L}_{\mathcal{L}}$$

where $h^{\hat{a}\hat{b}}$, a k×k (k= dim $\mathcal{K}$) matrix of constant real coefficients, is an inverse metric on G which is $H^{Left} \times G^{Right}$ - invariant (its Lie derivatives with respect to e_i^L and e_i^R vanish). In order to write down a simple formula for the projection of h^{-1} onto H\G we must choose a local cross-section σ: H\G→G. Then, since $\pi_*[e_\alpha^L(\sigma(y))] = {}^\sigma e_\alpha(y)$ and $\pi_*[e_i^L(\sigma(y))] = 0$, we find for the inverse of the projected metric the expression

$$(2) \quad \hat{h}^{-1} = h^{\hat{a}\hat{b}}\ {}^\sigma e_{\hat{a}} \otimes {}^\sigma e_{\hat{b}} + \lambda \delta^{ab}\ {}^\sigma e_a \otimes {}^\sigma e_b \ .$$

Since the projected vector fields ${}^\sigma e_\alpha$ are not G-invariant, it is not at all obvious from the above formula that $\hat{h}$ is G-invariant. This follows however from the following simple observation : although the fields ${}^\sigma e_\alpha$ are not G-invariant, they transform, under the action of G, by the adjoint representation of H (cf. 3.3-8). The first term of (2) is then invariant owing to the fact that H acts trivially on $\mathcal{K}$ (cf. 3.2 - 12), while the second term is invariant owing to the fact that it is proportional to the restriction to $\mathcal{L}$ of a biinvariant metric g. For a similar reason the metric (2) does not depend on a choice of section σ: H\G→G; indeed any two such choices are related by an H-*valued* "gauge" transformation : $\sigma'(y) = \sigma(y)\lambda(y)$, $\lambda(y) \in H$ so that the bases ${}^\sigma e_\alpha$ and ${}^{\sigma'} e_\alpha$ are connected by an (y-dependent) Ad(H) transformation.

The formula (2) gives the most general G-invariant metric on H\G in the case when $\mathcal{L}$ is irreducible under H (thus there is a 1 + k(k+1)/2 family of such metrics in this case). If $\mathcal{L}$ is not irreducible then we can write $\mathcal{L} = \mathcal{L}_1+\mathcal{L}_2+..+\mathcal{L}_p$, with $\mathcal{L}_1,...,\mathcal{L}_p$ carrying (real) irreducible representations of $\mathcal{H}$. It is then clear that a metric

$$\hat{h}^{-1} = h^{\hat{a}\hat{b}}\ \sigma_{e_{\hat{a}}}\otimes\sigma_{e_{\hat{b}}} + \lambda_1\delta^{a_1b_1}\ \underbrace{\sigma_{e_{a_1}}\otimes\sigma_{e_{b_1}}}_{\mathcal{L}_1} +..+ \lambda_p\delta^{a_pb_p}\ \underbrace{\sigma_{e_{a_p}}\otimes\sigma_{e_{b_p}}}_{\mathcal{L}_p}$$

is again G-invariant. Even more generally, if $\mathcal{L}_1$ and $\mathcal{L}_2$ carry two copies of the same irreducible representation of H, then the term $\lambda_{12}\delta^{a_1b_2}\ \sigma_{e_{a_1}}\otimes\sigma_{e_{b_2}}$ is again invariant. The dimension d of the manifold of G-invariant metrics on H\G is therefore equal to

$$d = \Sigma_{i=0..p}\ r_i(r_i + 1)/2$$

where $r_0 = k = \dim \mathcal{K}$ and where r_i are the multiplicities of nontrivial (non-singlets) representations of H occuring on $\mathcal{L}$. Notice that when $\mathcal{S}$ is irreducible (we say then that H\G is isotropy irreducible) then d = 1 i.e. up to a scale there is only one G-invariant metric on H\G. As we shall see later this number d is also the number of scalar fields which arise from the process of dimensional reduction of Einstein's gravitation.

Let us give a few examples:

SO(7)\SO(8) = S^7 , d=1 since the space is irreducible;

SU(3)\SU(4) = S^7 , then the reduction of Ad G with respect to H.K is

$[15]=[8\otimes1]+[1\otimes1]+[3\otimes1+\bar{3}\otimes\bar{1}]$

therefore $\mathcal{S}$ = $\underbrace{[1]}_{\mathcal{K}}$ + $\underbrace{[6]}_{\mathcal{L}}$ and d = 1+1 = 2.

U(2,$\mathbb{H}$)\U(1,$\mathbb{H}$) = S^7 , then, in the same way,

$[10]=[3\otimes1]+[1\otimes3]+[2\otimes2]$

$$\text{therefore } \mathbf{S} = \underbrace{1+1+1}_{K} + \underbrace{[4]}_{L} \quad \text{and} \quad d = (3\times 4)/2 + 1 = 7.$$

We have expressed a G-invariant metric on H\G in terms of the locally defined section-dependent vector fields ${}^{\sigma}\mathbf{e}_{\alpha}$ which were images by projection of right-invariant vector fields $\mathbf{e}_{\alpha}{}^{L}$ on G. We can also express such a metric in terms of the fundamental fields $\mathbf{e}_{\alpha}$ which are projections of left-invariant vector fields $\mathbf{e}_{\alpha}{}^{R}$ on G. Let $\mathbf{g}$ now denote such a metric. Then, since we work in a moving frame of fundamental fields, the metric components in this frame will be y-dependent. Explicitly, if $\mathbf{g}_{\alpha\beta}(y)$ are defined by the formula

$$(3) \qquad \mathbf{g}_{\alpha\beta}(\mathbf{y}) = \mathbf{g}(\mathbf{e}_{\alpha}(\mathbf{y}),\mathbf{e}_{\beta}(\mathbf{y}))$$

then $\mathbf{g}_{\alpha\beta}(\mathbf{y})$ propagate along H\G according to

$$(4) \quad \partial_{\alpha}\mathbf{g}_{\beta\gamma} = \mathbf{f}_{\alpha\gamma\beta} + \mathbf{f}_{\alpha\beta\gamma}$$

where

$$(5) \qquad \mathbf{f}_{\alpha\beta\gamma}(y) = \mathbf{g}_{\beta\delta}(y)\ \mathbf{f}_{\alpha}{}^{\delta}{}_{\gamma}(y)$$

and the structure functions $\mathbf{f}_{\alpha}{}^{\delta}{}_{\gamma}(y)$ are defined (in a neighborhood of the origin) by 3.3-3. knowing also the derivatives of the structure functions as derived in 3.3-5 one can calculate the curvature tensor $R_{\alpha\beta\gamma\delta}$ of the metric. When calculated at the origin the result reads (see [104])

$$
\begin{aligned}
R_{\alpha\beta\delta\gamma} = & -1/4\ \{C_{\alpha\beta}{}^{i}\, C_{i\gamma,\delta} - C_{\alpha\beta}{}^{i}\, C_{i\delta,\gamma} + C_{\gamma\delta}{}^{i}\, C_{i\alpha,\beta} - C_{\gamma\delta}{}^{i}\, C_{i\beta,\alpha}\ \} + \\
& +1/4\ \{C_{\alpha\gamma}{}^{\chi}\, C_{\beta\delta,\chi} + 2\, C_{\alpha\beta}{}^{\chi}\, C_{\gamma\delta,\chi} - C_{\beta\gamma}{}^{\chi}\, C_{\alpha\delta,\chi}\ \} + \\
(6) \qquad & +1/4\ \mathbf{g}^{\chi\lambda}\{(C_{\chi\alpha,\lambda} + C_{\chi\gamma,\alpha})(C_{\lambda\beta,\delta} + C_{\lambda\delta,\beta})\} \\
& -1/4\ \mathbf{g}^{\chi\lambda}\{(C_{\chi\beta,\gamma} + C_{\chi\gamma,\beta})(C_{\lambda\alpha,\delta} + C_{\lambda\delta,\alpha})\}
\end{aligned}
$$

For the Ricci tensor $R_{\beta\gamma} = \mathbf{g}^{\alpha\delta}\, R_{\alpha\beta\delta\gamma}$ one then gets

$$R_{\beta\gamma} = + 1/4\, C_{\alpha\chi,\beta}\, C_{\alpha\chi,\gamma} - 1/2\, C_{\beta\alpha,\chi}\, C_{\gamma\alpha,\chi} - 1/2\, C_{\beta\alpha,\chi}\, C_{\gamma\chi,\alpha}$$

(7) $$- 1/2\, C_{\alpha\beta}{}^{i}\, C_{\gamma i}{}^{\alpha} - 1/2\, C_{\gamma\alpha}{}^{i}\, C_{\beta i}{}^{\alpha} +$$

$$-1/2\, (C_{\chi\beta,\gamma} + C_{\chi\gamma,\beta})\, C_{\chi\alpha}{}^{\alpha}$$

where we use the convention that the summation over repeated one the same level is performed with $g^{\alpha\beta}$. Thus, for example

$$C_{\alpha\chi,\beta}\, C_{\alpha\chi,\gamma} = g^{\alpha\alpha'}\, g^{\chi\chi'}\, g_{\beta\beta'}\, g_{\gamma\gamma'}\, C_{\alpha\chi}{}^{\beta'}\, C_{\alpha'\chi'}{}^{\gamma'}$$

For the scalar curvature $R = g^{\beta\gamma} R_{\beta\gamma}$ we obtain

(8) $R = -1/4\, C_{\alpha\beta,\gamma}\, C_{\alpha\beta,\gamma} - 1/2\, C_{\alpha\beta,\gamma}\, C_{\alpha\gamma,\beta} - C_{\beta\alpha}{}^{i}\, C_{\beta i}{}^{\alpha} - C_{\chi\alpha}{}^{\alpha}$
$C_{\chi\beta}{}^{\beta}$

Comments

a) The last terms of the formula (7) and (8) are included for the sake of completeness only. For a compact group, and more generally, for a unimodular group, we have (see e.g. [51, Ch.19,16 Prob.14]) $C_{\alpha i}{}^{i} = 0$, and therefore (since $C_{\alpha\chi}{}^{\chi} = 0$ by reductiveness of $\mathfrak{G} = \mathfrak{H} + \mathfrak{S}$) the last terms in (7) and (8) vanish. In the following these terms will be omitted.

b) H\G is called naturally reductive homogeneous space if $C_{\alpha\beta,\gamma}$ are antisymmetric also in the last two indices. Then $\nabla_X Y = 1/2\, [X,Y]$ and

$$R_{\beta\gamma} = + 1/4\, C_{\beta\chi}{}^{\alpha}\, C_{\gamma\alpha}{}^{\chi} + 1/2\, k_{\beta\gamma}$$

$$R = + 1/4\, C_{\beta\chi}{}^{\alpha}\, C_{\beta\alpha}{}^{\chi} + 1/2\, g^{\beta\gamma}\, k_{\beta\gamma}$$

where $k_{\beta\gamma} = - C_{il}^{m} C_{jm}^{l}$ is the Killing metric (non-negative, minus the Killing form !)

c) In particular if H\G is a symmetric space (i.e. ; $C_{\alpha\beta,\gamma} = 0$) , then

$R_{\beta\gamma} = 1/2\ k_{\beta\gamma}$

$R \ = 1/2\ g^{\beta\gamma} k_{\beta\gamma}$

d) H\G is called a normal space if $g_{\alpha\beta}$ is a restriction to S of a biinvariant metric g_{ij} on G. Clearly d) implies b). Conversely, if G is connected then b) implies d) but g_{ij} may be, in general, semidefinite. The subparticular cases are

d_1) H = {e}, (the group case) and $g_{\alpha\beta} = k_{\alpha\beta}$. Then

$R_{\beta\gamma} = 1/4\ k_{\beta\gamma}$

$R \ = 1/4 \dim G$

d2) H\G is a symmetric homogeneous space and $g_{\alpha\beta} = k_{\alpha\beta}$. Then

$R_{\beta\gamma} = 1/2\ k_{\beta\gamma}$

$R \ = 1/2 \dim H\backslash G$

e) The Ricci tensor $R_{\beta\gamma}$ considered as a symmetric bilinear form on S is also $Ad(H)$-invariant. If H\G is *isotropy irreducible* i.e. if $Ad(H)$ acts irreducibly on S, then, by the Shur's Lemma $R_{\beta}{}^{\alpha} := g^{\alpha\gamma}R_{\beta\gamma}$, must be a multiple of $\delta_{\beta}{}^{\alpha}$ and therefore $R_{\beta\gamma} = \lambda\ g_{\beta\gamma}$. Thus H\G is an *Einstein* space.

f) We considered the space of *right* cosets $H\backslash G$. For a *left* coset space G/H the fundamental vector fields satisfy $[\tilde{\xi},\tilde{\tau}] = -[\xi,\tau]^{\sim}$ with the effect that everywhere in our formulas $C_{ij}{}^k$ is to be replaced by $-C_{ij}{}^k$. Such a change has no effect on the curvature formulae which are all quadratic in structure constants.

g) The formulae 6, 7, 8 hold at the origin $y_0 \in H\backslash G$ and, more generally, at any other point $y \in H\backslash G$ for which the isotropy group G_y is H. To go to an arbitrary point $[a]$ one has to transform the indices by the adjoint representation $Ad(a)$ (and then reexpress the vector fields $e_i(y)$ in terms of the basis $e_\alpha(y)$).

h) It is sometimes useful to express the Ricci tensor Ric, with $(Ric)^\alpha{}_\beta = R^\alpha{}_\beta$ in terms of the following 2×2 matrices: F, T, S, M, N where,

$F^\alpha{}_\beta = C_{\gamma\alpha}{}^\delta C_{\beta\delta}{}^\gamma$ $\quad$ $T^\alpha{}_\beta =$ $\quad$ $S^\alpha{}_\beta =$

$M^\alpha{}_\beta = -\, g^{\alpha\gamma} C_{\gamma\delta}{}^i C_{i\delta'}{}^{\beta'} g^{\delta\delta'} g_{\beta\beta'}$, $\quad$ $N^\alpha{}_\beta = -\, C_{\delta i}{}^a C_{\delta'\beta}{}^i g^{\delta\delta'}$,

then

$$Ric = 1/2\,(F - T + S/2 + M + N)$$

Assume now (till the end of this subsection) that G simple and $H\backslash G$ is endowed with a normal metric g (i.e. inherited from a biinvariant metric g^G on G), then

$$M = N;\ F = T = S \text{ and } Ric = F/4 + M.$$

We can also write the metric itself as $g = F + 2M$, indeed, using an orthonormal frame on G and full antisymmetry of the $C_{ij,k}$, we have

$$g_{\alpha\beta} = -\, C_{\alpha i}{}^j C_{\beta j}{}^i = +\, C_{\gamma\alpha}{}^\delta C_{\beta\delta}{}^\gamma + 2\, C_{\delta\alpha,i} C_{\beta i,\delta}$$

Therefore we can also write $Ric = 1/4\, g + M/2$; the trace of this last expression will give us a nice expression for the scalar curvature $R = tr\, Ric$. Notice first that $tr\, g = \dim G/H = \dim G - \dim H = n-h$, and that $tr\, M = h - c.h$ where c is the index of H in G (cf. Ch. 2), indeed, using again full antisymmetry of the $C_{ij,k}$, we have

$$-\, C_{\hat{m}i}{}^j C_{\hat{n}j}{}^i = -\, C_{\hat{m}\hat{o}}{}^{\hat{p}} C_{\hat{n}\hat{p}}{}^{\hat{o}} - C_{\delta\alpha\hat{m}} C_{\beta\hat{m}\delta}$$, which can be read

$$g^G_{\hat{m}\hat{n}} = g^H_{\hat{m}\hat{n}} + M_{\hat{m}\hat{n}},$$ i.e. using the definition of c,

$$g^G_{\hat{m}\hat{n}} = c.\, g^G_{\hat{m}\hat{n}} + M_{\hat{m}\hat{n}},$$ whose trace gives

$$h = c.h + \mathrm{tr}\, M.$$

Using these two results for **tr g** and **tr M**, we get the following expression for the scalar curvature R

(8') $\quad R = n/4 + h/4 - c.h/2.$

This last expression could be gotten directly by writing G as a H bundle over H\G and using the formula for dimensionnal reduction (Sec 4.5.2); we would actually get

$R^G = R^{G/H} + R^H - 1/4\, F^2$, i.e., in this case

$n/4 = R + ch/4 - 1/4\, .\, h\,(1-c)$, hence (8').

3.5 Invariant metrics on H\G related to the fibration of H\G.

As we already mentioned in Sect. 3.0, H\G is always a left N|H principal bundle over N\G -and G/H is a right N|H principal bundle over G/N ([28], [165]).

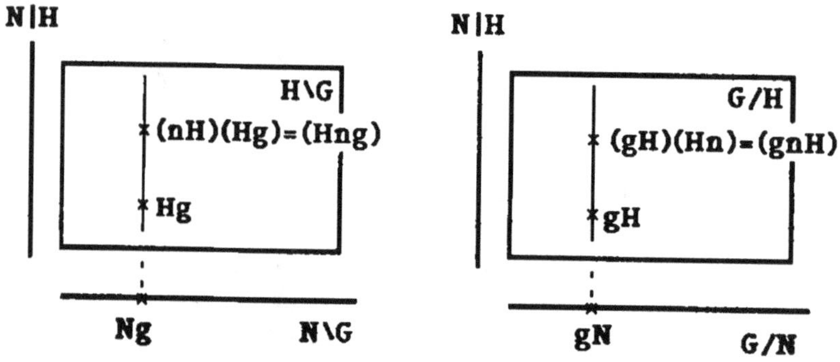

Remark: ∀n∈N, nH=Hn

The structure group N|H acts on the bundle space H\G from the left, while the group G acts on H\G from the right by bundle automorphisms. Among G-invariant metrics on H\G there is an interesting class of metrics which are also N|H-invariant. These metrics are dimensionally reducible to N\G and project onto (G-invariant) metrics on N\G. We already know that a G-invariant metric on H\G is uniquely determined by a matrix ($\mathbf{h}_{\alpha\beta}$) which gives the scalar product in $\mathbf{S}$-the space tangent to H\G at the origin. The only condition on $\mathbf{h}_{\alpha\beta}$ is that it is **Ad(H)**-invariant. One can ask then which **Ad(H)** matrices $\mathbf{h}_{\alpha\beta}$ determine G-invariant metrics on H\G which are also N|H invariant? The answer is that $\mathbf{h}_{\alpha\beta}$ must be, in addition, also **Ad(N)**-invariant. Notice that by eqs 3.2-4,6 we have not only $\mathbf{Ad(N)S} \subset \mathbf{S}$, but also separately $\mathbf{Ad(N)}\mathcal{K} \subset \mathcal{K}$ and $\mathbf{Ad(N)}\mathcal{L} \subset \mathcal{L}$. One should also notice that $\mathcal{K}$ and $\mathcal{L}$ are orthogonal with respect to any **ad(H)**-invariant scalar product on $\mathbf{S}$. This follows from the fact that representations of H on $\mathcal{K}$ and $\mathcal{L}$ are disjoint : $\mathcal{K}$ consists of **Ad(H)**-singlets while there are no **Ad(H)**-singlets in $\mathcal{L}$.

Let $\mathbf{g}$ be a normal metric on H\G. It is defined by the projection of a biinvariant metric from G, and it is given by the restriction to $\mathbf{S}$ of an **Ad(G)**-invariant scalar product on **Lie** G. Since $\mathcal{K}$ and $\mathcal{L}$ are orthogonal, we may write

$$(9)\quad \mathbf{g} = \mathbf{g}\{\mathrm{proj}.\mathcal{L}\} + \mathbf{g}\{\mathrm{proj}.\mathcal{K}\}$$

each term being **Ad(N)**-invariant separately. Therefore the following one-parameter family

$$(10)\quad \mathbf{g}^t = \mathbf{g}\{\mathrm{proj}.\mathcal{L}\} + t^2\,\mathbf{g}\{\mathrm{proj}.\mathcal{K}\}$$

consists of K × G metrics on H\G. Using the examples given in Sec.3.2 we obtain for instance U(2,$\mathbb{H}$)×SU(2) invariant metrics on S^7 or $E_8 \times SU(3)/Z_3$ invariant metrics on E_8/E_6 (here, for example, we would write E_8/E_6 as a $SU(3)/Z_3$ bundle over the irreducible space

$E_8/((SU(3)\times E_6)/Z_3)$. Although we could use the formulae 3.4-6-8 to calculate the curvatures of the above metrics, it is more convenient to use the dimensional reduction formula of Ch.4 (see also the remark h) at the end of 3.4) applied now to the case where the base space space is N\G, and the internal space is a group K=N|H, taking also into account the fact that the induced Yang-Mills connection is G-invariant. It will be shown in Sect. 5.9.1 that the scalar curvature of this squashed metric (10) reads

$$R_S(t) = R_L + 1/t^2 \; R_K - t^2/4 \; F^2$$

where R_L and R_K are the curvatures of L : = N\G and K=N|H endowed with the metrics given by corresponding terms of (10) (for t = 1) while $F^2 = g^{aa'} g^{bb'} g_{\hat{c}\hat{c}'} C_{ab}{}^{\hat{c}} C_{a'b'}{}^{\hat{c}'}$. Taking, in particular, for g the normal metric of S = H\G and assuming, for simplicity L symmetric and K simple, we find R_L = (dim L)/2, R_K = c (dim K)/4 and F^2 = (dim K)(1-c), where c = c(K:G) is the index of K in G (cf Ch.2). We get

$$R_S(t) = (\dim L)/2 + c\,(\dim K)/(4\,t^2\,) + (\dim K)(c-1)t^2/4$$

(notice that from its very definition the index refers to the embedding of the Lie algebras rather than groups).

3.6 - Einstein metrics on homogeneous spaces

From the calculations given at the end of sect. 3.4, we see that, if G is simple, the normal metric on a symmetric space H\G is Einstein; this is also true if H\G is isotropy irreducible (symmetric or not). If H\G is not irreducible, the normal metric is usually not an Einstein metric, however, by considering the one parameter family of metrics (sect.3.5) obtained from the normal metric by "squashing" it in the

N|H direction, it is usually possible to find values of the parameter for which the metric is Einstein .

When t varies, the volume of the homogeneous space S varies; to keep it fixed, we consider the family of metrics

$\bar{h}(t) = (1/t^2)^{k/s}\, h(t)$, with $k = \dim K$, $s = \dim S = k+L$ and $L = \dim L$.

Then the scalar curvature is

$$R_S(t) = t^{2k/s}\, [\, L/2 + c\, k/(4\, t^2\,) + k(c-1)t^2/4\,]$$

We now vary this expression with respect to t and find

$$dR_S(t)/dt = 1/2.k(k/s+1)(c-1)\, t^{2k/s-3}\, [t^2-\alpha_+]\,[t^2-\alpha_-]$$

with

$$\alpha_\pm = \{-1 \pm [1+c(2k/L\, +1)(c-1)]^{1/2}\}/\{(2k/L\, +1)(c-1)\}$$

which gives us two Einstein metrics for the values α_+ and α_- of the parameter t^2.

For example, for $G = U(2,\mathbb{H})$, $H = U(1,\mathbb{H})$, $H\backslash G = S^7$, we have $N|H=SU(2)$, then $\dim K = 3$, $\dim L = 4$, we find $c = 2/3$ in the tables [!!]. Therefore

$$dR/dt = -5/7\, .\, t^{-15/7}\, .\, (t^2-2)(t^2-2/5)$$

These two Einstein metrics obtained for $t^2 = 2$ and $t^2 = 2/5$ are at least $U(2,\mathbb{H}) \times SU(2)$ invariant; actually the one corresponding to $t^2 = 2$ is the usual $SO(8)$-invariant metric on S^7. Notice that $t^2 = 1$ (the normal metric) is not Einstein (which should not be surprising since $H\backslash G$ is not irreducible).

Let us see what is the common volume of this one parameter family of metrics $\bar{h}(t)$ on S^7. The $t^2=2$ sphere corresponds to the usual sphere $\Sigma x_i^2=\rho^2$, (ρ is to be determined), its volume is therefore $2\pi^4/3!$

ρ^7 and its scalar curvature (computed by the above formula for **R** with $t^2 = 2$) is $= 2^{3/7}.7/4$; but we know also that the scalar curvature of a standard sphere S^n is $n(n+1)/\rho^2$, i.e. $42/\rho^2$ in our case. Therefore $42/\rho^2 = 2^{3/7}.7/4$ and $\rho = \sqrt{3}\ 2^{9/7}$. The volume of the sphere (with $t^2 = 2$) is therefore $2\pi^4/3! \times 2^9 3^{7/2}$ and all the other spheres of the $\overline{\mathbf{h}}(t)$ family have the same volume by construction.

It is also sometimes possible to obtain new Einstein metrics by distorting a normal metric in a direction parallel to a *non principal fibration* (for some precise values of the parameter t^2). Let us just mention (as an example) that besides the usual Einstein metric on S^{15} - which is $SO(16)$ invariant - and besides the non standard Einstein metric on S^{15} which is $U(4,\mathbb{H}) \times SU(2)$ invariant (and is obtained as in the above example of S^7 by writing S^{15} as a $SU(2)$ principal bundle over HP^3 and "squashing" appropriately), there is a last homogeneous Einstein metric on S^{15} which is obtained for some value of t^2 by writing S^{15} as a *non principal* S^7 bundle over S^8 and "squashing" in the S^7 direction; this last metric is **Spin**(9) invariant. [242]

Let us mention another non principal fibration:

The complex projective space $CP^{2n+1} = U(n+1,\mathbb{H})/(U(n,\mathbb{H}) \times U(1))$ is a S^2 bundle over $\mathbb{H}P^n = U(n+1,\mathbb{H})/(U(n,\mathbb{H}) \times SU(2))$. Here again, by squashing the normal metric on CP^{2n+1} in the direction of the fiber S^2, one gets new metrics which are $U(n+1,\mathbb{H})$ invariant. Notice that $U(n+1,\mathbb{H})/U(n,\mathbb{H}) \times U(1)$ is <u>not</u> an irreducible space (indeed $[4n+2] = [4n] + [2]$) although $SU(2n+2)/S(U(2n+1) \times U(1))$, also defining CP^{2n+1} is an irreducible space.

Let us conclude this section by mentionning that there exists also usually <u>many</u> Einstein metrics on a homogeneous space S which are not homogeneous (they are not invariant under a group transitive on S)

3.7 Dimensionally reducible metrics

All metrics on homogeneous spaces discussed so far are "dimensionally reducible to a point" in the sense that they are completely determined by their expression at the origin of H\G, however they are not the only ones to share this property.

3.8 Pointers to the litterature

Differential geometry on homogeneous spaces (general) 127
Classification 89, 225, 234, 235, 236
Curvatures for invariant metrics 9, 10, 25
Einstein metrics on homogeneous spaces 12', 241, 242

IV

RIEMANNIAN GEOMETRY OF A (RIGHT) PRINCIPAL BUNDLE

4.0 Summary section

In this chapter, we consider spaces P which can be written locally as M×G, M being some m-dimensional manifold (to be a model of space-time, for example) and G being isomorphic with a compact Lie group (also called G). We assume the product M×G to be only "local", for example a three-sphere P can be written locally as a product of a two-sphere M and a circle G, but clearly S^3 is not equal to $S^2 \times U(1)$. Using a physicist's terminology, we say that at each point x of the "external" space M, there exists an "internal" space G_x; although isomorphic with G, G_x differs from G in the sense that we do not know where the origin is (a circle becomes a group U(1) when we know where the origin of the group is, and this choice of an origin is arbitrary). The "abstract" group G is assumed to know how to act on P but usually not from both sides, we will assume it acts from the right: if $\mathbf{p} \in P$ and $\mathbf{g} \in G$ then $\mathbf{pg} \in P$. It is convenient to choose for all $\mathbf{x}$ in $\mathbf{M}$ an origin in the internal space (the fiber over x) G_x, this choice (called choice of a section, or gauge choice) cannot be done globally when P is not a product; it is then possible to parametrize points $\mathbf{z}$ of P by a couple $(\mathbf{x},\mathbf{y})$, $\mathbf{x} \in \mathbf{M}$, $\mathbf{y} \in G$, we have then "above" each point of $\mathbf{M}$ a copy of the group G.

We are interested in particular metrics on such spaces P: those which are "dimensionally reducible" to $\mathbf{M}$, i.e. those which can be built out of quantities defined only on $\mathbf{M}$. Among them, the G-invariant metrics are particularly important. In section 4.4 we will prove the following "reduction theorem":

There is a one to one correspondence between G*-invariant metrics on* P *and the triples* $(\mathbf{\gamma},\mathbf{A},\mathbf{h})$ *where* $\mathbf{\gamma}_{\mu\nu}(x)$ *is an arbitrary metric on* $\mathbf{M}$, $\mathbf{A}_\mu{}^i(x)$ *is a Yang-Mills field valued in* Lie(G) *and* $\mathbf{h}_{ij}(x)$, *with* $\mathbf{h}_{ij}=\mathbf{h}_{ji}$, *is a set of scalar fields which geometrically correspond to the choice of a* G*-invariant metric in each internal space* G_x.

The physical interpretation of $\mathbf{\gamma}$ and $\mathbf{A}$ is well-known, the one of $\mathbf{h}$ remains to be found (the usual situation corresponds to the case where $\mathbf{h}_{ij}$ is x-independent and describes a bi-invariant metric -written δ_{ij} in some particular basis-); however the appearance of the $\mathbf{h}_{ij}$ is very natural from the geometrical point of view.

To each element T_i in the Lie algebra of G, one can associate a vector field $\mathbf{e}_i{}^R$ in P since G acts on P from the right: $\mathbf{e}_i{}^R$ at $\mathbf{p} \in P$ is the image of T_i under the jacobian (matrix of partial derivatives) of the map $\mathbf{a} \in G \rightarrow \mathbf{pa} \in P$. These vector fields are called fundamental vector fields and they turn out to be Killing vector fields for G-invariant metrics. By choosing a (local) gauge σ, we can represent P (locally) as M×G, this allows one to define a left action of the group G on P and also the corresponding vector fields ${}^\sigma\mathbf{e}_i{}^L$.

These vector fields (that we call "invariant") are, however, not globally defined since they depend upon the choice of σ. One can use these vector fields to write (locally) the inverse of the G-invariant metric $\mathbf{g}$ at the point $(x,a) \in \mathbf{M}\times\mathbf{G}$ as

$$\mathbf{g}^{-1} = \gamma^{\mu\nu}(x)\,(\partial_\mu - {}^\sigma\mathbf{A}_\mu{}^i(x)\,{}^\sigma\mathbf{e}_i{}^L(a)) \otimes (\partial_\nu - {}^\sigma\mathbf{A}_\nu{}^j(x)\,{}^\sigma\mathbf{e}_j{}^L(a)) + {}^\sigma\mathbf{h}^{ij}(x)\,({}^\sigma\mathbf{e}_i{}^L(a) \otimes {}^\sigma\mathbf{e}_j{}^L(a))$$

and the metric itself as

$$\mathbf{g} = \gamma_{\mu\nu}(x)\,(dx^\mu \otimes dx^\nu) + {}^\sigma\mathbf{h}_{ij}(x)\,(\,{}^\sigma\mathbf{A}_\mu{}^i(x)\,dx^\mu + {}^\sigma\omega_i{}^L(a)) \otimes (\,{}^\sigma\mathbf{A}_\nu{}^j(x)\,dx^\nu + {}^\sigma\omega_j{}^L(a))$$

where dx^μ and ${}^\sigma\omega_i{}^L$ are the one forms dual to the ∂_μ and ${}^\sigma\mathbf{e}_i{}^L$. This way of writing clearly exhibits G-invariance (the Lie derivative of $\mathbf{g}^{-1}$ with respect to $\mathbf{e}_i{}^R$ vanishes automatically since $[\mathbf{e}_i{}^R, {}^\sigma\mathbf{e}_i{}^L] = 0$ -cf. sect. 2.3 -); however it destroys the explicit gauge invariance since the quantities entering the right hand side of this expression are gauge-dependent. The metric can also be written in a manifestly gauge independent way, cf. sect. 4.4.5.

We then study the curvature tensors of P associated with a G-invariant metric and we find that they can always be written in terms of quantities defined on M; in particular the scalar curvature reads:

$$R^P = R^M + R^G - 1/4\, F_{\mu\nu,i}F_{\mu\nu,i} - 1/4\, D_\mu \mathbf{h}_{ij} D_\mu \mathbf{h}_{ij} - 1/4\, \mathbf{h}_\mu \mathbf{h}_\mu - D_\mu \mathbf{h}_\mu$$

where $R^M(x)$ is the scalar curvature of M for the metric γ. $R^G(x)$ is the scalar curvature of G for the metric $\mathbf{h}_{ij}(x)$ and is given in the case of a unimodular group by

$$R^G(x) = 1/2\, C_{ijk}C_{jki} - 1/4\, C_{ijk}C_{ijk}$$

$F^i_{\mu\nu}$ is the curvature of the connection, the Yang Mills field, we have also used the notation h_μ to denote $h^{ij}D_\mu h_{ij}$, finally,

$$D_\mu h_{ij} = \partial_\mu h_{ij} + C_{ik}{}^l A_\mu{}^k h_{lj} + C_{jk}{}^l A_\mu{}^k h_{il} .$$

Summations over repeated indices on the same level are always carried out using $\gamma_{\mu\nu}$, h_{ij} or their inverse (for instance $C_{ijk}C_{jki} = C_{ij}{}^k C_{j'k}{}^i h^{jj'}$).

To the particle physicist, the above expression is reminiscent of the usual Einstein-Yang-Mills lagrangian describing coupled gravity and (non-abelian) gauge fields, strong forces for example. However it differs from it by the presence of the h_{ij} field (one could call it "internal gravity field"). Moreover, we should decide what to do with the integration over the internal coordinates: one possibility is to take $\int(R^P(x)\,\gamma^{1/2}\ d^4x)$ as a candidate for the piece of the four dimensional action describing the coupling between gravity and non-abelian gauge fields. However, this choice is very unnatural since after worrying with "extra dimensions", we decide to forget them completely; the most natural choice is of course to choose

$$\int(R^P(x)\,\gamma^{1/2}h^{1/2}\ d^4x\ d^ny = \int R^P(z)dvol(P)$$

i.e. the Einstein lagrangian in P, integration over the internal coordinates y leads then to an effective four dimensional action which differs from the previous one by the presence of an overall factor

$$V(x) = \int h^{1/2}(x)d^ny = \int dvol(G_x).$$

This volume factor describes physically a change of the unit scale at each point x of M; it is therefore natural to re-absorb it by a conformal transformation of the metric on **M** (cf. sect. 4.7): $\gamma' = \gamma \times$ (conformal factor). In terms of new variables $\gamma', A_\mu{}^i, h'_{ij}$, the resulting action functional reads

$$\int R^P \, dvol(P) = K \,.\, \int \{ R^M[\gamma'] + R^G[h'] - 1/4.\, F_{\mu\nu}{}^i F_{\mu'\nu'}{}^j \,.\gamma'^{\mu\mu'} \gamma'^{\nu\nu'}.h'_{ij} - 1/4\ h'^{ij} h'^{kl} D_\mu h'_{ik} D_\nu h'_{jl} \gamma'^{\mu\nu} + 1/(4(d-2))\ h'_\mu .\, h'_\nu \gamma'^{\mu\nu} \} \gamma'^{1/2}\, d^m x + \text{a boundary term}$$

where $d = m + n = \dim(M) + \dim(G)$ and K is a constant equal to the volume of G determined by the chosen basis $T_i \in Lie(G)$ (for instance the standard volume).

Possible contact with experiment can now be done provided we restore units, coupling constants etc. This is done in 4.8 where we also discuss the behaviour of the scalar potential.

Another point that we discuss in sect 4.6 is the problem of "consistency": suppose that we have a lagrangian in m+n dimensions and that we impose an ansatz (constraints) leading after integration over extra dimensions to a dimensionally reduced lagrangian in m dimensions. It is clear that every solution of the (m+n)-dimensional field equations satisfying the constraints will also be a solution of the m dimensional theory. However the converse is not necessarily true: "consistency" is obtained if, by definition, every solution of the resulting m-dimensional field equations can also be interpreted (using the ansatz) as a solution of the original (m+n)-dimensional theory. Choosing for example a pure gravity theory with cosmological constant in m+n dimensions (whose saddle points are Einstein metrics, cf. sect. 1.3.2), we show in 4.6 that dimensional reduction based on the G-invariant ansatz for the metric tensor is indeed consistent in the above sense; a similar discussion will also appear in chapter 5 (spaces which are locally a product $M \times G/H$).

In 4.9 we discuss geodesics in **P** for G-invariant metrics and study their projection on **M** (the physical idea is that test particles in **P** follow geodesics and what we see is just a shadow...); the equations of motion for particles of mass m and color charge q^i are

$$D\dot{x}_\mu/dt = q_i F^i{}_{\mu\nu}\dot{x}^\nu + 1/2.\ q^i q^j D_\mu(h_{ij})$$

$$Dq_i/dt = C_{ij,k}\, q^j q^k$$

Observe that the form of these equations will change after the rescaling discussed above).

The first equation generalizes the usual Lorentz equation of electromagnetism (notice the appearance of the "internal gravity force"), the second one describes charge non-conservation (charge is here interpreted as the component of the velocity in the "internal" direction).

In sect. 4.10, we show that the G-invariant ansatz is not the only possible one which leads to dimensional reduction from **P** to **M**: one can get an m-dimensional theory by starting for example with the scalar curvature associated with the following (inverse) metric on **P** and integrating over the "internal" coordinates.

$$g^{-1} = \gamma^{\mu\nu}(x)(\partial_\mu - B_\mu{}^i(x)\, e_i{}^R(a) - {}^\sigma A_\mu{}^i(x)\, {}^\sigma e_i{}^L(a)) \otimes (\partial_\nu - B_\nu{}^j(x)\, e_j{}^R(a)\, {}^\sigma A_\mu{}^i(x)\, {}^\sigma e_i{}^L(a)) + {}^\sigma h^{ij}(x)\, (e_i{}^R(a) \otimes e_j{}^R(a)) \quad ,$$

This metric usually has no isometries at all but leads after dimensional reduction to an effective theory on **M** incorporating Yang-Mills fields valued in G×G. The geometrical meaning (and gauge invariance) of this ansatz may be unclear at first sight in the case where **P** is not a product. We will see however that such a metric can be gotten from a G×G invariant metric on the "tautological" space **P**×**M** by passing to the quotient through the action of G_{diag}. We argue however that such metrics are usually "inconsistent" -in a sense defined previously-, the source of the inconsistency is that the scalar curvature associated to such metrics at a point (x,y) of P is usually not independent of $y \in G$, contrarily to what happens in the case of a G-invariant metric. An analoguous situation exists in the case where G is replaced by G/H (cf. next chapter). There, we argue in the same way that the "popular" ansatz, leading to gauge group G, is as a rule inconsistent for the same reasons ("our" ansatz, i.e. the G-invariant one, will lead to a gauge group N(H)/H).

The above dimensionally reducible but non G-invariant metrics constitute actually a particular case of a very general geometrical situation that we discuss in chapter 8 (action of a bundle of groups); for this reason, we give in sect. 4.11 the definition of several (infinite dimensional) groups of diffeomorphisms related to a given principal bundle P (group of automorphisms, group of vertical automorphisms -the gauge group-, etc.) and discuss some of their properties. In 4.12, we show how the non G-invariant metrics of 4.9 can be discussed in terms of the action of the bundle of groups AdP.

The next subsection begins with a study of the example P=M×SU(2)

4.1 Example of M×SU(2)

Let P be the product of M (a model of space-time) and of the three dimensional sphere S^3. Remember that S^3 may be endowed with a Lie group structure (the Lie group SU(2)) and that there exists many metrics on S^3 which are invariant under SU(2)(cf. ch. 2): there is a one parameter family of bi-invariant metrics (isometry group is SU(2)×SU(2)) and a $3\times4/2= 6$-parameters family of (let us say right) invariant metrics (isometry group is only SU(2)). At each point (x,y) in $M\times S^3$, there is a tangent space which naturally splits in the sum of the tangent space to M at x and the tangent space to S^3 at y. For each x in M let $(h_{ij}(x))$ be an SU(2)-invariant metric on S^3 = SU(2), let $(\gamma_{\mu\nu}(x))$ be a metric on M and let $A_\mu{}^i(x)$ be a Yang-Mills field on M valued in the Lie algebra of SU(2). It is intuitively clear that this data indeed specifies a metric on P and this metric is far from being arbitrary (for instance the scalar curvature associated with such a metric at the point (x,y) will only depend upon x).

We will analyse below the situation when P is not necessarily a product of M and of the Lie group G but a principal bundle. The particular case of $G = S^3$ will be continued in sec. 5.9.2.

4.2 Fundamental fields on P

4.2.1 Definition

Let P be a (right) principal bundle with base M, projection π and structure group G, we set dim(M)=m, dim(G)=n, then dim(P)=m+n. G acts on P from the right, let $p \in P$, call R_p the map $a \in G \rightarrow R_p(a)=pa \in P$; the tangent map R_{p*} at the identity of G associates with every element T_i of Lie(G) a vector tangent to P at p, we call $e_i^R(p)=R_{p*}(T_i)$. These fundamental vector fields e_i^R on P corresponding to infinitesimal transformations of P generated by T_i are explicitly given by

$$e_i^R(p) = d/dt(p \exp(tT_i)) |_{t=0}, \quad p \in P.$$

4.2.2 Properties

Let $\{T_i\}$ be a basis in Lie(G), and C_{ij}^k the corresponding structure constants

$$[T_i, T_j] = C_{ij}^k T_k$$

Since G acts on P from the right we get for the associated fundamental fields:

$$[e_i^R, e_j^R](p) = C_{ij}^k e_k^R(p) \quad p \in P$$

Notice that we would get a minus sign for a left principal bundle. It is clear from their definition that the fields e_i^R are vertical i.e. $e_i^R(p)$ is tangent to the fiber at p; moreover the family $\{e_i^R(p)\}$ is a basis in the vertical subspace V_p at p. For the moment, we have no way of defining a horizontal supplementary subspace H_p at p; this will be done when we introduce a connection in P (or a metric). We will sometimes make a slight notational abuse and write $e_i^R(p) = pT_i$.

From the definition of fundamental fields,or using the above notation, we find immediately how they propagate along the fiber and we get the relation

$$e_i^R(pa)=Ad(a)^j{}_i\, e_j^R(p)\, a, \quad p \in P,\ a \in G,$$

where $Ad(a)^j{}_i$ is the matrix of the adjoint representation of G

$$Ad(a)T_i = aT_i a^{-1} = Ad(a)^j{}_i\, T_j.$$

4.3 Local product representation of P

4.3.1 Local trivialization of P

Let σ: M $\rightarrow$ P be a local cross-section of the principal bundle P. Then σ determines a local trivialization ϕ: M×G $\rightarrow$ P of P by $\phi(x,a)=\sigma(x)a$, $a \in G$; reciprocally, given σ, any point p of P can be parametrized by a pair $(x,a) \in M\times G$, where $x=\pi(p)$ and a such that $p=\sigma(x)a$. Above every point x of M we now have a "copy" of the group G with origin $e=\sigma(x)$.

4.3.2 Definition of invariant fields (via the choice of a local section)

Invariant vector fields on P are fields which are left invariant by the right of G. On the infinitesimal level (or assuming G connected), there are fields commuting with the fundamental fields e_i^R:

$[\xi, e_i^R] = 0$, $i=1,...,n$.

As we shall see later, invariant vector fields generate automorphisms of the bundle (gauge transformations). Typically, however, they arise as a result of choosing some local trivialization (or section) of P.

With such a choice of local trivialization, it becomes possible to act with G on the left of P:

$b(x,a) = (x,ba)$ $x \in P$; $b,a \in G$.

We can then define fundamental vector fields for this left action, in a way similar to what we did previously with the right fundamental ones; we will call ${}^\sigma e_i{}^L$ these vector fields. Notice that the left action being local only (it depends upon the choice of σ), these fields are not globally defined (unless the principal bundle P admits a global section, in which case it would be trivial). Notice that the ${}^\sigma e_i{}^L$ are right-invariant ($[{}^\sigma e_i{}^L, e_j{}^R] = 0$); for this reason, and also because we study here a right principal bundle we will drop the adjective "right" in the following and will simply call "fundamental fields" the (globally defined) right fundamental fields $e_i{}^R$ and "invariant fields" the right-invariant fields. Notice that the $\{{}^\sigma e_i{}^L\}(p)$ also span the vertical subspace at $p \in P$, but in this case:

$$[{}^\sigma e_i{}^L, {}^\sigma e_j{}^L](p) = - C_{ij}{}^k\, {}^\sigma e_k{}^L(p).$$

Another set of invariant vector fields are the fields $\partial_\mu = \partial/\partial x_\mu$; we have

$$[\, \partial_\mu, e_i^R \,] = 0 \,.$$

Although these fields too depend on the trivializing section σ (otherwise we do not know what are the x-coordinate lines on P), and they should be, in principle, written as ${}^\sigma\partial_\mu$; nevertheless we shall consequently omit the superscript σ in this case. We shall also denote by the same symbols ∂_μ the vector fields on M, and their invariant lifts to P; this abuse of notation is eased by the fact that they satisfy the same commutation relations. Finally, let us mention that $({}^\sigma e_i^L)$ and (∂_μ) together form a (local) basis of vector fields on P.

4.3.3 Relation between fundamental fields and (local) invariant fields

By choosing a local section $x \in M \rightarrow \sigma(x)$ P, the fiber G_x can be identified with G and we have the usual relation between left and right fundamental fields on a Lie group (cf. sect. 2.2.). Let Ad a be the adjoint representation of G, $Ad(a)T_i = aT_i a^{-1}$, we get in the vertical subspace at $p \in P$

$${}^\sigma e_i^L(p) = (Ad(a)^{-1})^j{}_i(p) \; e_j^R(p) \quad ,$$

where $p=(x,a)$ in the local section σ.

4.4 G-invariant metrics and the reduction theorem

4.4.1 The reduction theorem

Let P be a principal bundle with base M, projection π and structure group G, then every G-invariant metric on P determines and is determined by a triple consisting of

i)for each $x \in M$, a G-invariant metric in G_x -the copy of G over x-

ii)a principal connection in the principal bundle (P,π,M)

iii)a metric on M

4.4.2 Proof of the reduction theorem

Let g now be a G-invariant metric on P, we prove first the direct proposition in i'),ii'),iii'), then the converse in iv').

i') g being a metric on P we know a fortiori, how to compute the scalar product of two vertical vectors. Since g is G-invariant, we obtain a G-invariant metric h_x on each fiber G_x of P.

ii')For each $p \in P$ let V_p be the *vertical subspace* (i.e. the subspace of the tangent space T_p consisting of vectors tangent to the orbits of G). Define H_p to be the orthogonal complement of V_p in T_p. This *horizontal distribution* (H_p), $p \in P$ is G-invariant and therefore determines a principal connection on P (the reader should remember that the connection form on P is then fully determined on T_p by the requirement that the value of the connection form on $e_i^R(p) \in V_p$ is $T_i \in Lie(G)$ and that it vanishes on H_p).

iii')The scalar product $\gamma_x(v,w)$ of two vectors tangent to M at a point x is obtained as follows: Choose an arbitrary point p in the orbit labelled x, and let $\hat{v}$ and $\hat{w}$ be the vectors in H_p which project onto v and w respectively (i.e. the *horizontal lifts* of v and w at p), then define $\gamma_x(v,w)=g_p(\hat{v},\hat{w})$. The result is independent of the choice of p in the orbit because of the G-invariance of the metric g. We therefore get a metric on M (a simpler way of defining γ is to set $\gamma := (\pi_* g^{-1})^{-1}$, cf. Sec. 4.4.5).

iv')To prove the converse theorem, it is easy to see that, given a metric γ on M, a principal connection on P (therefore a horizontal distribution (H_p), $p \in P$) and G-invariant metrics on the fibres of P, one construct G-invariant metric on P. Indeed, given $w,v \in T_p$, let $\pi(w),\pi(v)$ denote their projection on M, $\hat{w},\hat{v}$ their projection on H_p (also the horizontal lifts of $\pi(w),\pi(v)$), then $v - \hat{v}$, $w-\hat{w}$ are vertical, and the scalar product of v, w is defined by:

$$g_p(v,w) = \gamma_x(\pi(v),\pi(w)) + h_x(v-\hat{v}, w-\hat{w}), \quad x=\pi(p).$$

4.4.3 Comments on scalar fields

Two of the three ingredients entering the Reduction Theorem are quite standard and will not be discussed here. We want to make a few comments about the family of metrics (h_x), $x \in M$. A coordinate representation of h_x is obtained as follows: choose a local section of P, σ: $M \to P$, then define $h_{ij}(x)=g_{ij}(\sigma(x))$, where $g_{ij}(p)=g(e_i^R(p),e_j^R(p))$, $p \in P$, are numerical functions on P; it is enough to know these functions at $\sigma(x)$, since, using G-invariance, h_{ij} can be unambiguously propagated all over the orbits P. We can interpret h_x locally as a right invariant metric in the "copy" of the group G at $x \in M$ (cf. chapter 2); physically h_x describes the "shape" of the "internal space" at x. The field $x \to h_x$ of G-invariant metrics on fibres of P can be identified with a cross-section of a bundle associated with P; indeed, from the law of transformation of fundamental fields, we find

$$g_{ij}(pa)=Ad(a)^{i'}{}_i\, Ad(a)^{j'}{}_j\, g_{i'j'}(p), \quad p\in P,\ a \in G,$$

so that the numerical functions $g_{ij}(p)$ on P can be identified with sections of associated bundle (denoted $P \times_{Ad} (Lie(G)^* \otimes_{Symm} Lie(G)^*)$). As for the notation for the scalar field, we choose to be inconsequent in the notation: sometimes it will be written g_{ij}, another time h_{ij}, while its argument can be x, or p, or (x,a), depending on the situation.

4.4.4 Adapted moving frame

Each tangent space T_p decomposes into $T_p=V_p + H_p$, where V_p, the vertical space at p is the space tangent to the orbit of G through p, and H_p, the horizontal space at p, is the orthogonal complement of V_p in T_p. We fix a local coordinate system (x^μ) in M and denote by e_μ the horizontal lifts of the vector fields ∂_μ; notice that without loss of generality we could take a non-holonomic moving frame in M, but we keep the notation ∂_μ in order not to introduce too many symbols. We already know that the fundamental fields $(e_i{}^R(p))$ span the vertical subspace at p, then $(e_A)=(e_\mu, e_i{}^R)$ is a moving frame in a neighbourhood of a point $p \in P$. The structure functions $f_{AB}{}^C$ of this moving frame are given by the following commutation relations

$$[e_i{}^R, e_j{}^R] = C_{ij}{}^k\, e_k{}^R,$$
$$[e_i{}^R, e_\mu] = 0,$$
$$[e_\mu, e_\nu] = -F_{\mu\nu}{}^i\, e_i{}^R + f_{\mu\nu}{}^\sigma\, e_\sigma,$$

where $f_{\mu\nu}{}^\sigma$ are the structure functions of the moving frame ∂_μ on M, F being the curvature of the connection ω induced from the G-invariant metric g on P. The first relation has already been given and comes from the fact that the $e_i{}^R$ are (right) fundamental fields; the second relation is obvious because the e_μ fields are invariant by construction; from the definition of the curvature F of a connection, we get $F(e_\mu, e_\nu) = F_{\mu\nu}{}^i\, T_i = d\omega(e_\mu, e_\nu) = -\omega([e_\mu, e_\nu])$, thus $-F_{\mu\nu}{}^i\, e_i{}^R$ is the vertical part of the commutator $[e_\mu, e_\nu]$. Its horizontal part is equal to the horizontal lift of $[\partial_\mu, \partial_\nu]$ since the fields e_μ and ∂_μ are π-related, i.e. $\partial_\mu = \pi_* e_\mu$.

To summarize, we get

$$f_{ij}{}^k = C_{ij}{}^k\,, \qquad f_{i\mu}{}^j = f_{i\mu}{}^\nu = 0\,, \qquad f_{\mu\nu}{}^i = -F_{\mu\nu}{}^i,$$

while $f_{\mu\nu}{}^\sigma$ are the same as these of ∂_μ on M. One should notice here that while $C_{ij}{}^k$ are constants and $f_{\mu\nu}{}^\sigma(x)$ depend on $x \in M$ only, the $F_{\mu\nu}{}^i$ in the above formulae are not constant along the fibers. This is clear from the fact that e_μ, e_ν and therefore also $[e_\mu, e_\nu]$ are (right-) invariant vector fields, while e_i^R are not; therefore $F_{\mu\nu}{}^i$ must be functions along the orbits to compensate for this difference. We have

$$F_{\mu\nu}{}^i(pa) = Ad(a^{-1})^i{}_j\, F_{\mu\nu}{}^i(p).$$

The following picture summarises the situation:

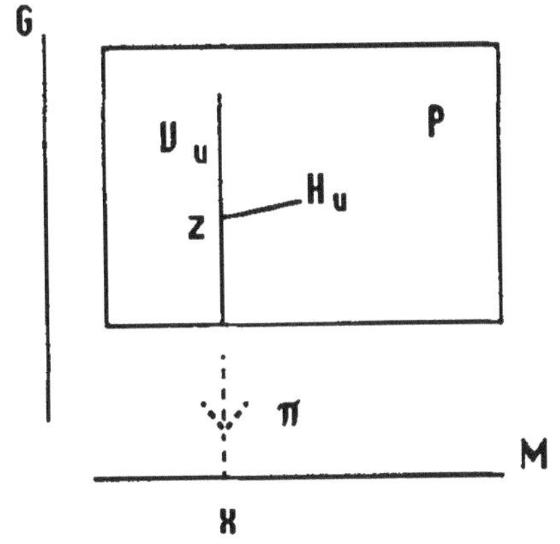

4.4.5 Several expressions for the metric and for the connection

The three ingredients of a G-invariant metric $g = (g_{AB}) = [g(e_A, e_B)]$ can be constructed as follows

$$\gamma_{\mu\nu}(x) = g(e_\mu(u), e_\nu(u)),\quad \pi(y)=x \in M$$
$$g_{ij}(p) = g(e_i^R(p), e_j^R(p)),\quad p \in P,$$
$$\omega^i(v(p)) = v^i, \text{ with } v(p)= v^i(p)\, e_i^R(p) + v^\mu(p)\, e_\mu(p) \;\in T_p$$

In order to become more familiar with the geometrical constructions presented in this chapter we will now express the metric and its inverse in several bases. Let $p = (x,a) \in M{\times}G$ be a local product representation of P in a given gauge $\sigma: M \to P$. The connection form can be written

$$\omega(p) = {}^{\sigma}A_{\mu}(x,a)\, dx^{\mu} + a^{-1}da\ ,$$

with

$${}^{\sigma}A_{\mu}(x,a) = a^{-1}\, {}^{\sigma}A_{\mu}(x)a \;=\; {}^{\sigma}A_{\mu}{}^{i}(x)\, a^{-1}T_i\, a,$$

where $a^{-1}da$ is the (left-invariant) Maurer-Cartan form discussed in 2.2. The horizontal lift $e_{\mu}(p)$ can be written

$$e_{\mu}(u) = \partial_{\mu} - {}^{\sigma}A_{\mu}{}^{i}(x,a)\, e_i{}^{R}(a) \qquad \text{(in terms of fundamental fields)}$$

$$= \partial_{\mu} - {}^{\sigma}A_{\mu}{}^{i}(x)\, {}^{\sigma}e_i{}^{L}(a) \qquad \text{(in terms of invariant fields).}$$

Notice again that while ω, $e_i{}^{R}$ are gauge independent quantities, ${}^{\sigma}e_i{}^{L}$, ${}^{\sigma}A_{\mu}(x)$, ${}^{\sigma}A_{\mu}{}^{i}(x,a)$ and also ∂_{μ} depend upon the choice of a local section σ. In the same way, for the scalar field $x \to h_x$, we set

$$h(p) = {}^{\sigma}h^{ij}(x)\, {}^{\sigma}e_i{}^{L}(a) \otimes {}^{\sigma}e_j{}^{L}(a) = h^{ij}(x,a)\, e_i{}^{R}(a) \otimes e_j{}^{R}(a)$$

The reader should be reminded here that the fact that the physical fields -like the field h- are usually sections of associated bundles, and that any such section can be described either by a locally defined gauge-dependent function on M, or by a globally defined equivariant function on P (cf. sect 6.1).

In term of fundamental fields, the inverse metric g^{-1} in P reads

$$g^{-1} = \gamma^{\mu\nu}\, e_{\mu} \otimes e_{\nu} + h^{ij}\, e_i{}^{R} \otimes e_j{}^{R}$$

$$= \gamma_{\mu\nu}(x)\, (\partial_{\mu} - {}^{\sigma}A_{\mu}{}^{i}(x,a)e_i{}^{R}(a)) \otimes (\partial_{\nu} - {}^{\sigma}A_{\nu}{}^{j}(x,a)e_j{}^{R}(a)) + h^{ij}(x,a)\, (e_i{}^{R}(a) \otimes e_j{}^{R}(a))$$

Indeed, by their very definition the vector fields $\mathbf{e}_\mu$ are orthogonal to the $\mathbf{e}_i$ and the metric itself takes the form

$$\mathbf{g} = \gamma_{\mu\nu}(x)\ (dx^\mu \otimes dx^\nu) + \mathbf{h}_{ij}(x,a)\ (\omega^{iR}(a) + {}^\sigma\mathbf{A}_\mu{}^i(x,a)dx^\mu) \otimes (\omega^{jR}(a) + {}^\sigma\mathbf{A}_\nu{}^j(x,a)dx^\nu),$$

$\{dx^\mu, \omega^{iR}\}$ being the dual basis of $\{\partial_\mu, \mathbf{e}_i{}^R\}$. In matrix notations, we get

$$\mathbf{g}_{AB} = \begin{pmatrix} \gamma_{\mu\nu}(x) & 0 \\ & \\ 0 & \mathbf{h}_{ij}(x,a) \end{pmatrix}$$

in the vertical-horizontal frame $\mathbf{e}_A$ while in the frame $\{\partial_\mu, \mathbf{e}_i{}^R\}$ we have

$$(\mathbf{g}) = \begin{pmatrix} \gamma_{\mu\nu}(x) + \mathbf{h}_{ij}(x,a)\ {}^\sigma\mathbf{A}_\mu{}^i(x,a)\ {}^\sigma\mathbf{A}_\nu{}^j(x,a)\ , & \mathbf{h}_{ij}(x,a)\ {}^\sigma\mathbf{A}_\mu{}^i(x,a) \\ & \\ \mathbf{h}_{ij}(x,a)\ {}^\sigma\mathbf{A}_\nu{}^i(x,a) & ,\ \mathbf{h}_{ij}(x,a) \end{pmatrix}$$

This simple block diagonal form of the metric in the canonical moving frame is of course compensated by the more complicated form of the structure functions. Notice that the vertical fields $\mathbf{e}_i{}^R$ are not necessarily orthogonal between themselves (this will have to be taken care of when we study spinors on **P** in chapter 10). In terms of the (locally defined) invariant fields ${}^\sigma\mathbf{e}_i{}^L$, the inverse metric takes the form

$$\mathbf{g}^{-1} = \gamma^{\mu\nu}(x)(\partial_\mu - {}^\sigma\mathbf{A}_\mu{}^i(x)\ {}^\sigma\mathbf{e}_i{}^L(a))\otimes(\partial_\nu - {}^\sigma\mathbf{A}_\nu{}^j(x)\ \ {}^\sigma\mathbf{e}_j{}^L(a)) + {}^\sigma\mathbf{h}^{ij}(x)\ ({}^\sigma\mathbf{e}_i{}^L(a)\otimes\ {}^\sigma\mathbf{e}_j{}^L(a))\ ,$$

and the metric itself

$$\mathbf{g} = \gamma_{\mu\nu}(x)(dx^{\mu}\otimes dx^{\nu}) + {}^{\sigma}h_{ij}(x)({}^{\sigma}A_{\mu}{}^{i}(x)\, dx^{\mu} + {}^{\sigma}\omega_{i}{}^{L}(a))\otimes({}^{\sigma}A_{\nu}{}^{j}(x)\, dx^{\nu} + {}^{\sigma}\omega_{j}{}^{L}(a))$$

The expression of $\mathbf{g}^{-1}$ in terms of ${}^{\sigma}e_j{}^{L}$ fields clearly exhibits G-invariance since $[e_i{}^R, {}^{\sigma}e_j{}^L] = 0$ and $[e_i{}^R, \partial_\mu] = 0$, therefore the Lie derivative of $\mathbf{g}^{-1}$ with respect to the $e_i{}^R$ vanishes: the fundamental fields $e_i{}^R$ appear as Killing vector fields for the G-invariant metric $\mathbf{g}$. Sometimes the Killing vectors $K_i = e_i{}^R$ at $y \in G$ are written as $K_i(y) = K_i{}^m(y)(\partial/\partial y^m)$ in a co-ordinate basis (y^m) on G and we could write

$$\mathbf{g}^{-1} = \gamma_{\mu\nu}(x)(\partial_\mu - A_\mu{}^i(x,y)K_i{}^m(y)\partial/\partial y^m)\otimes(\partial_\nu - A_\nu{}^j(x,y)K_j{}^n(y)\partial/\partial y^n) + h_{ij}(x,y)K_i{}^m(y)K_j{}^n(y)\,(\partial/\partial y^m \otimes \partial/\partial y^n).$$

Notice the explicit dependence of A and h on y.

Examples

We already gave an example with $P = M \times SU(2)$ in sect. 4.1, but we do not have to restrict ourselves to a product case: Let us construct all possible SU(2) invariant metrics on the seven sphere S^7; first we notice that S^7 is an SU(2) bundle over S^4 -the Hopf fibration, cf. sect. 3.6- and that S^7 is precisely the $k=1$ instanton bundle. Therefore a direct use of the above theorem tells us that in order to construct a general SU(2)-invariant metric on S^7 one must choose

1) an arbitrary metric $\gamma(x)$ on $M = S^4$,

2) an arbitrary Yang-Mills field $A = (A_\mu{}^i(x))$ defined on S^4 with values in Lie(SU(2)),with instanton number $k=1$, and

3) an arbitrary $SU(2)$-right-invariant metric $h = (h_{ij}(x))$ on each copy of $SU(2)$ above the points x of S^4 (there is a 6-parameter family of such metrics).

Notice that, in the principal bundle case studied in this chapter, the dimension of the space of scalar fields, being right-invariant metrics on G -cf.sect 2.6- is $n(n+1)/2$, with $n=\dim(G)$.

4.5 Curvature of G-invariant metrics on a principal bundle P

4.5.1 Levi-Civita connection in the adapted moving frame

The structure functions of the moving frame e_A have already been given; from G-invariance of the metric, we get

$$e_\mu(g_{ij}) = D_\mu g_{ij},$$
$$e_i^R(g_{jk}) = C_{ij,k} + C_{ik,j}$$

where $C_{ij,k} = g_{kl}\, C_{ij}^{\;l}$, and $D_\mu g_{ij}$ denotes the covariant derivative of g_{ij} with respect to the connection ω. When a local section $\sigma : M \rightarrow P$ is chosen, it can be explicitly written as

$$D_\mu g_{ij} = \partial_\mu g_{ij} + C_{ik}^{\;l} A_\mu^{\;k}\, g_{lj} + C_{jk}^{\;l} A_\mu^{\;k}\, g_{il}$$

where $A = \sigma^*\omega$ is the Yang-Mills potential.

We now express the coefficients $\Gamma_{AB,C}$ of the Levi-Civita connection associated with the metric g -cf. sect.1.3.1-in terms of M-based quantities

$$\Gamma_{ij,k} = 1/2\ (f_{ij,k} - f_{ki,j} + f_{jk,i})$$
$$\Gamma_{\mu i,j} = \Gamma_{i\mu,j} = -\Gamma_{ij,\mu} = 1/2\ D_\mu g_{ij}$$
$$\Gamma_{\mu\nu,i} = -\Gamma_{\mu i,\nu} = -\Gamma_{i\mu,\nu} = -1/2\ F_{\mu\nu,i}$$

$\Gamma_{\mu\nu,\rho}$ = (the Christoffel symbols of $\gamma_{\mu\nu}$ on M).
Notice that in the present chapter $f_{ij,k} = C_{ij,k}$.

4.5.2 Ricci tensor and scalar curvature

We give below the formulae for Ricci and scalar curvature of P endowed with a G-invariant metric g. In the adapted moving frame (e_A) defined previously, we obtain

$$R_{ij} = R_{ij}(G) + 1/4\ F_{\mu\nu,i}\ F_{\mu\nu,j} + 1/2\ h^{kl}\ D_\mu h_{ik}\ D^\mu h_{jl}$$
$$- 1/4\ \ D_\mu h_{ij}.h^{kl} D^\mu h_{kl}\ -1/2\ D_\mu(D^\mu h_{ij})$$

$$R_{\mu\nu} = R_{\mu\nu}(M) - 1/2\ F_{\mu\sigma,i}\ F_{\nu\sigma,i} - 1/4\ h^{ij}\ h^{kl}\ D_\mu h_{ik}\ D_\nu h_{jl}$$
$$- 1/2\ D_\mu(h^{ij}\ D_\nu h_{ij})$$

$$R_{\mu i} = 1/2\ D^\sigma\ F_{\mu\sigma,i} + 1/4\ F_{\mu\sigma,i}\ h^{jk}\ \ D^\sigma\ h_{jk} - 1/2\ C_{ij}{}^k\ h^{il}\ D_\mu h_{kl}$$

$$R\ \ = R(M)+ R(G) - 1/4\ F_{\mu\nu,i}\ F_{\mu\nu,i} -1/4\ D_\mu h_{ij} D_\mu h_{ij}$$
$$-1/4 h_\mu h_\mu - D_\mu h_\mu$$

Comments

a) $R_{ij}(G)$ and $R(G)$ are given in sect. 2.5 with understanding that h_{ij} is now a function of x.

b)The derivative D_μ in the above formulae acts both on internal and space-time indices with $A_\mu{}^i$ and $\Gamma_{\mu\nu}{}^\rho$ respectively.

c)The summation over repeated indices on the same level is performed with h_{ij} and $\gamma_{\mu\nu}$. For example the term $F_{\mu\nu,i}\ F_{\mu\nu,i}$ should be read as $g^{\mu\mu'} g^{\nu\nu'} g_{ii'} F_{\mu\nu}{}^{i} F_{\mu'\nu'}{}^{i'}$ (notice that we use h_{ij} and g_{ij} to denote the same object).

d)We call $h = \det(h_{ij})$ and $h_\mu = h^{ij} D_\mu h_{ij} = h^{ij} \partial_\mu h_{ij} = D_\mu \ln(h)$, where we made use of the unimodularity of G ($C_{ij}{}^{i} = 0$).

e) The last term of R can be written in several ways, for example,

$$D_\mu (h^{ij} D_\mu h_{ij}) = - h^{ij} h^{kl} D_\mu h_{ik} D_\mu h_{jl} + h^{ij} D_\mu D_\mu h_{ij} \quad .$$

Later we will get field equations from an action principle on P, the quantity of interest is then $R[g]\, \gamma^{1/2}\, h^{1/2}$ where $\gamma = \det(\gamma_{\mu\nu})$. The last term of R gives

$$-D_\mu (h^{ij} D_\mu h_{ij})\, \gamma^{1/2}\, h^{1/2} = - (D_\mu h_\mu)\, \gamma^{1/2}\, h^{1/2}$$
$$= 1/2\ h_\mu h_\mu\, \gamma^{1/2}\, h^{1/2} - \partial_\mu (h_\mu\, \gamma^{1/2}\, h^{1/2})$$

where appears a total divergence, and we get

$$R^P[g]\ g^{1/2} = (R^M + R^G - 1/4\ F_{\mu\nu,i} F_{\mu\nu,i} - 1/4\ D_\mu h_{ij} D_\mu h_{ij}$$
$$+ 1/4\ h_\mu h_\mu)h^{1/2}\, \gamma^{1/2} - \partial_\mu (h_\mu\, \gamma^{1/2}\, h^{1/2})$$

4.6 Action principle and the consistency requirement

4.6.1 The (dimensionnally reduced) scalar curvature as a Lagrangian ?

There are several possibilities of using the results of 4.5.2 to determine field equations. One possibility is to take an action $S = \int_M R\, \gamma^{1/2}\, dx$. With this choice the last term of the Lagrangian is already a divergence; but such a choice is quite unnatural since it does not take the "extra dimensions" seriously, the natural choice is to take

$$S = \int_P R\, g^{1/2}\, d^{(n+m)}x = \int_P R\, \gamma^{1/2}\, h^{1/2}\, d^{(n+m)}x.$$

The scalar curvature R for G-invariant metrics being independent of $y \in G$, we get in this way an effective m-dimensionnal theory, with

$$S = \int_M R(x)\, V(x)\, \gamma^{1/2}\, dx$$

where V(x) is the volume of the fiber at x. As we shall see below, it is useful to make a conformal rescaling on the space-time metric $\gamma_{\mu\nu}$ in order to re-absorb the internal volume factor; before that we want to analyse the "consistency requirement".

4.6.2 Validity of the consistency requirement

We already explained in the summary section of the present chapter what the consistency requirement was; first we have to specify what the model is: in any case, the Yang-Mills, gravitational and scalar fields will be described by the Einstein action on P, however one could add other terms (source terms for the gravitationnal field). Since we want our theory to admit a ground state solution of the field equations, we will consider the simple case of a Lagrangian $(R-2\Lambda)\, g^{1/2}$ with cosmological constant in d=n+m dimensions. We consider this model beeing fully aware of its non-realistic features, and we consider it just for the reason that it is the simplest model (or rather class of model) which exhibits a consistent dimensional reduction. The model admits "spontaneous compactification" on $E = M \times G$ with M being, for instance a de Sitter space and G a Lie group endowed with a G-invariant Einstein metric (the Einstein constants have also to agree); the field equations resulting from the above Lagrangian are

$$\hat{E}_{AB} + \Lambda\,\hat{g}_{AB} = \hat{T}_{AB} \quad ,$$

where $\hat{E}_{AB} = \hat{R}_{AB} - 1/2\,\hat{R}\,\hat{g}_{AB}$ is the Einstein tensor on P and where $\hat{T}_{AB}$ describes the contribution of matter fields. In this simple model we consider $T_{AB} = 0$, and so the field equations become

$$\hat{R}_{\mu\nu} = 2\,\Lambda/(d-2)\,\hat{g}_{\mu\nu}$$
$$\hat{R}_{\mu i} = 0$$
$$\hat{R}_{ij} = 2\Lambda/(d-2)\,\hat{g}_{ij}$$

i.e. they describe an Einstein space with Einstein constant $2\Lambda/(d-2)$, $d=m+n$. (The case $\Lambda=0$ would correspond to spaces which are Ricci flat and among compact Lie groups, only tori would be allowed). To check the consistency of the ansatz, we have to compare the above equations, which are now equations for $\gamma_{\mu\nu}(x)$, $A_\mu{}^i(x)$, $h_{ij}(x)$, with the ones obtained from the m-dimensional action which results from integration of the original Lagrangian over the internal coordinates. Modulo a constant proportionality factor (related to the standard volume on G), this m-dimensional action is

$$S_{eff} = \int_M (\hat{R} - 2\Lambda)\,\gamma^{1/2}\,h^{1/2}\,d^m x \,.$$

(It is worthwhile to notice here a simple but important and useful, fact: if the Lie algebra basis is taken to be orthonormal with respect to a bi-invariant metric on G which gives G a unit volume, then $h^{1/2}(x)$ is equal to the volume of the internal space at x determined by the metric $g_{AB}(p)$).

This action should now be varied with respect to $\gamma_{\mu\nu}(x)$, $A_\mu{}^i(x)$ and $h_{ij}(x)$ in order to obtain a set of m-dimensional equations for those fields. One gets, by explicit calculation, the following set of equations

$$\hat{R}_{\mu\nu} - 1/2\,(\hat{R} - 2\,\Lambda)\,\gamma_{\mu\nu} = 0,$$
$$\hat{R}_{\mu i} = 0$$
$$\gamma^{1/2}\,h^{1/2}\,[\,\hat{R}_{ij} - 1/2\,(\hat{R} - 2\,\Lambda)\,h_{ij}\,] = 0$$

where $\hat{R} = \hat{g}^{\mu\nu}\hat{R}_{\mu\nu} + h^{ij}\,\hat{R}_{ij}$ (notice that $\gamma_{\mu\nu} = g_{\mu\nu}$). This set of equations is evidently the same as the other set given previously.

We want to warn the reader that the discussion of consistency is slighty more involved when we come to discuss (in chapter 5) the case where the internal space is no longer a group but a homogeneous space (several constraints will have to be taken into account by introducing Lagrange multipliers).

This ends the proof of the consistency of the G-invariant ansatz in the principal bundle case. Observe that it is essential in this statement that every solution which is an extremum of the effective m-dimensional action is an extremum of the original d-dimensional action. The inverse statement, that is , that every constrained solution of the original field equations is a solution of the m-dimensional ones follows from the fact that the effective m-dimensional action is defined as an integral of the original one over the internal variables.

4.7 Conformal rescaling and the effective Lagrangian

4.7.1 Conformal rescaling

The integrand of the d-dimensional action was $L = R^P[g]g^{1/2}$, and could be written, with obvious notations,

$$L = \{R^M[\gamma] + R^G[h] - 1/4\;\; F\,F\,[\gamma^{-1}][\gamma^{-1}][h]$$

$$- 1/4\, [\gamma^{-1}]\, [h^{-1}][h^{-1}]\, [Dh][Dh]\ - 1/4\, h.h\ - D\,h\}\ h^{1/2}\,\gamma^{1/2}$$

where $h_\mu = h^{ij} D_\mu h_{ij}$.

We rescale $\gamma_{\mu\nu}$ and h_{ij}, i.e, we define $[\gamma'] = h^x\, [\gamma]$ and $[h'] = h^y\, [h]$ and determine the numbers x and y by the following requirements:

i) $[\gamma^{-1}][\gamma^{-1}][h]\, h^{1/2}\,\gamma^{1/2} = [\gamma'^{-1}][\gamma'^{-1}][h']\,\gamma'^{1/2}$

ii) $[\gamma^{-1}]\, h^{1/2}\,\gamma^{1/2} = [\gamma'^{-1}]\,\gamma'^{1/2}$

These requirements traduce the fact that we want to "remove" the global $h^{1/2}$ factor as much as possible; notice that ii) removes it both from the kinetic energy energy term for the h_{ij} field and from the gravitational piece R^M. Indeed the Ricci tensor is invariant under a -constant- conformal rescaling therefore the scalar curvature transforms as the inverse metric itself. Actually, we are not making a constant conformal rescaling, but this simple reasoning is enough to find the correct powers x and y and we easily find

$$x = y = 1/(m-2) \text{ with } m = \dim \mathbf{M}.$$

In order to compute the transformed Lagrangian as a function of the new variables $[\gamma']$, $[h']$, $F'=F$, we need to know how R^M transforms under a non-constant rescaling; we have the general formula

$$\gamma'_{\mu\nu} = h^y.\gamma_{\mu\nu}$$

$$R[\gamma'].h^y = \{R[\gamma] - (m-1)\, y\, \Delta h/h$$
$$- (m-1)y\{(m-2)y-4\}\, D_\nu h\, D_\mu h\, g^{\mu\nu}/(4\, h^2)\,\}$$

with $\Delta h = \gamma^{\mu\nu} D_\mu D_\nu h$.

We need then to transform the other terms of L as well. A tedious but straightforward calculation leads to the effective lagrangian

$$L = \{R^M[\gamma'] + R^G[h'] - 1/4\, FF[\gamma'^{-1}][\gamma'^{-1}][h']$$
$$- 1/4\, [h'^{-1}][h'^{-1}][\gamma'^{-1}]\, [Dh'][Dh']$$
$$+ 1/(4(d-2))\ [h'].[h']\ \}\, \gamma'^{1/2},$$

where d = dim P = dim M + dim G = m + n, and we omitted the surface terms . It is useful in the calculation to notice that

$$h^{1/2}\ \gamma^{1/2} = (h')^{(-1/(d-2))}\, \gamma'^{1/2}\ , \quad h = h'^{(m-2)/(d-2)}$$

and that

$$h_\mu = h^{ij} D_\mu h_{ij} = D_\mu \ln h$$
$$= (m-2)/(d-2) D_\mu \ln h' = (m-2)/(d-2)\ h'_\mu.$$

It is striking to observe that the above lagrangian only differs from the first by the absence of the global $h^{1/2}$ factor and by a simple modification of the coefficient of the internal volume factor $h_\mu\, h_\mu$.

4.7.2 The particular case G=U(1)

This is the original Kaluza-Klein theory with Jordan-Thirry scalars (i.e a scalar field $\sigma(x)$). We take the metric in the 5th dimension to be $h_{55} := \sigma^2$. Then, using the results of sect. 4.5.2, we get (with the original metric, i.e., before the conformal rescaling) the following Ricci tensor

$$R^P_{\mu\nu} = R^M_{\mu\nu} - 1/\sigma\ \nabla_\mu(\partial_\nu\sigma) - \sigma^2/2\ F_{\mu\rho}\, F_\nu{}^\rho$$
$$R^P_{\mu 5} = 3/2\ F_\mu{}^\rho \sigma \nabla_\rho \sigma + 1/2\ \sigma^2 \nabla_\rho F_\mu{}^\rho = 1/(2\sigma)\ \nabla_\rho(\sigma^3\, F_\mu{}^\rho)$$

$$R^P{}_{55} = 1/4\ \sigma^4 F^2 - \sigma\ \nabla_\mu(\partial^\mu\sigma).$$

The curvature scalar is

$$R^P = R^M - 1/4\ F^2\sigma^2 - 2\ \nabla_\mu(\partial^\mu\sigma)/\sigma.$$

The original Lagrangian is

$$L = R^P.\sigma.(\det \gamma_{\mu\nu})^{1/2}$$

With this formulation, the "gravity constant" is not constant because of the factor σ in front of the expression. The conformal rescaling $\gamma'_{\mu\nu} = \sigma.\ \gamma_{\mu\nu}$ (which should be considered as a field redefinition), along with $h'_{55} = \sigma.\ h_{55}$ leads to the following expression for L

$$L = [\ R^M - 1/4\ F_{\mu\nu}F^{\mu\nu}\ L^2 e^{2\varphi} + (-2/3)\ L^2\ \partial_\mu\varphi\ \partial^\mu\varphi]\ \ (\det \gamma_{\mu\nu})^{1/2}$$

where we have set $h'_{55} = \ = L^2 \exp(2\varphi(x))$, L beeing a constant. Notice that φ (and not σ) enters the lagrangian with a "conventional" kinetic energy term.

After this rescaling it is possible to define a "usual" gravitational constant.

4.8 On the Interpretation of the scalar fields. Physical units.

4.8.1 Comment on scalar fields

The potential for the scalar fields h_{ij} when there is no cosmological constant in the 'big' space P is $\mathcal{V} = -R^G[h(x)]$; it has therefore a clear geometrical interpretation since its value at $x \in M$

measures the scalar curvature of the internal space (which, by G-invariance, is constant on the internal space at x). It is clear that $\mathfrak{V}$ is in general unbounded from below; indeed if for example, we "deflate" the internal space G_x by making a scaling $\mathbf{h}_{ij} \longrightarrow \rho^2\, \mathbf{h}_{ij}$, its scalar curvature may become arbitrarily big. It should be stressed here that it is probably not possible to construct a "physical" theory out of the Einstein Lagrangian of pure gravity -even in higher dimensions!-, one should add terms of the higher order in the curvature, other kind of matter fields, spinor fields for example and maybe also a cosmological constant; this may allow for further constraints like freezing the volume of the internal space and considering only for "squashing" deformations. The internal space in this chapter being a compact group, we recover the usual Einstein-Yang-Mills theory as the lowest order of an expansion of $\mathbf{h}_{ij}(x)$ around the Killing metric, which indeed corresponds to a saddle point of $\mathfrak{V}(\mathbf{h}(x))$ if we consider only metrics with fixed volume -remember that the critical points of the functional $[\mathbf{h}] \longrightarrow \int R[\mathbf{h}]$, when $\mathbf{h}$ varies in the space $\mathfrak{M}_0$ of metrics of fixed volume,are precisely the Einstein metrics . The Killing metric or one proportional to it, is a critical point of $\mathfrak{V}[\mathbf{h}]=-R[\mathbf{h}]$, $\mathbf{h} \in \mathfrak{M}_0$, and is not usually the only one although it is usually a minimum, for example the following curve gives the behaviour of $\mathfrak{V}[\mathbf{h}]$ when $\mathbf{h}$ varies in the set of normalised $SU(4)\times SO(4)$ invariant metrics on $G=SU(4)$; the general expression was given in sect. 2.9 which in this case reads

$$\mathfrak{V} = -R^G = -\,(t^2)^{2/5}\,[9/2 + 3/(2t^2) - 9/8\,.\,t^2\,]$$

and its derivative with respect to t is:

$$d\mathfrak{V}/dt = 63/20\;\; t^{-11/5}\,(t^2 - 1)\,(t^2 - 1/7)$$

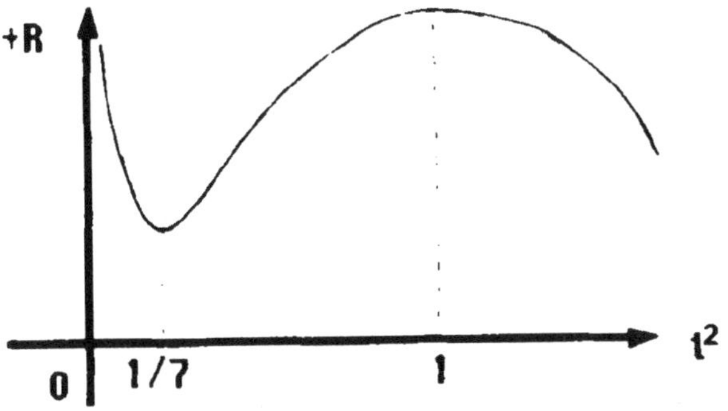

The point A corresponds to a bi-invariant metric on SU(4) (the standard Einstein metric) and the point B to a non standard squashed Einstein metric. It is remarkable that if we expand **h** around another critical point, we are in a situation similar to the Higgs mechanism and we get a mass spectrum for the gauge fields. A straightforward calculation using the results of section 2.9 (supposing that we expand **h** around a G×K invariant Einstein metric on G, and that the pair (G,K) is symmetric) leads formally to $k = \dim(K)$ massless bosons and $l = n-k$ massive bosons with mass $m^2 = 2l^2/(4k^2-l^2)$ in dimensionless units (for example 5 massive bosons with mass $m^2=18/7$ in the last example). Unfortunately, in the present case it is not legitimate to perform such an expansion around this position since it corresponds to an unstable point of the potential; however it may be that in a realistic theory, such a point is stabilized by the effect of other interactions. Before ending this paragraph, let us mention that in order to interpret $h_{ij}(x)$ and $A_\mu{}^i(x)$ as usual scalar fields and Yang-Mills fields, it may be necessary to "redefine" them (to make a change of field variables), in such a way that they appear with a usual kinetic energy term in the Lagrangian: for instance, we saw in 4.7.2 that a kinetic term

$(\partial_\mu \sigma\, \partial^\mu \sigma)/\sigma^2$ may be written as $\partial_\mu \varphi\, \partial^\mu \varphi$ via the field redefinition $\sigma = \exp(\varphi)$.

4.8.2 Cosmological constant and units

As already discussed in 4.6.1, we choose $R^P \sqrt{g}$ as the original d-dimensional lagrangian; however, we allow for the presence of a possible cosmological constant Λ^P already in the "big space" P. Also, in order to make contact with the usual formulation of physical laws, we have to resore units (scale of measurement). In the system where (Planck's constant) = (Speed of light) = 1, there is only one unit (for example the centimeter) and this unit can be fixed by choosing arbitrarily the mass of some elementary particle (for instance, if we choose the electron and set its mass to m_e = 6.7657 10^{-59} cm^{-1}, this defines what the centimeter is). In this system, a given quantity has dimension $[L^p]$ where L is a length; traditionally one takes R^M (scalar curvature of space-time) $\tilde{=}$ $[L^{-2}]$ and A_μ (Yang-Mills field) $\tilde{=}$ $[L^{-1}]$, therefore the Lagrangian density L is homogeneous to $[L^{-4}]$ and the four- dimensional action $\int L\, d^4x$ is dimensionless. We will choose dimensionless scalar fields : $h_{ij}(x)$ $\tilde{=}$ $[L^0]$.

Let therefore ρ denote some unit scale homogeneous with a length, we get the following effective 4-dimensional lagrangian (here, we take dim M = 4 hence d=4+n).

$$L = \{ R^M/\rho^2 + R^G/\rho^4 - 1/4\, F_{\mu\nu}{}^i F_{\rho\sigma}{}^j h_{ij} - 1/(4\rho^2)\, h^{ij} h^{kl} (D_\mu h_{ik} D^\mu h_{jl} - D_\mu h_{ij} D^\mu h_{kl}) - 2\, \Lambda^P/\rho^2 \} \sqrt{h}\, \sqrt{\gamma}$$

$$L = \{ R^M/\rho^2 + R^G/\rho^4 - 1/4\, F_{\mu\nu}{}^i F_{\rho\sigma}{}^j h_{ij} - 1/(4\rho^2)\, h^{ij} h^{kl} (D_\mu h_{ik} D^\mu h_{jl} - 1/(n+2)\, D_\mu h_{ij} D^\mu h_{kl}) - 2\, \Lambda^P \cdot h^{-1/(n+2)}/\rho^2 \}\, \sqrt{\gamma}$$

These two expressions describe the same dynamics since the second system is gotten from the first via the (conformal) change of variables discussed in 4.7. $R^M\sqrt{\gamma}$ is usually interpreted as the part describing pure gravity, then $\rho^2 = 16\pi G = 131.193\ 10^{-66}\ cm^2$ (but maybe we could keep in mind another possibility: the suggestion of using this term as a description of a spin-2 strong force; in this case the numerical value of ρ could be quite different).

If **M** is a four-dimensional space-time with signature -+++, then the Einstein Lagrangian with cosmological term is $(R^M - 2\Lambda^M)/(16\pi G)$. For reasons of positivity of the energy, the signature of the "internal" metric has to be spacelike, i.e. of positive sign with the above conventions. If we choose the signature +--- on **M**, the signature of the "internal" metric should be taken negative (but then, with our definition of the curvature tensor, the scalar curvature on a standard two sphere S^2 would also be negative). In what follows we assume that the choice (-+++) has been made on **M**. The potential for the scalar field $\mathbf{h}_{ij}(x)$ is then (write the lagrangian as "kinetic energy"-"potential")

$$\mathfrak{V} = - R^G(\mathbf{h})/\rho^4 + 2\Lambda^P/\rho^2 (\det \mathbf{h})^{-1/(n+2)}$$

where $R^G(\mathbf{h})$ was given in 4.0 and where n = dim G. When $\Lambda^P = 0$, the Killing metric $\mathring{\mathbf{h}}_{ij}$ is not even a stationary point of $\mathfrak{V}$: one has to "freeze" the volume. But, if $\Lambda^P \neq 0$, there exist a metric $\mathbf{h}_{ij} = L^2\, \mathring{\mathbf{h}}$ with a real constant L which is a stationary point for $\mathfrak{V}$; this value L is determined by Λ^P : assume G compact and simple, then $R^G(\mathbf{h}) = n/4L^2$ and $\det(\mathbf{h}) = L^{2n} \det(\mathring{\mathbf{h}})$; we may absorb $\det(\mathring{\mathbf{h}})$ in the definition of Λ^P and get:

$$-\mathfrak{V} = n/(4L^2\rho^4) - 2\Lambda^P/(L^{2n/n+2}\,\rho^2)$$
$$-\partial\mathfrak{V}//\partial L = 1/\rho^2\,(-n/(2L^3\rho^3) + 4n\,.\,\Lambda^P/(n+2)\,.\,1/L^{(2n/n+2)+1}\,)$$

Therefore, $\partial\mathfrak{V}//\partial L = 0 \Leftrightarrow L = L_0$, with

$\operatorname{Log} L_0 = -(n+2)/4\ \operatorname{Log}(\ 8\Lambda^P\rho^2/(n+2))$; we call $\mathfrak{V}_0 := \mathfrak{V}(L_0) = 1/(2L_0{}^2\rho^4)$.

If we expand the internal metric h_{ij} around this background L_o^2 $\mathring{h}_{ij}$, we get a constant to be identified with the cosmological constant in **M**

$$\mathcal{V}_0 = + 2\ \Lambda^M/\rho^2 \qquad \text{therefore } \Lambda^M = 1/(4L_o^2\rho^2)$$

Notice that Λ^M is then positive (if we had not added a term Λ^P in the lagrangian, besides the fact that the potential would not have had a saddle point for $h = L^2\ \mathring{h}$, a direct identification of $\mathcal{V}_0 = -n/4L^2\rho^4$ with the cosmological term $2\ \Lambda^M/\rho^2$ would have led to a negative cosmological constant $\Lambda^M = -n/(8L^2\rho^2)$). If we now perform an expansion of h_{ij} around $L_0^2\ \mathring{h}_{ij}$, it is convenient to write $h_{ij}(x) = [\ \varphi_i^{i'}(x)\varphi_j^{j'}(x)\ \mathring{h}_{i'j'}]\ L_o^2$ and also $\varphi_i^{i'}(x) = \delta_i^{i'} + \sigma_i^{i'}(x)$. With this change of variables the Yang-Mills term becomes

$$-1/4\ F_{\mu\nu}{}^i\ F_{\mu\nu}{}^j\ h_{ij} = -1/4\ \ F_{\mu\nu}{}^i\ F_{\mu\nu}{}^j\ L_o^2\ \mathring{h}_{ij}\ + \ ....$$

The length L_o^2 is then a measure of the Yang-Mills coupling constant (cf. sec. 2.11); if g_σ denotes the coupling constant defined in the representation σ of G, and if i_σ, i_G denote the indices of the representations (cf. sec. 2.11), we get

$$1/g_\sigma^2\ .\ i_\sigma/i_G = L_o^2/2 = 1/(8\Lambda^M\rho^2)$$

Let us choose for example G=SU(2) and σ the fundamental representation: $i_\sigma = 1/2$, $i_G = 2$; we can use the relation linking g^2 and Λ^M in two possible ways: either we choose a reasonable value for g^2 (for instance 1/10) and find an enormous cosmological constant, or we use the fact that, experimentally $\Lambda^M \leq 4.\ 10^{-56}\ cm^{-2}$ and find that $g^2 \leq 10^{-119}$ which is not acceptable from a phenomenological point of view.

4.9 Color charges, scalar charges and the particule trajectories

Consider a geodesic C: $t \longrightarrow C(t)$ in P. Since g_{AB} is G-invariant, it follows that for each $a \in G$ the path Ca: $t \longrightarrow C(t)a$ is a geodesic too. Observe that both C and Ca have the same projection $\pi(C)=\pi(Ca)$ on M, and it is this projection that an observer "living on M", that is blind to the extra dimensions, sees. He does not distinguish between the individual elements of the whole bundle $[C]=\{Ca/a \in G\}$ of geodesics with the same projection on M. Our aim in this section is to identify the set of data which is adequate for a description of the projected motion i.e. the set of quantities which characterize the equivalence class [C] rather than one of its representatives [62], [103]. In the present case where P is a principal bundle, this problem does not cause any difficulty since trough any point $p \in \pi^{-1}(\pi(C(t))$ there passes exactly one representative of the class [C]; we shall see in chapter 5 that it is not the case when the internal space is a homogeneous space rather than a group, in which case a subtler analysis is necessary.

4.9.1 Geodesics in P

Consider a class [C] of geodesics in P with $C' \in [C]$ if $C' = Ca$ for some $a \in G$. and let p a point in P above x_0, $\pi(p)=x_0$. Given $C \in [C]$, let $\dot{C}(t) = z(t) + v(t)$ be the (orthogonal) decomposition of the vector $\dot{C}(t)$, tangent to C at t into its horizontal $z(t)=z^{\mu}(t)e_{\mu}(C(t))$ and vertical $v(t)=v^{i}(t)e_{i}^{R}(C(t))$ components. Owing to G-invariance of vector fields e_{μ} it follows that the components $z^{\mu}(t)$ are the same for all members of the equivalence class [C]; in fact the $z^{\mu}(t)$ are the components of the vector tangent to the projection $\pi[C]$ in M. Now, v(t) being vertical can be identified with a vector q(p) in Lie(G): $q(p)=v^{i}(p)T_{i}$, and q(p) does not depend on the choice of C in the class [C], this because through a given p there passes only one member of the class. Therefore q(p), considered as a vector q(x)

$= p.q(p)$ in the fiber of the vector bundle **adP** associated to **P** via the adjoint representation of G on Lie(G), is physically interpreted as the coloured charge, at x, of the particle described by the geodesic class [C].

4.9.2 Projection in M : equations of motion

According to the previous dicussion, the adequate set of initial data describing the motion of a particle in external fields, $A_\mu{}^i, \gamma_{\mu\nu}, h_{ij}$ consists of $x_0{}^\mu, \dot{x}_0{}^\mu, q_0$. It remains to find differential equations governing the time evolution of x^μ and q; we only sketch the method here since we will do it explicitely in chapter 6. Denoting by $u^A=(z^\mu,v^i)$ the components of the vector $\dot{C}$ tangent to C, the geodesic equations are

$$du^A/dt + \Gamma^A{}_{BC}\, u^B u^C = 0 ,$$

we then substitute the connection coefficients found in sect.4.5 and get an equation for dz^μ/dt and another one for dv^i/dt; it is then handy to make a convenient gauge choice (i.e. to choose a local section of P) in order to identify z^μ with $\dot{x}(t)$ and $v^i(t)$ with $q^i(t)$. A short calculation leads to:

$$D\dot{x}_\mu/dt = q_i\, F^i{}_{\mu\nu}\, \dot{x}^\mu + 1/2.\, q^i q^j\, D_\mu(h_{ij})$$
$$Dq_i\,/dt = C_{ij,k}\, q^j q^k$$

where $Dq_i\,/dt$ and $D\dot{x}_\mu/dt$ are the covariant derivative of q_i and $\dot{x}_\mu$ along $x_\mu(t)$:

$$Dq_i\,/dt = dq_i\,/dt - \dot{x}_\mu(t)\, A_\mu{}^j(x(t))\, C_{ji}{}^k q_k$$
$$D\dot{x}_\mu/dt = d\dot{x}^\mu/dt - \Gamma_{\mu\nu}{}^\sigma\,(x(t))\dot{x}^\mu \dot{x}_\sigma$$

The first equation describes the rate of change of the velocity of a colored test-particle in the external field $(\gamma_{\mu\nu}, A_\mu{}^i, h_{ij})$, the second describes the colour charge non-conservation. Notice that, contrary to the case where h_{ij} is a constant bi-invariant metric, for a general configuration of h_{ij} neither $q^2 := h^{ij}q_i q_i$ nor $\dot{x}^2 := \gamma_{\mu\nu}\,\dot{x}^\mu\,\dot{x}^\nu$ are separately conserved, but it is their sum that is a constant of motion. Notice also that what we call charge here means actually charge/mass. At present, physical signifiance of the above equations is far from clear.

4.10 Generalised Kaluza-Klein metrics (action of a bundle of groups)

4.10.1 Motivations

We discuss here metrics which are not G-invariant but are, in some sense "dimensionally reducible". What is commonly refered to as "the Kaluza-Klein ansatz" in the physics articles of the 80s -even when the internal space is a homogeneous space and not a group-does not describe a G-invariant metric; in its more general setting, this non-invariant "full-scale ansatz" leads to a gauge group G×G if the internal space is G (cf. 4.10.2) and to a gauge group G × N(H)/H when the internal space is G/H . As we shall see, this most popular ansatz, which is not G-invariant, should be considered as guilty of inconsistency unless explicitly proven innocent for a specific model.

We give now a particular example of such a non-invariant ansatz: let us again choose $\gamma_{\mu\nu}(x)$, a metric on **M**, $B_\mu = B_\mu{}^i\,T_i$ a Yang-Mills field on **M** and a family h_{ij} of scalar fields as before. We consider the following inverse metric at the point (x,y) of **P**:

$$g^{-1}(x,y) = \gamma_{\mu\nu}(x)\,(\partial_\mu - B_\mu{}^i(x) e_i{}^R(y)) \otimes (\partial_\nu - B_\nu{}^j(x) e_j{}^R(y)) + h_{ij}(x)\; e_i{}^R(y) \otimes e_j{}^R(y)\,.$$

By expressing the fundamental fields $e_i{}^R$ in terms of the (locally defined) ${}^\sigma e_i{}^L$, i.e. writing $e_i{}^R(y)=K_i{}^j(y)\, {}^\sigma e_j{}^L(y)$, we can express the previous metric as

$$g^{-1}(x,y) =$$
$$\gamma_{\mu\nu}(x)\, (\partial_\mu - B_\mu{}^i(x)\, K_i{}^{i'}(y)\, {}^\sigma e_{i'}{}^L(y)) \otimes (\partial_\nu - B_\nu{}^j(x)\, K_j{}^{j'}(y)\, {}^\sigma e_{j'}{}^L(y))$$
$$+ \quad h_{ij}(x)\, K_i{}^{i'}(y) K_j{}^{j'}(y)\ \ {}^\sigma e_{i'}{}^L(y) \otimes {}^\sigma e_{j'}{}^L(y).$$

The result of the mixture of left and right invariant fields in the above expression is that $g(x,y)$ is neither left nor right invariant. The reader willing to see the geometrical meaning of $g(x,y)$ in terms of "gauge invariant" quantities on the principal bundle P may wonder...indeed, the $B\mu$ field does not come from a principal connection on P! However the above ansatz has a precise geometrical meaning that we discuss in the next subsection.

4.10.2 G×G invariant metrics on the tautological bundle P×G

A very general geometrical construction giving the full-scale ansatz will be given elsewhere (chapter 10). This construction applies to all cases where internal spaces have transitive isometry groups. Here, we restrict our discussion to the case of P=P(M,G) being a right principal bundle that we write locally as M×G. The following "recipe" then gives the full-scale ansatz:

1) Artificially enlarge P to $\overline{P}$=P×G ($\overline{P}$ is sometimes called the tautological bundle)

2) Now G×G acts from the right on $\overline{P}$ by $(x,a,b)(c,d)=(x,ac,d^{-1}b)$

3) In particular, $G_{diag} \subset G\times G$ acts on $\overline{P}$ by $(x,a,b)(c,c)=(x,ac,c^{-1}\, b)$

4) Therefore $\overline{P}/G_{diag}$ is isomorphic with P; indeed, this isomorphism is given by $(x,a,b) \mapsto (x,ab)$. The original G action on P defined by $(x,a)b=(x,ab)$ comes from the following G-action on $\overline{P}$=M×G×G: $(x,a,b)(c)=(x,a,bc)$. This last action is not killed during

the projection $\overline{P} \mapsto P=\overline{P}/G_{diag}$, whereas no other part of the action of $G\times G$ on $\overline{P}$ introduced in step 2) survives at the level of P (assuming G is semi-simple); this owing to the fact that, in general, the centralizer (commutant) of G_{diag} in $G\times G$ is trivial.

5) Consider all metrics on P which are projections from $\overline{P}$ to $P = \overline{P}/G_{diag}$ of $G\times G$ invariant metrics on $\overline{P}$ for the action 2).

Indeed those metrics, being $G\times G$ invariant for the action 2) project on P on metrics which have no remaining invariance in general (because of 4)). Those very particular metrics of $\overline{P}$ which are $(G\times G)\times G$ invariant [for the actions of 2) and 4)] or, at least, $G_{diag}\times G$ invariant, project onto G-invariant metrics on P. From the point of view of set theory, we have taken the product by G and then the quotient by G, we are again on P. But from the point of view of the field content, we have produced a large class of metrics on P which contain all G-invariant metrics, but also contains much more. This is the full scale ansatz. It gives rise to gauge fields of $G_L\times G_R$. It should be noticed that the scalar curvature of a given $G\times G$-invariant metric on $\overline{P}$ [for the action of 2)] is only a function of $x \in M$ and can be "dimensionally reduced " to M by using the general technique of sect. 4.4, 4.5 (the so-called G-invariant ansatz,here $G\times G$ rather than G !). However the scalar curvature of the projected metric on P has no reason to be independent of $\mathbf{a} \in G$, since it will not be G-invariant in general.

4.10.3 Dimensional reduction and inconsistency of this ansatz

One can now perform calculations of the scalar curvature associated with this more general ansatz. For the horizontal basis at $(x,\mathbf{a}) \in P = M\times G$ one has now:

$$e_\mu(x,\mathbf{a}) = \partial/\partial x_\mu - A_\mu{}^i(x,\mathbf{a})e_i{}^R(\mathbf{a}) - B_\mu{}^i(x)e_i{}^R(\mathbf{a})$$

We denote by $F_{\mu\nu i}$ and $G_{\mu\nu j}$ the field strengths of gauge fields A_μ and B_μ respectively. This time also we have at our disposal not $n(n+1)/2$ but $2n(2n+1)/2$ scalar fields, $n=\dim(G)$. The piece of scalar curvature which interests us has the form

$$\hat{R} = R(M)-1/4\ g_{ij}(x,a)(F_{\mu\nu}{}^i(x,a)+G_{\mu\nu}{}^i(x))(F_{\mu\nu}{}^j(x,a)+G_{\mu\nu}{}^j(x)) + \dots$$

where $F_{\mu\nu i}(x,a)$ depends on a and $g^{ij}(x,a)$ is a function of the scalar fields and a. For consistency, the solutions of equations of motion obtained from the Lagrangian

$$L[A_\mu,B_\mu,\gamma_{\mu\nu},h_{ij},\dots] = \int_G [\hat{R}(x,a) - 2\ \Lambda]\ dvol_G(a)$$

should also be solutions of the set of equations (in P):

$$\hat{R}_{MN} = (2\Lambda/(d-2))\ g_{MN}.$$

This last set of equations is a dependent in general, whereas the first is not. With a special choice of scalar fields, one can make the dependence of g_{ij} in a reasonable, but even then, it is impossible to make the piece of $\hat{R}$ that contains the field F and G to be a-independent unless $h_{ij}(x)=f(x)k_{ij}$, k_{ij} being the Killing metric of G; here we assume G simple and use the relation

$$\int_G K^i{}_j(a)K^k{}_l(a)\ dvol(G)= (\delta_{jl}\ \delta^{ik}/\dim(G))\ vol_G.$$

But this will be generally incompatible with the field equation containing R_{ij}.

4.11 <u>The group of automorphisms of a principal bundle P</u>

4.11.1 - The structure group and associated bundles

In all this section 4.11, P will denote a principal bundle with base M, structure group G, and projection $\pi : P \to M$ (thus looking locally as M×G), on which G acts from the right. In particular G acts transitively on the fibers, and every fiber is isomorphic (as a homogeneous space) to G itself. Now if F is any space on which G acts from the left (F need not be a vector space and the action of G on F need not be a linear one), then one can build the associated bundle $\mathfrak{F} := P \times_G F$, which looks locally like M×F, and is constructed out of M and F by identifying the pairs (p,f) and $(pa, a^{-1}f)$ in P×F. The equivalence class of (p,f) is denoted p.f and is interpreted as "an object having in frame p coordinate f". By the very definition of the equivalence relation the same object has, in a frame **pa**, coordinate $a^{-1}f$. In case there are more than one action of G on F it is better to give such an action a name, say ρ, and write ρ(a)f instead of a and $P \times_\rho F$ instead of $P \times_G F$. Then $\mathfrak{F}$ is the bundle of objects tranforming under the representation (not necessarily linear) ρ of G. Notice that the group G does not act on the associated bundle, since we already divided P×F by the action of G to get $\mathfrak{F}$. To give an example : the group GL(n) does not know how to act on vectors tangent to an n-dimensional manifold ; it knows however how to act on their components in any given frame. Below we will be particularly interested in two associated bundles Ad $P := P \times_{Ad} G$ and **ad** $P = P \times_{Ad} Lie(G)$ with typical fibers G and Lie(G) respectively.

4.11.2 The group Aut P of automorphisms of P

A diffeomorphism Φ of P is called an automorphism of P if it commutes with the G-action. It maps fibers onto fibers and induces a diffeomorphism, called Φ_M, on the base M of the bundle. Schematically we have the following figure

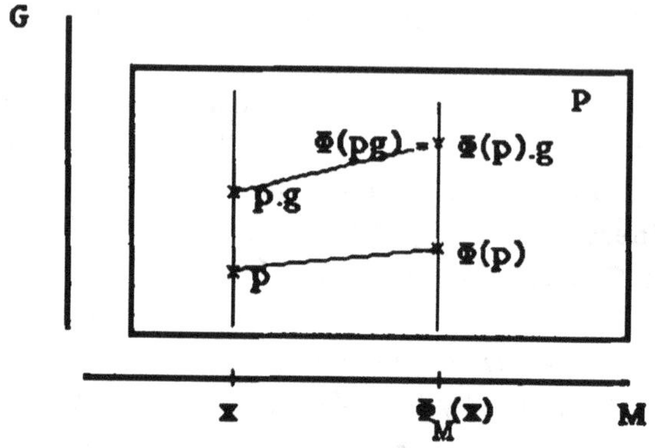

The automorphism group Aut P is an infinite-dimensional group ; its elements act not only on P but also on every bundle $\mathcal{F}$ associated to P : if $\mathcal{F} = P\times_G F$ and $u = p.f$ then $u \to \Phi(u) := \Phi(p).f$ is a well-defined diffeomorphism of $\mathcal{F}$ which maps every fiber F_x of $\mathcal{F}$ diffeomorphically onto the fiber $F_{\Phi_M(x)}$.

The group Aut P also acts on the space $\mathfrak{A}$ of all connections of P ; indeed, let $\omega \in \mathfrak{A}$ and $\Phi \in$ Aut P , we define then ω^Φ by

$$\omega^\Phi{}_p(v) = \omega_{\Phi(p)}(\Phi_*(v))$$

This action has a straightforward geometrical interpretation: horizontal spaces of the transformed connection ω^Φ are images under Φ of the horizontal spaces of the original connection ω.

Let $\sigma : M \to P$ be a local section of P, then $\Phi(\sigma(x))$ is in the same fiber as $\sigma(\Phi_M(x))$ and these two elements must therefore differ by a unique element, denoted ${}^\sigma\Phi(x)$, of G

$$(1) \qquad \Phi(\sigma(x)) = \sigma(\Phi_M(x))\ {}^\sigma\Phi(x)$$

Therefore Φ is locally characterized by a pair $(\Phi_M, {}^{\sigma}\Phi)$ consisting of a diffeomorphism of M and a G-valued function on M.

4.11.3 The group Int P of vertical automorphism P (the gauge group)

First definition : We call $\mathfrak{G}$ = **Int P** the set of all automorphisms Φ of P which induce the identity map on the base space : $\Phi_M = id_M$. In other words $\mathfrak{G}$ does not move the points of M at all, it acts vertically along the fibers. Notice that the map $\varphi \rightarrow \varphi_M$ from **Aut P** to **Diff M** is a group homomorphism whose kernel is precisely $\mathfrak{G}$; $\mathfrak{G}$ is therefore a normal subgroup of **Aut P**; however the above map **Aut P**$\rightarrow$**Diff M** is not necessarily onto. A diffeomorphism $\varphi \in M$ is called "liftable" if there exists an automorphism Φ of P such that $\varphi = \Phi_M$. An example of a principal bundle for which each diffeomorphism φ of M is liftable is the frame bundle of M; in this case every φ has a natural lift to the bundle, so that we have not only that **Diff M** = **Aut P/Int P**, but also **Aut P** = **Int P**$\times_s$**Diff M** (semi-direct product). Thus, in this particular case of the frame bundle, **Aut P** is a trivial **Int P** - principal bundle over **Diff M**, although it is not the direct product from the point of view of its group structure.

There exist two other, equivalent to the first one, definitions of the gauge group $\mathfrak{G}$ that are very useful, and which we will discuss now.

Second definition : To each vertical automorphism Φ we associate a map φ from P to G via the equation

$$(1)\quad \Phi(p) = p\,\varphi(p), \qquad p \in P,\ \varphi(p) \in G.$$

This function φ : P$\rightarrow$G is equivariant in the sense that

$$(2)\quad \varphi(pa) = a^{-1}\varphi(p)a,$$

which follows from (1) and the property $\Phi(pa) = \Phi(p)a$ of the automorphism Φ. Equation (1) sets up a one-to-one correspondence between the elements Φ of $\mathfrak{G}$ and the maps $\varphi = P \to G$ satisfying the property (2) ; we could therefore define $\mathfrak{G}$ directly as the space of these maps. Observe that to the composition $\Phi_3 = \Phi_1 \circ \Phi_2$ of automorphisms of P there corresponds the pointwise multiplication of the representing functions (which have values in G) : $\varphi_3(p) = \varphi_1(p)\varphi_2(p)$.

Third Definition There exists a third definition of $\mathfrak{G}$ which involves the bundle of groups **Ad P** $= P\times_{Ad}G$. This associated bundle is built by letting G act on itself by the adjoint action : we consider the right action of G on P×G given by $(p,a) \to (pb, Ad(b^{-1})a) = (pb, b^{-1}ab)$, and we quotient P×G by this action. An element $p.a$ of **Ad P** is the equivalence class $\{(pb, b^{-1}ab) : b \in G\}$. **Ad P** is locally M×G but the group G does not act on it ; it is the bundle **Ad P** that acts on itself. Indeed, the fibers of **Ad P** have a natural group structure; one defines $(p.a)(p.b) := p.ab$, and this definition does not change if $p.a$ and $p.b$ are replaced by other representatives $p'.a'$ and $p'.b'$ of the equivalence classes - this follows from the fact that the adjoint transformations act on G by group automorphisms. The bundle **Ad P** is therefore an example of a bundle of groups; we will discuss more general bundles of groups in Ch. 10 It is now clear that there is again a one-to-one correspondence between sections of the bundle **Ad P** and the equivariant maps $\varphi : P \to G$ as discussed in the second definition. This correspondence is a particular case of a generally valid correspondence between sections of an associated bundle and equivariant functions on the principal bundle [96,127] . Explicitly it is given as follows : let $\varphi : P \to G$ satisfy (2); to this φ we associate section $\tilde{\varphi}$ of **Ad P** which at a point $x \in M$ takes the value $\tilde{\varphi}(x) = p.\varphi(p)$, where p is any point of P over x. The equivariant property of φ assumes that $\tilde{\varphi}(x)$ defined that way does not depend on the particular choice of p. We could therefore define directly $\mathfrak{G} = \Gamma(\mathrm{Ad}\,P)$, the letter Γ denoting the space of sections. As already observed above (4.11.2.1), locally the elements of $\mathfrak{G}$ can be considered as maps $M \to G$ (this is even true globally if P is trivial and therefore admits a global section). Indeed, given a local section $\sigma: M \to P$, and $\tilde{\varphi}$, we define ${}^{\sigma}\varphi: M \to G$ by the relation

$$4.11.3.4 \qquad {}^{\sigma}\varphi(x) = \tilde{\varphi}(\sigma(x))$$

The formulae 4.11.3.1, 4.11.3.3., 4.11.3.4 and 4.11.2.1 are consistent in the sense that ${}^{\sigma}\varphi(x) = {}^{\sigma}\tilde{\varphi}(x)$, i.e. we have one local description of a vertical automorphism (i.e. of a global gauge transformation).

The group $\mathfrak{G} = \mathrm{Int}\,P$ is a subgroup of Aut P, and therefore it acts on any associated bundle $\mathfrak{F} = P \times_G F$. If $\Phi \in \mathfrak{G}$ is represented by an equivariant section $\varphi : P \to G$ and if $u \in \Gamma F$ is a section of $\mathfrak{F}$ considered as a map $u: P \to F$ then the action of φ on u is given by a simple formula $(\varphi u)(p) = \varphi(p)u(p)$.

The group $\mathfrak{G}$ also acts on the space $\mathfrak{A}$ of connections on P :

$$\omega^{\varphi}{}_p(\vec{v}) = \varphi(p)^{-1}\, \omega_p(\vec{v})\, \varphi(p) + \varphi(p)^{-1}\, d\varphi_p(\vec{v})$$

4.11.4 The pointed gauge group

The action of $\mathfrak{G}$ on the space $\mathfrak{A}$ is not free ; indeed it is well known that the stability subgroup of $\mathfrak{G}$ leaving a given connection ω invariant is isomorphic to the centralizer of the holonomy of ω (see [104]). Therefore, even if G is simple, this stability group will be nontrivial unless the connection is irreducible. In order to have free action of gauge transformations on connections, in which the space of orbits is a manifold (of course infinite-dimensional), one has to restrict either the space of connections or the space of gauge transformations. We will follow here the second route by introducing the so called "pointed gauge group" $\mathrm{Int}_x\,P := \mathfrak{G}_x$. Let us fix $x \in M$, then $\mathfrak{G}_x$ can be defined in one of the following three equivalent ways :

i) $\mathfrak{G}_x$ is the set of all elements Φ of $\mathfrak{G}$ whose restriction to the fiber G_x of P is an identity transformation

$\mathfrak{G}_x = \{\Phi \in \mathfrak{G} : \Phi(p) = p , \quad \forall p \in \pi^{-1}(x) \}$

ii) When $\mathfrak{G}$ is considered as the space **Map Ad**(P,G) of equivariant maps φ φ: $P \rightarrow G$, then we can equivalently define $\mathfrak{G}_x$ as the set of all those φ for which $\varphi(p) = e$ for all (equivalently : for at least one) $p \in \pi^{-1}(x)$. In other words $\mathfrak{G}_x$ can be defined as the kernel of the *evaluation map* $h_p : \varphi \in \mathfrak{G} \rightarrow \varphi(z) \in G$. This map being onto, it follows that $\mathfrak{G}_x$ is a normal subgroup of $\mathfrak{G}$, and that G is isomorphic to $\mathfrak{G}/\mathfrak{G}_x$.

iii) Suppose now $\mathfrak{G}$ is considered as the space of sections Γ(Ad P) of the bundle Ad P. We already noticed that this bundle is a bundle of groups ; denoting e_x the identity of the fiber G_x of Ad P (in the notation introduced above $e_x = p.e$, where $p \in \pi^{-1}(x) \subset P$) we can now characterize $\mathfrak{G}_x$ as consisting of those sections $\tilde{\varphi} \in$ Γ(Ad P) which take the value e_x at the point x of M.

If M is pathwise-connected, the action of $\mathfrak{G}_x$ on the space $\mathfrak{A}$ of all connections is free. Indeed, let ω be a connection invariant under an element $\Phi \in \mathfrak{G}_x$. To see that Φ is an identity transformation take any $p \in \pi^{-1}(y)$, connect x and y by a path $t \rightarrow \gamma(t)$, $\gamma(0)=x$, $\gamma(1)=y$, and let $\tilde{\gamma}$ be the horizontal lift of γ through p. Then $t \rightarrow \Phi(\tilde{\gamma}(t))$ is also a horizontal lift of γ, and since both coincide at $\gamma(0)=x$, they must coincide everywhere, thus $\Phi(p) = \Phi(\tilde{\gamma}(1)) = \tilde{\gamma}(1) = p$. It follows from the above that $\mathfrak{A}$ is an infinite dimensional principal bundle with structure group $\mathfrak{G}_x$ and base $\mathfrak{A}/\mathfrak{G}_x$ - the space of "physical" gauge fields i.e. gauge potentials modulo gauge transformations. It is interesting to notice that the dimensional reduction scheme discussed in sect. 4.4 works also in this case, and is even necessary for understanding the functional integral over the gauge potential and the Fadeev-Popov trick.

Notice that if $\mathfrak{F} = P\times_G F$ is an associated bundle then $\mathfrak{A}\times_{\mathfrak{G}_X}\mathfrak{F}$ is a vector bundle over $\mathfrak{A}/\mathfrak{G}_X$ with an infinite dimensional fiber $\mathfrak{F}$. The same space can be obtained by constructing first the associated bundle $\mathfrak{Q} = \mathfrak{A}\times_{\mathfrak{G}_X}P$ whose fibers are G-spaces P, and which can also be considered a G-principal bundle over $M\times\mathfrak{A}/\mathfrak{G}_X$, and then by building the associated bundle $\mathfrak{Q}\times_G F$, this time with a finite-dimensional fiber F, but a bigger base $M\times\mathfrak{A}/\mathfrak{G}_X$. One can also build the associated bundle $\mathfrak{A}\times_{\mathfrak{G}_X}\Gamma\mathfrak{F}$ whose sections are the physical configurations of the joint system consisting of gauge fields and matter fields, and can be equivalently considered as sections of the bundle $\mathfrak{Q}\times_G F$.

4.11.5 **The infinite-dimensional Lie algebras : aut P = Lie(Aut P), int P = Lie (Int P) = Lie($\mathfrak{G}$) = Lie($\mathfrak{G}_X$).**

The infinite-dimensional Lie groups **Aut P** and **Int P** are not Lie groups in the usual spaces as they are not modelled on Banach spaces [149], however they are "diffeological groups" (or "Souriau groups", [206]) thus possessing a well defined Lie algebra and exponential map. $\mathfrak{G}$ is even a Campbell - Baker - Haussdorft Lie group [149] ; the easiest way to define the vector space **Lie($\mathfrak{G}$)** is to consider the bundle of Lie algebras **ad P** : = $P\times_{Ad}\mathrm{Lie}(G)$ and define **Lie($\mathfrak{G}$)** as **Γ(ad P)**. An element $\mathfrak{X} \in \mathrm{Lie}(\mathfrak{G})$ can be therefore considered as an equivariant function $\mathfrak{X}: P \rightarrow \mathrm{Lie}(G) : \mathfrak{X}(pa) = \mathrm{Ad}(a^{-1})(p)$. Locally, with respect to a given section $\sigma: M \rightarrow P$, $\mathfrak{X}$ can be represented by a function ${}^{\sigma}\mathfrak{X} : M \rightarrow \mathrm{Lie}(G)$ given by ${}^{\sigma}\mathfrak{X}(x) := \mathfrak{X}(\sigma(x))$.

4.12 Pointers to the literature

Principal bundles: 97, 210
Kaluza-Klein theories with internal spaces as groups: 37, 38, 41, 48, 105, 114', 116, 122, 126, 134', 136', 161-163, 207, 208, 215'
Consistency problem : 45, 59, 61
Classical equations of motion (in Kaluza-Klein theories): 62,103
Groups of automorphisms of bundles: 3,149

V

RIEMANNIAN GEOMETRY OF A BUNDLE WITH FIBERS G/H AND A GIVEN ACTION OF A LIE GROUP G

5.0 Summary

In this chapter we consider spaces E which can be written locally as M×G/H, M being some m-dimensional manifold (to be a model of space-time, for example) and G/H being a homogeneous space for the Lie group G. This is a very common situation. Whenever we meet a Lie group G acting on a manifold E then E will decompose into several disjointed pieces (called "strata") $E = E_1 \cup E_2 \cup ..$ and each stratum E_i will look locally like $M \times G/H_i$, H_i being a subgroup of G. Thus the analysis and the geometrical constructions of this chapter can have as many applications as there are instances of Lie groups (even infinite-dimensional) acting on manifolds. Kaluza-Klein theory with a global group of isometries is one example, others are : all problems involving groups of isometries, analysis of symmetric solutions of Einstein and/or Yang-Mills equations, action of the group of diffeomorphisms on the space of metrics etc. (Of course when applying the constructions of the present chapter and of the following chapters to non-compact or infinite-dimensional cases, a appropriate care has to be taken.) As we have already said we consider on a space $E \cong M \times G/H$ the action of a (compact) Lie group G. We assume the product to be only "local" : the space E will be a collection of fibers (orbits of G), each of the type G/H, parametrized by M(i.e. $x \in M$), but E is not supposed to be either globally equal or canonically isomorphic to the product of the "external" space M and the "internal" space G/H (a simple example is $E = S^7$ globally acted upon by $G = SU(2) = S^3$ to give $M = S^4$ - the Hopf fibration - but clearly S^7 is not globally a product $S^4 \times S^3$). However, we will assume that G, which plays the role of an internal symmetry group, knows how to act on E ; this would be self-evident if M is $M \times G/H$ (the trivial case), but it is less so when E is only a "local" product. And indeed there exist two different kinds of spaces which, although both are local products $M \times G/H$, differ by the fact of existence or not of a **global** G-action. In the present chapter we assume that such an action exists and is given, while in Ch.10 we will study the other case (while the present case is best exemplified by the Hopf fibration as mentioned above, the simplest illustration of the second one is a U(1) local action on the Klein bottle - see Ch.10).

For technical reasons (and for consistency with the notation of Ch.3 and Ch.4) we will assume that G acts on E from the right and we will write therefore the typical fiber as $H \backslash G$ (rather than G/H). It is possible to parametrize points of E by a couple (x,y), where $y = [a]$ is a right coset $y = Ha$; such a parametrization cannot be done globally when E is not a product. The operation of G on E is, in this parametrization, given simply by right translations : if $u = (x,y) \in E$, then a group element $b \in G$ acting from the right on u transforms it into $ub = (x,yb) \in E$, with $yb = [ab] = Hab$. We will also introduce *local* transformations $(x,y) \to (x,f(x,y))$ which connect different *admissible* parametrizations, an admissible parametrization being one in which the group action is given as above. Thus the local transformations must commute with the global ones (those of G). It is shown in 5.1 that every local transformation is described by a function $x \to n(x)$, where $n(x)$ belongs to the group $N|H$, N being the normalizer of H in G : this result should not be too surprising in view of the definition and properties of the group N|H (cf. Ch.3). In the particular case where H = {e} then H\G = G and the internal space is a group, the global and local symmetry groups happen to be the same in this case ; the first acting on the internal space from the right and the other from the left: this is the situation already studied in Ch.4. When the internal space is a homogeneous space H\G, then the situation becomes more complicated and the local symmetry group is no longer isomorphic to the global one.

We are interested in particular metrics on E : those which are "dimensionally reducible" to M, in particular those which are G-invariant. We prove in 5.4 the following "reduction theorem". There is a one to one correspondance between G-invariant metrics on E and the triples ($\mathbf{\gamma}$,A,$\mathbf{h}$) where $\mathbf{\gamma}_{\mu\nu}(x)$ is an arbitrary metric on M, $A_\mu{}^i(x)$ is a Yang-Mills field valued in Lie(N(H)|H) and $\mathbf{h}_{\alpha\beta}(x)$ - with $h_{\alpha\beta} = \mathbf{h}_{\beta\alpha}$ - is a set of scalar fields which geometrically correspond to the choice of a G-invariant metric in each internal space $(H\backslash G)_x$.

Observe again that the Yang-Mills field which one gets out of the process of dimensional reduction is neither valued in Lie(G) nor in Lie(H) but in Lie(N(H)|H) - which can be considered as a subalgebra of Lie(G).

For example, we take as a model a space E which is a local product of space time M and the 16 dimensional complex Stiefel manifold SU(5)/SU(3). Then assuming that SU(5) acts globally on E, we find that an SU(5) invariant metric on the 20-dimensional space E can be constructed out of a gravity field on M, a Yang-Mills field valued in the Lie algebra of SU(2)×U(1) and a scalar field $\mathbf{h}_{\alpha\beta}(x)$ characterizing the shape of the internal SU(5)/SU(3) homogeneous space sitting above the point $x \in M$. Notice that the Lie algebra of the normalizer of SU(3) in SU(5) is indeed the Lie algebras of SU(3) × SU(2) × U(1). More examples can be constructed using the results of Ch.3 as well as Ch.4 (remember that the group N|H is discrete when G/H is an irreducible homogeneous space).

We will use often the following property: given two groups T,G, a subgroup H⊂ G and a homomorphism λ from H into T, we may consider the group $\overline{G}=T\times G$ and its subgroup $\overline{H}$ = **diag** $(\lambda(H),H) = \{(\lambda(h),h),h\in H\}$ then, calling $\overline{N}$ the normalizer of $\overline{H}$ into $\overline{G}$, we have the property that $\overline{N}|\overline{H}$ is locally isomorphic with $Z(\lambda(H))\times N|H$ where $Z(\lambda(H))$ is the centralizer of the image of H into T and N the normalizer of H into G. This property can be used here to build new examples (cf. sect. 5.2.5) and it will be used many times in Ch. 8 and 9.

Let us choose u_1 and u_2 as two points of E and call H_{u_1} and H_{u_2} the corresponding "little groups", (stabilizers) i.e. $u_1h_1 = u_1 \Leftrightarrow h_1 \in H_{u_1}$ and $u_2h_2 = u_2 \Leftrightarrow h_2 \in H_{u_2}$; although isomorphic (their isomorphism is assured by the fact that we deal with one stratum), H_{u_1} and H_{u_2} are usually two different subgroups of G. Choose u_0 in E and call $P = P_0$ the set of all points u of E which have the same little group $H_u=H_0$ as u_0; P is a submanifold of E (it is in general less-dimensional that E) and it plays an important role in the sequel (we will see in sect.5.2 that it is actually a *principal bundle* over M with structure group $N(H)|H$, to which E is associated).

To each element T_i in the Lie algebra of G, one can associate a (fundamental) vector field e_i^R on E (since G acts on E) : e_i^R at $u \in E$ is the image of T_i under the Jacobian of the map $a\in G \rightarrow ua\in E$. These vector fields will turn out to be Killing vector fields for the G-invariant metrics (they are sometimes denoted by $K_i = e_i^R$ and sometimes people also write $K_i = K_i{}^m\, \partial_m$, ∂_m being a coordinate basis in $(G/H)_x$). The vector fields K_i span the vertical (=internal) tangent space at $u\in E$ but they form an *overcomplete* system in this vertical tangent space (we have n = dim (G) of Killing fields but the vertical subspace has only dimension dim(S) = dim (G) - dim(H)). We will study the properties of these vector fields in sect. 5.3 and will see in particular that if we split a basis $\{T_i\}$ in Lie(G) into $(T_{\hat\alpha}, T_\alpha)$, $T_{\hat\alpha}$ being a basis in Lie(H) and T_α a basis in $\mathfrak{S} = T(S=G/H, e)$ then the K_α - the Killing vectors corresponding to T_α - form a basis of the vertical subspace at $p\in P\subset E$ (also when p is chosen in a neighborhood of P). In sect. 5.4, we will complete this vertical basis to a full basis in E and build an *adapted moving frame* at the points of E (not too far from P) by adding to the K_α a set of "horizontal fields" $e_\mu = \partial_\mu - A_\mu{}^\alpha e_\alpha$ constructed from a basis $\{\partial_\mu\}$ on **M** and the Yang-Mills field A_μ valued in **Lie(N(H)|H)**. We will then express any G-invariant metric on **E** in this moving frame and will compute in sect. 5.5 the Ricci and scalar curvature tensors (on **E**) associated to such G-invariant metrics. Not surprisingly, we will find that they can always be written in terms of quantities defined on **M**. In particular the scalar curvature reads

$$R^E = R^M + R^{G/H} - 1/4\, F_{\mu\nu,\hat a}\, F_{\mu\nu,\hat a} - 1/4\, D_\mu h_{\alpha\beta}\, D_\mu h_{\alpha\beta} - 1/4\, h_\mu h_\mu - D_\mu h_\mu$$

where R^M is the scalar curvature of **M** for the metric γ, $R^{G/H}(x)$ is the scalar curvature of **H\G** for the metric $h_{\alpha\beta}(x)$ it plays the role of a potential energy for scalar fields and is given by formula 3.4(8). As usual contraction of indices is carried out using $\gamma_{\mu\nu}$, $h_{\alpha\beta}$ or their inverse; for instance $C_{\alpha\gamma\beta}\, C_{\beta\gamma\alpha}$ means $C_{\alpha\beta}{}^\gamma\, C_{\beta'\gamma}{}^\alpha\, h^{\beta\beta'}(x)$.

$F_{\mu\nu}{}^{\hat a}$ is the curvature of the connection (the Yang-Mills field) : remember (cf. sect.3) that we split $\mathcal{S} = \mathcal{K} + \mathcal{L}$ with $\mathcal{S} = T_{[e]}(G/H)$, $\mathcal{K} = Lie(N(H))$, $\mathcal{L} = T(G/N)$ and correspondingly the basis T_α as $T_\alpha = (T_{\hat a} \in \mathcal{K}, T_a \in \mathcal{L})$.

Finally h_μ denotes $h^{\alpha\beta} D_\mu h_{\alpha\beta}$ (it describes the gradient of the volume of the internal space), and

$$D_\mu h_{\alpha\beta} = \partial_\mu h_{\alpha\beta} + C_{\alpha\hat a}{}^\gamma A_\mu{}^{\hat a}\, h_{\beta\gamma} + C_{\beta\hat a}{}^\gamma A_\mu{}^{\hat a}\, h_{\alpha\gamma}$$

As we shall see in sect. 5.4, G-invariance of the family $h_{\alpha\beta}(x)$ is expressed by the vanishing of the following Lie derivative :

$$L_{\hat a} h_{\beta\gamma} := C_{\hat a\beta}{}^\delta\, h_{\delta\gamma}(x) + C_{\hat a\gamma}{}^\delta\, h_{\beta\delta}(x) = 0.$$

This is a constraint that the " Thirry " fields (first introduced by Thirry in the context of five dimensional relativity) have to satisfy.

We analyse several examples in sect. 5.6 and then investigate in sect. 5.7 the properties of the Einstein lagrangian in E, namely

$$\int R^E(u)\, dvol_E(u) = \int R^E(x)\, \gamma^{1/2}\, h^{1/2}\, d^m x\, d^n y$$

Integration over the internal coordinates y introduces then a factor $V(x) = \int h^{1/2}(x) d^n y = \int dvol_{S_x}$ and leads then (as in Ch.4) to an effective four dimensional action (if m=4). The volume factor describes physically a change of the unit scale at each point x of M and it is therefore natural to re-absorb it by a conformal rescaling of the fields $(h, \gamma \rightarrow h', \gamma')$, (cf. sect. 5.8). In terms of the new variables γ', h', A the resulting action functional reads :

$$\int R^E dvol_E = V \int \{R^M[\gamma'] + R^{G/H}[h'] - 1/4\, FF[\gamma'^{-1}][\gamma'^{-1}][h]$$
$$-1/4\, [h'^{-1}]\, [h'^{-1}][\gamma'^{-1}]\, [Dh'][Dh'] + 1/(4(d-2))\, [h'].[h']\ \}\, \gamma'^{1/2}\, d^m x$$

where d = m + s = dim M + dim S , S = H\G and V is a constant equal to the volume of H\G determined by the choosen basis $T_i \in$ Lie G (for instance the standard volume on H\G). The reader will have already noticed that the above expression is similar to the one obtained in Ch.4 with the difference that the range of indices for the Yang-Mills term corresponds to the gauge group K = N|H. We also discuss in sect. 5.8 the behaviour of the "scalar potential" $R^{G/H}(x)$ for the fields $h_{\alpha\beta}$. Another point to be discussed in sect. 5.7 is the problem of "consistency", the problem was already stated in sect. 4.0 and 4.6 and is solved here in an analogous way : we show that in a model of pure gravity with cosmological constant in E (saddle points are Einstein metrics, cf. sect. 1.3.2.), all solutions of effective field equations on M gotten from the effective action corresponding to the G-invariant are also solutions of the original field equations in E (here it is important to take into account the constraint $L_{\hat{a}} h_{\alpha\beta} = 0$ by introducing Lagrange multipliers); the G-invariant ansatz is therefore consistent.

In sect. 5.10 we discuss geodesics on **E** for G-invariant metrics and study their projection on **M** ; in this case, we have to introduce two kinds of non abelian charges and two kinds of scalar-field forces (compare with sect. 4.0 and sect. 4.9). A complete description of the projected trajectory, for particles of charge $(q_{\hat{a}}, \lambda_a)$ with $q_{\hat{a}} \in$ Lie(N/H), $\lambda_a \in \mathcal{L}$, is given by

$$D\dot{x}_\mu/dt = q_{\hat{a}}\, F_{\mu\nu}{}^{\hat{a}}\, \dot{x}^\nu + 1/2\; q^{\hat{a}}\, q^{\hat{b}}\, D_\mu(g_{\hat{a}\hat{b}}) + 1/2\; \lambda^a \lambda^b D_\mu(g_{ab})$$
$$Dq_{\hat{a}}/dt = C_{\hat{a}\hat{b}\hat{c}}\, q^{\hat{b}}\, q^{\hat{c}} + C_{\hat{a}bc}\, \lambda^b \lambda^c$$
$$D\lambda_a/dt = C_{abc}\, \lambda^b \lambda^c$$

In sect. 5.11, we show that the G-invariant ansatz is not the only possible one which leads to dimensional reduction from E to M ; we show how to build geometrically the so-called "popular ansatz" (leading to a gauge group at least equal to G): it can be gotten from a G × G invariant metric on $\overline{E}$ = E × G by passing to the quotient E = $\overline{E}/G_{diag}$. through the action of G_{diag} . After integration over the internal coordinates of **E**, we are led to a theory on **M** incorporating a gauge field valued in the Lie algebra of G × N(H)|H. This method of dimensional reduction is usually "inconsistent"; one reason for that being that the scalar curvature on **E** at a point (**x**, [y]) is usually not [y]-independent as it was in the case of a G-invariant metric.

Finally, in sect. 5.12 we generalize the dicussion to the study of G-spaces: we no longer assume that the "multidimensional universe" E has only one kind of orbits (G/H). In the case where we have a global group action of G on E, E can be "stratified" into subspaces ("strata") which have the same type of internal space : one of these strata (if G is not compact, then one has usually several open strata) is then open dense in E (in other words, the internal space is "almost everywhere" of the same type !). We indicate then how to apply the construction of G-invariant metrics to this more general situation.

5.1. The structure of a simple G-space

5.1.0. Summary

Here, we study in some detail the bundle structure of a simple G-space E. The problem has several aspects (or perspectives) among which are the following:

- Starting from E with a given global action of G one can *construct* a principal bundle P (here M = E/G and structure group N|H) as a *submanifold* of E.
- Then one can show that E can be considered as a bundle associated to P
- Conversely, if one starts with a principal bundle P(M,N|H) then, using the natural action of N|H on H\G one can construct an associated bundle E (base M, fiber H\G).
- Then one can define a global G-action on E.
- One can also identify the original principal bundle P with a submanifold of the associated bundle.
- Quite a different fibration of E is possible : E can be considered as a bundle over G/N and with typical fibre P. The use of this second fibration in physical applications is not clear.

5.1.1. Bundle structure of E

Let G be a compact group of transformations of a manifold E. For each $u \in E$, let G(u) denote the orbit of G through **u** :

$$G(u) = \{ua : a \in G\}$$

Then G(u) is a compact submanifold of E, called also a fibre or an "internal space". When u_1 and u_2 belong to the same orbit, then their isotropy groups (or "stability group", or "little groups"), denoted by H_{u_1} and H_{u_2} respectively, are conjugate. But isotropy groups associated to points in different orbits need not be conjugate - then E decomposes into "strata". By the "principal orbit theorem" [27] the stratum consisting of orbits with maximal dimension is an open dense submanifold of E. In our case it is natural to assume from the very beginning that E, being a model of an extended space-time, consists of one stratum only, i.e., that all isotropy groups H_u ($u \in E$) are mutually conjugate. With the above in mind, we state a theorem that we will comment on and explain later in this section.

Let E be a manifold with a right action of a compact Lie group G, and suppose that all isotropy groups H_u ($u \in E$) are conjugate to a standard one, say $H_{u_0} = H$. Let M be the set of all orbits, $H\backslash G$ the coset space of right classes Ha along H, and let N be the normalizer of H in G. Then M is a manifold and $E(M,H\backslash G)$ is an associated bundle with base M, fiber $H\backslash G$ and group $N|H$.

Because of its importance for us we shall give in sect. 5.1.2, 5.1.3, 5.1.4, 5.1.5 some more information about several ingredients of the above result.

5.1.2. The role of the normalizer

We already defined what the normalizer is in Ch.3 and studied its properties, in particular we have shown that there exists a one to one correspondance between elements of the group $N|H$ and automorphisms of $H\backslash G$ (by automorphism of $H\backslash G$ we mean an invertible mapping $H\backslash G \rightarrow H\backslash G$, which commutes with the right action of G).

5.1.3. - From $P(M,N|H)$ to $E(M,H\backslash G)$

Let us consider a principle bundle $P(M,N|H)$ with a base M and structure group $N|H$. We shall discuss here the geometry involved in the constructions of the associated bundle $E(M,H\backslash G)$. The procedure described is nothing but a particular case of the standard construction of an associated bundle from a principal one (see, e.g[127]).

In the direct product $P{\times}H\backslash G$, define the following relation

$$(p,[a]) \sim (p',[a']) \Leftrightarrow \exists [n]\in N|H \text{ so that } p'=p[n] \text{ and } [a']=[n^{-1}][a]$$

where $p[n]$ is obtained from p by using the right action of $n\in N|H$ on the principal bundle P, and $[n^{-1}][a] = [n^{-1}a]$. In other words, N|H acts on the typical fibre $H\backslash G$ by automorphisms, those discussed in 3.2 : $[a] = \alpha_n([a'])$. It is easy to see that the above relation is an equivalence relation. We shall denote an equivalence class by the symbol $p.[a]$. Let us recall what is the intuitive meaning of this : writing $u = p.[a]$ means that the "geometrical object" u has "coordinate" [a] in the "frame" p. Of course, $u = p.[a] = (p[n^{-1}]).([n][a])$, so that the group N|H plays the role of the group of transformations of "frames". The space of equivalence classes (quotient of $(P{\times}H\backslash G)$ by the equivalence relation) is, by definition, the associated fibre bundle $E = E(M,G\backslash H)$ with "geometrical objects" u in the fibre; M is the basis and transition functions are valued in N|H.

In a local trivialization determined by a local cross-section (gauge) $\sigma: M\to P$, the element p of P can be represented as $p = (x,[n])$, i.e. , $p = \sigma(x)n$, where $x\in M$ and $n\in N|H$. The element $u\in E$ can be written as (x,y), where $y = [a] \in H\backslash G$ and $u = \sigma(x).a$.

5.1.4. Global action of G on E(M,H\G)

Let us ask what are the transformations $\beta : H\backslash G\to H\backslash G$ of the typical fibre which pass through the equivalence relation defining E, to induce transforamtions of E itself. If, in some local trivialization (i.e. a local parametrization of E), $u = (x,y)$ and we want to define $\beta(u)$ as $(x,\beta(y))$, we must check that this definition is gauge independent, i.e., that $p.\beta[a] = p[n^{-1}].\beta([n][a])$ or $\beta([n][a]) = [n]\beta([a])$. In other words, the left action of the structure group on the typical fibre must commute with β. Since our structure group is, in general, non-abelian, it is clear that β cannot be the left multiplication by N|H. But it can well be the right multiplication by an element of G so that $ug = (p.[a])g=(p.[ag])$, since we have shown that $N|H$ is precisely the set of all transformations of $H\backslash G$ which commute with such β - s. The situation is now the following : we have an associated fibre bundle $E(M,H\backslash G)$ with structure group $N|H$, with G acting on E from the right and operating transitively on each fibre, so that the fibres of E coincide with the orbits of G. However, this action is not free; indeed, if $u = p.[a] \in E$, then the isotropy group of u is $H_u = a^{-1}Ha$. In particular, all isotropy subgroups are mutually conjugate. (Here and below e denotes the identity of G).

5.1.5. Embedding of P in E.

We will show now how P can be identified with a submanifold of E. Let P and E be as above (E associated to P) and consider, for each $p \in P$, the point $u(p) \in E$ defined by $u(p) = p.[e]$, i.e. $u(p)$ is the geometrical object uniquely defined by the requirement that it has "coordinates" $[e]$ in the "frame" p (e denotes, as usual, the identity of G). It is easy to see that (owing to the fact that the action of $N|H$ or $H\backslash G$ is clearly free) the mapping $p \to u(p)$ from P to E is injective ; it also satisfies $u(p[n]) = u(p).n$ for all $p \in P$, $n \in N$. In particular $u \in E$ is of the form $u = u(p)$ if and only if the isotropy group H_u of u is precisely H. Conversely, let E be a simple G-space, then all the isotropy groups H_u are conjugate to a standard one $H = H_{u_0}$, u_0 being some fixed point in E. When $u_2 = u_1 a$ then $H_{u_2} = a^{-1}H_{u_1}a$ but we get $H_{u_1} = H_{u_2}$ iff $a \in N$; let P be defined as the set of all $u \in E$ such that $H_u = H$. We already observed that if $p \in P$, $a \in G$ then $pa \in P$ if and only if $a \in N$, and since H acts trivially on the points of P, it is the quotient group $N|H$ which freely acts on the fibres of P. Therefore P has the algebraic structure of a principal bundle with base M (the space of orbits) and with structure group $N|H$. It is also clear that E can be identified with the bundle associated to P via the natural action of $N|H$ on $H\backslash G$. Indeed, we have a natural surjection $P \times H\backslash G \to E$ given by

$$(p,[a]) \to p.[a] := pa, \qquad p \in P \subset E,\ a \in G.$$

Analytically, to assure that **P** defined as above is a smooth fiber bundle, one needs the so-called "slice theorem" [152] [167] for the non compact case.

The resulting structure is represented as follows:

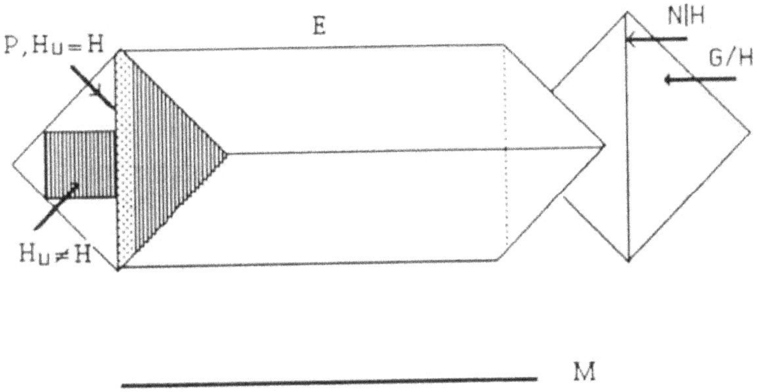

5.1.6. **The "other" fibration of E.**

Previously, we have seen that **E** can be considered as a collection of fibres of the type **H\G**, glued together and parametrized by the manifold of orbits **M** ; moreover this bundle **E** was associated to a principal bundle **P** which has same base and structure group **N|H**. We will now show that there exists another way of looking at **E**, another fibration which corresponds to the freedom we had of choosing the submanifold **P** in **E** (remember that **P** can be embedded in **E** via the choice of a particular $u_0 \in \mathbf{E}$ for which we set $\mathbf{H} = \mathbf{H}_{u_0}$). Actually we will never need to use this other fibration (and the reader may skip it) but it is nevertheless interesting to consider : remember that **H\G** is a **N|H** principal bundle over **N\G** (cf. sect.3.2.6) , but **N|H** acts on **P**, as we know, therefore we can build the associated bundle $\mathbf{H\backslash G} \times_{\mathbf{N|H}} \mathbf{P}$ which is a non-principal fiber bundle of base N\G and typical fiber **P** (neither a group nor a homogeneous space!) ; it is clear that the total space of the bundle is **E** itself. In other words **E** can be considered as a collection of spaces of the type **P** (itself a local product of **M** and N|H) glued together and parametrized by the space of orbits N\G. Schematically , we have

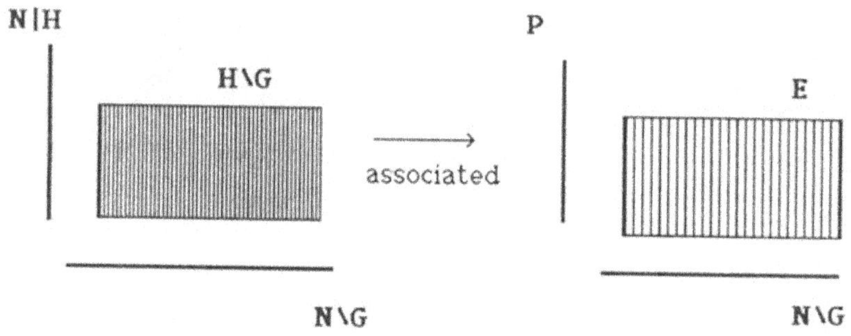

It is interesting to notice that what remains of the structure of **P** after passing to the quotient by the action of N|H is the action of the infinite-dimensional group of gauge transformations (cf. sect. 4.11 for definition and properties of this group).

5.2. Examples

By "examples" here, we mean "general classes of examples" : particular cases with particular groups G and H will be given later (sect 5.6.3).

5.2.1. Trivial examples

Take an arbitrary manifold **M** together with a compact group G and a (closed) subgroup **H** of G, then consider the trivial bundle **E** = M×H\G. E is then a G-universe in the previous sense.

5.2.2. Less trivial examples

We may consider G-spaces of the kind **E**=(H\G×H\G)/(N(H)|H) considered as bundles over N\G with typical fiber H\G. Indeed, we use first the fact that H\G is a N(H)|H bundle over N\G. Therefore we have a right action : p∈H\G, a∈N|H pa∈H\G := S of N|H := **K** over H\G and we define therefore the following diagonal right action of N(H)|H on H\G×H\G : (p,p')∈ S×S , a∈K, (pa,p'a)∈S×S and define **E** as the coset space obtained via the diagonal action. It is clear that **E** can also be gotten as a S-bundle over N\G and build the associated bundle **E** via the following left action of **K** on H\G : **ap** := **pa**$^{-1}$.

Schematically

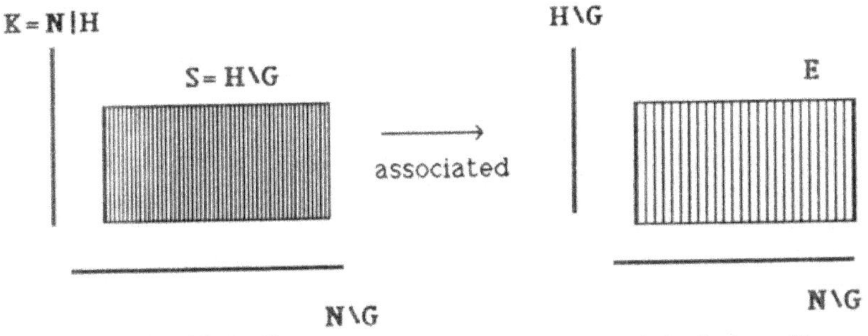

Principal bundle   Associated bundle

The points of **E** are equivalence classes :

class(p,p')={(pa,a^{-1}p')/ a∈K} = {(pa,p'a)/ a∈K}.

We can choose for example :

a) $G = Sp(2)$, $H = Sp(1)$ therefore $N = Sp(1) \times SU(2)$ and E becomes an S^7 bundle over S^4 (which is not trivial) with a global $Sp(2)$ action.

b) $G = SU(3)$, $H = SU(2)$ therefore $N : SU(2) \times_{Z_2} U(1)$ and E becomes an S^5 bundle over $\mathbb{C}P^2$ with a global $SU(3)$ action.

etc.

A similar construction starting with G writen as an H bundle over G/H and letting H act on G from the right allow us to write $E=(G \times G)H$ as a non-trivial G bundle over G/H (but this bundle is then principal for the G-action (extension of H to G).

5.2.3. Even less trivial examples

We start with two different homogeneous spaces $H_1 \backslash G_1$ and $H_2 \backslash G_2$ which are such that $N_1 | H_1$ is isomorphic with $N_2 | H_2$: we can then use $N_1 | H_1$ to act on $H_2 \backslash G_2$ and $N_2 | H_2$ to act on $H_1 \backslash G_1$, thereby building two associated bundles. Schematically:

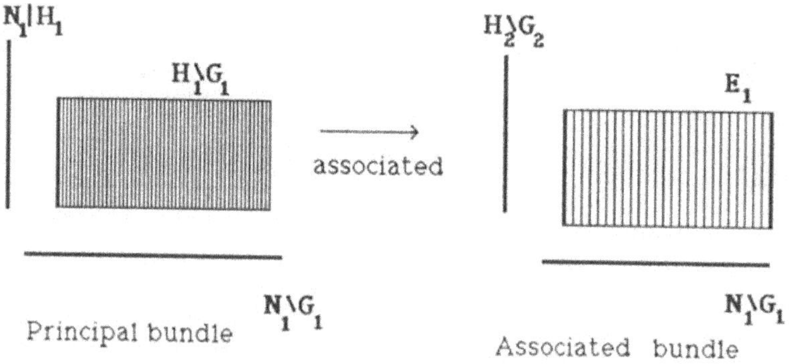

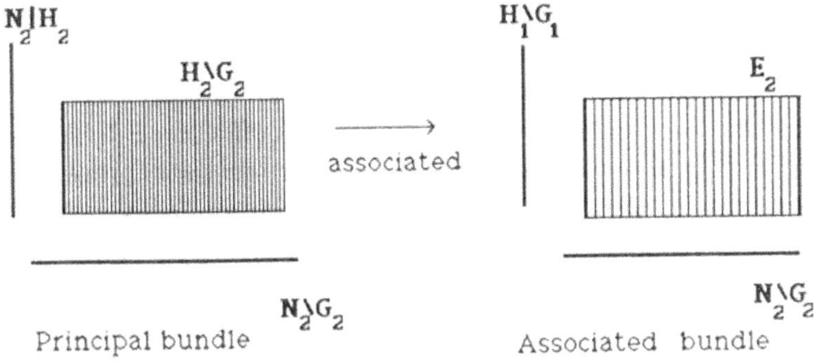

But it is easy to see that E_1 and E_2 have the same total space E, fibrated differently; actually $E = (H_1\backslash G_1 \times H_2\backslash G_2)/(N|H)$, where $N|H := N_1/H_1 = N_2/H_2$ and we divide by the diagonal action. For instance take

$G_1 = SU(3)$, $H_1 = SU(2)$, $N_1 = SU(2)\times_{Z_2}U(1)$, $N_1|H_1 \cong U(1)$,

$G_2 = SU(4)$, $H_2 = SU(3)$, $N_2 = SU(3)\times_{Z_2}U(1)$, $N_2|H_2 \cong U(1)$,

then build $E = (S^5\times S^7)/U(1)$, an 11-dimensional manifold which may be viewed either as a S^7 bundle over $\mathbb{C}P^2$ with a global SU(4) action or as a S^5 bundle over $\mathbb{C}P^3$ with a global SU(3) action.

5.2.4. **Remark**

More generally the principal orbit theorem (cf. section 5.12) provides many examples of G-spaces.

5.2.5. **Unexpected examples**

Take as a model a space E which is a local product $M \times G$, G being a Lie group, but suppose we are not interested in the G action but in a larger action on E (supposing that such an action exists! For example we take $E = M \times G$, a trivial bundle)for instance $G\times G$ action or $G\times H$ action, H being a subgroup of G. More precisely, we consider the situation where there exists a global $G\times H$ action on E and where the typical fiber is

$S=(G\times H)/\overline{H}$, with H being the diagonal embedding of H, i.e.$\overline{H}$ = $\{(h,h)\in H\times H,\ h\in H \subset G\}$. Notice that S is homeomorphic to G itself but does not usually carry a quotient group structure. E is associated with a principal bundle with structure group $\overline{N}(\overline{H})|\overline{H}$ where $\overline{N}(\overline{H})$ is the normalizer of $\overline{H}$ in $\overline{G}$ = G×H, and using then a theorem recalled in the introduction, we see that the structure group is locally isomorphic with $Z(i(H))\times N(H)|H$, where Z(i(H)) is the centraliser of H in G and N(H) the normaliser of H in G. For instance, if E is a local product of H and the three sphere S^3, and assuming that there exists at least an SU(2) action, we will be in one of the following situations :

-The typical fiber is (SU(2)×SU(2))/SU(2) and the structure group $\overline{N}|\overline{H}$ is just the center of SU(2),

-The typical fiber is (SU(2)×U(1))/U(1) and the structure group $\overline{N}|\overline{H}$ is U(1),

-The typical fiber is (SU(2))/{e} and we are in the principal bundle case; the structure group $\overline{N}|\overline{H}$ is SU(2).

Notice that there exists many fiber bundles with typical fiber S^3 and on which there is no SU(2)-action at all whatsoever. Take for instance $E = P\times_{Ad}SU(2)$ where P is a principal bundle with structure group SU(2) (cf. 4.11).

5.3. Fundamental and invariant vector fields

Whenever a Lie group G acts on a manifold E (we assume right action) we have two distinguished families of vector fields on E: the fundamental and the invariant vector fields. The first family is finite-dimensional (if dim(G) < ∞) and, in fact isomorphic to Lie(G) , while the second family is infinitely dimensional (unless M=E/G is discrete and finite). Roughly speaking the two families correspond to gauge transformations of the first (global) and of the second (local) kind respectively. We now proceed to define these two families and to discuss their essential properties.

5.3.1 Fundamental vector fields

As in sect. 3.3, to each element $\xi \in Lie(G)$ there corresponds the fundamental vector field $\tilde{\xi}$ on **E** given by the formula

(0) $\tilde{\xi}(y) = d/dt(y\, e^{t\xi})|_{t=0}$

with the property

(1) $[\tilde{\xi},\tilde{\tau}] = [\xi,\tau]^{\sim}$.

From the definition we find out how the group G acts on the fundamental fields :

$$\tilde{\xi}(y)a = d/dt(y\, e^{t\xi} a) = d/dt(ya\, e^{ta^{-1}\xi a}) = \tilde{\xi}_1(ya),$$

with $\xi_1 = Ad(a^{-1})\xi$.

Denoting T_i as a basis in Lie(G) and $e_i(y)$ the fundamental vector fields corresponding to these basic vectors : $e_i := \tilde{T}_i$, we have

(2) $e_i(y)\, a = Ad(a)^j{}_i \; e_j(ya)$

and

$[e_i,e_j] = C_{ij}{}^k\, e_k$, (we would get a minus sign for a left action).

We already studied in sect. 3.2 the decomposition of the Lie algebra of G. Let us recall a few notations

$$\mathcal{G} = \underbrace{\overbrace{\mathcal{H} + \mathcal{K}}^{\mathcal{N}} + \mathcal{L}}_{},\quad \text{with } \mathcal{S} = \mathcal{K} + \mathcal{L},$$

$[\mathcal{H},\mathcal{H}] \subset \mathcal{H}$, $[\mathcal{K},\mathcal{K}] \subset \mathcal{K}$, $[\mathcal{H},\mathcal{K}]=0$,

$[\mathcal{H},\mathcal{L}] \subset \mathcal{L}$, $[\mathcal{K},\mathcal{L}] \subset \mathcal{L}$

$\mathcal{G}$ = Lie(G), $\mathcal{H}$ = Lie(H), $\mathcal{N}$ = Lie(N), $\mathcal{K}$ = Lie(N|H),

$\mathcal{S} = T_0(S=H\backslash G)$, Z_H = centraliser of $\mathcal{H}$ in $\mathcal{G}$, C_H = centre of $\mathcal{H}$,

N is the normaliser of H in G, it is locally a direct product of two normal subgroups $N \cong HK$ (it may be disconnected).

The above decomposition of Lie(G) = $\mathcal{G}$ may be, for instance, made by using some chosen Ad G invariant scalar product in Lie(G).

Also, remember that we split accordingly the generators of the Lie algebra: T_i span Lie(G), $T_{\hat\alpha}$ span Lie(H), T_α span $\mathcal{S}$; moreover we split $\{T_\alpha\}$ in $\{T_a, T_{\hat a}\}$ where T_a span $\mathcal{L}$ and $T_{\hat a}$ span $\mathcal{K}$. We split the fundamental fields $\{e_i\}$ in the same way as the $\{T_i\}$, i.e. we write $\{e_i\}=\{e_{\hat\alpha}, e_\alpha\}$ and $\{e_\alpha\}=\{e_a, e_{\hat a}\}$.

It should be noticed that, at a given $u \in E$, the family $\{e_i(u)\}$ is over complete in the vertical tangent space at u. It is already obvious, from a dimensional argument, that we get n = dim G vertical fundamental fields corresponding to a basis in Lie(G) but the vertical subspace at u has only dimension s = dim G/H = n-h . Something special happens, however, on the submanifold $P \subset E$: the vector fields $\{e_{\hat\alpha}\}$ corresponding to Lie(H) vanish identically on P, whereas the s vector fields $\{e_\alpha\}$ are linearly independent at every point of P (and therefore also in some open neighbourhood U of P). we remind the reader that the principal bundle P was defined as the set of all points invariant under H (besides, this property is already obvious in the trivial case where $E = M \times H \backslash G$). The commutator of two vertical fields being again vertical[1], we have

$$[e_\alpha, e_\beta](u) = f_\alpha{}^\gamma{}_\beta(u)\, e_\gamma(u) , \quad u \in U,$$

where the $f_\alpha{}^\gamma{}_\beta$ are the structure functions of the vertical moving frame $\{e_\alpha\}$; notice that they depend generally on the point u at which they are calculated, however they are constant on P and the structure functions coincide there with the structure constants of G (cf. sect. 3.3 and also footnote 2 in sect. 1.3):

$$f_\alpha{}^\gamma{}_\beta(p) = C_{\alpha\beta}{}^\gamma \qquad \text{for } p \in P.$$

---

[1]This property follows from fact that vertical fields are "π-related" to the zero vector field on the base, and the property that π_*, when defined, preserved commutators.

Notice also that the property for $\{e_\alpha\}$ to constitute a (vertical) moving frame may fail if one goes too far from **P** : the vectors $\{e_\alpha\}$ may become linearly dependent there. We will use the vector fields $\{e_\alpha\}$ in the next section in order to build a moving frame on **E**.

5.3.2 Invariant vector fields

We discuss now invariant vector fields. An invariant vector field is a vector field **X** on **E** which commutes with the action of G :

(3) $\quad \mathbf{X}(y)\,\mathbf{a} = \mathbf{X}(y\mathbf{a})$

It is clear from (2) that fundamental fields are never invariant, unless they correspond to central generators. An infinitesimal version of (3) reads: invariant fields are those commuting with the fundamental ones.

(4) $[\mathbf{X},\tilde{\xi}] = 0 \qquad \forall\ \xi \in G$

or

(4') $[\mathbf{X},\mathbf{e}_i] = 0 \qquad i=1,2,\ldots,\dim G$

It is clear from the very definition that fundamental fields form a finite-dimensional space, which in fact (assuming G-action on E effective) is isomorphic to **Lie**(G). To see that invariant vector fields form usually an infinite-dimensional space observe that, by (4), if **X** is an invariant vector field and if $f : \mathbf{E} \rightarrow \mathbb{R}$ is a G-*invariant function* on **E**, then $y \rightarrow f(y)\mathbf{X}(y)$ is again an invariant vector field. Thus invariant vector fields form a *module* over the algebra of invariant functions. As we will discuss later, (Ch. 10) the algebra of invariant functions on E coincide with the algebra of *all* functions on **M** = E/G - which is an infinite-dimensional algebra for dim **M** > 0. In practical applications one meets usually two kinds of invariant vector fields. Fields of the first kind are vertical invariant vector fields. These correspond to local left actions of N(H)|H on **E** (see sect. 4.11 and Ch. 10). Invariant vector fields of the second kind come from a local trivialization of **E**. If $\sigma : \mathbf{M} \rightarrow \mathbf{P} \subset \mathbf{E}$ is a local section of the principal bundle P then σ induces a local trivialization of E (since E is an associated bundle). We can write this trivialization $u=(x,y)$, $x\in M$, $y\in H\backslash G$ for $u=\sigma(x).y$. This local parametrization of the points of E by $x\in M$ and $y\in H\backslash G$ allows us to introduce vector fields $\partial_\mu = \partial/\partial x^\mu$ on E. These vector fields are invariant by their very definition; we have:

$$(5) \qquad [\partial_\mu, e_i] = 0.$$

5.4. G-invariant metrics on E

5.4.1. The reduction theorem

Let us state a theorem that generalizes the one proved in section 4.4.1.

Let E be a simple G-space, then every G-invariant metric on E determines and is determined by a triple consisting of

i) for each $x\in M$, a G-invariant metric in S_x - the copy of $H\backslash G$ over x,

ii) a G invariant horizontal distribution (Z_u) $u\in E$ on E or, equivalently a principal connection in $P(N|H,M)$,

iii) a metric γ on M.

5.4.2. Proof of the reduction theorem

A given manifold may not always admit a non-degenerate pseudo-Riemannian metric of given signature; in the following we shall always assume that problems of this type do not arise.

In subparagraphs i) ii) and iii) we prove the direct proposition, while in iv) we will show the converse.

i) g being a metric on E, we know, a fortiori, how to compute the scalar product of two vertical vectors. Since g is G invariant, we obtain a G invariant metric h_x, on every fibre E_x of E.

ii) For every $u \in E$, let V_u denote the internal, or vertical, tangent space i.e., the subspace of $T_u(E)$ consisting of all vectors tangent to the fibre at u. Let Z_u be the orthogonal complement of V_u in $T_u(E)$ with respect to our given G invariant metric g. Since g is G-invariant, we have $(Z_u)a = Z_{ua}$, $a \in G$, i.e., we have a G-invariant horizontal distribution on E. In order to show that the distribution determines a principal connection, we must prove that, at $p \in P$, Z_p is *tangent to* P. This follows by observing that Z_p is orthogonal to the vectors $e_a(p)$ which span the orthogonal complement to P at p. More precisely the argument goes through the following steps:

1) With $p \in P$ the tangent space T_pE contains the subspaces T_pP and $\mathcal{L}_p$, where T_pP consists of vectors tangent to P and $\mathcal{L}_p$ consists of fundamental vectors from $\mathcal{L} \subset Lie(G)$.

2) The group H leaves the point p invariant and therefore rotates the vectors in T_pE.

3) Since H does not move the points of P at all, its action on T_pP is trivial.

4) On the other hand, by eq. 3.2(11) and the following discussion, we know that $\mathcal{L}_p$ does not contain vectors invariant under H.

5) Therefore T_pP and $\mathcal{L}_p$ must be orthogonal (w.r. to the H-invariant scalar product induced by the metric g in T_pE.

6) By counting the dimensions we must have $T_pE = T_pP \oplus \mathcal{L}_p$.

7) Then, since Z_p is orthogonal to *all* vertical vectors, it is *a fortiori* orthogonal to vectors from $\mathcal{L}_p$ and thus by 6) must be a part of T_pP.

It should be observed that the above argument may fail if the metric g is pseudo-Riemannian and the induced metric on $\mathcal{L}_p$ is degenerate ! This however should be considered as an exception worth noticing but neglected in the analysis of a generic case.

Now, $(Z_p)_{p \in P}$ is an N and therefore also an N|H invariant horizontal distribution on P, i.e. a principal connection.

iii) The scalar product $\gamma_x(v,w)$ of two vectors, tangent to M at a point x, is obtained as follows : choose an arbitrary point u in the orbit labelled by x, and let v^* and w^* be the vectors in Z_u which project onto v and w respectively. Then define $\gamma_x(v,w)=g_u(v^*,w^*)$. The result is independent of the choice u on the orbit because of the G invariance of the metric g (cf. the principal bundle case 4.4.2).

iv) Conversely, it is easy to see that given a metric γ on M, a principal connection $(Z_p)_{p\in P}$ on P, and G-invariant metrics h_x on the fibres of E, one constructs a G-invariant metric on E. Indeed, given $\xi,\tau \in T_uE$, let $p\in P$ be such that $u = pa$ for some $a\in G$, and let $\pi(\xi)$ and $\pi(\tau)$ denote the projections of ξ and τ on M. Denote by ξ^* and τ^* the horizontal lifts of $\pi(\xi)$ and $\pi(\tau)$ to Z_p. Then the vectors $\xi-\xi^*a$ and $\tau-\tau^*a$ are vertical and the scalar product of ξ and τ in E can be defined as

$$g_u(\xi,\tau) = \gamma_x((\pi(\xi)),(\pi(\tau))) + h_x(\xi-\xi^*a,\tau-\tau^*a).$$

5.4.3. Comments on scalar fields

Constraints on the fields $h_{\alpha\beta}(x)$

Notice that a co-ordinate representation of h_x is obtained as follows : choose a local cross section (gauge) of P, $\sigma : M \to P$ -such a section "marks the origin" on each fibre- then define $h_{\alpha\beta}(x) := g_{\alpha\beta}(\sigma(x))$, where $g_{\alpha\beta}(p) := g_p(e_\alpha(p),e_\beta(p))$, $p\in P$ are numerical functions on P. The matrix $h=(h_{\alpha\beta})$, being associated with a G invariant metric on a space isomorphic to $H\backslash G$ is automatically Ad H invariant, and therefore satisfies the constraints

(1) $$R(a)^T h \; R(a) = h, \quad a\in H,$$

where $R(a)$, $a\in H\subset N$, is given by $R(n)^\alpha{}_\beta=(\Lambda(n))^\alpha{}_\beta$ and

$$\Lambda(n) = \begin{pmatrix} (\Lambda(n))^{\hat\alpha}{}_{\hat\beta} & 0 \\ 0 & ((\Lambda(n))^{\alpha}{}_{\beta} \end{pmatrix}$$

$(\Lambda(a))^i{}_j$ denotes the matrix of the adjoint representation of G:

$(\mathrm{Ad}\ a)T_i = aT_ia^{-1} = \Lambda(a)^i{}_j\, T_i$.

According to (1) the matrix $h(x)$ defines an Ad H invariant scalar product in $\mathcal{S}$. We have defined a splitting of $\mathcal{S}$ into the direct sum $\mathcal{K}$ and $\mathcal{L}$ of two subspaces, using an auxiliary bi-invariant metric on G. We now realize that $\mathcal{K}$ and $\mathcal{L}$ are also orthogonal with respect to the scalar product induced by $h_{\alpha\beta}(x)$. Indeed, $\mathcal{K}$ and $\mathcal{L}$ carry disjoint representations of H and therefore they are orthogonal with respect to every Ad H invariant scalar product:

$$h = \begin{pmatrix} h_{\hat a\hat b} & 0 \\ 0 & h_{ab} \end{pmatrix}$$

Owing to formula 1, we also have

$$R(a) = \begin{pmatrix} \delta^{\hat a}{}_{\hat b} & 0 \\ 0 & R^a{}_b \end{pmatrix}$$

The field $x \to h_x$ of G-invariant metrics on fibres of E can be identified with a cross-section of a bundle associated to P. Actually, we find

$$g_{\alpha\beta}(pn) = R(n)^{\alpha'}{}_{\alpha} R(n)^{\beta'}{}_{\beta}\, g_{\alpha'\beta'}(p)\ , \qquad n \in N$$

and

$$g_{\alpha\beta}(p) = R(h)^{\alpha'}{}_{\alpha}\, R(h)^{\beta'}{}_{\beta}\, g_{\alpha'\beta'}(p)\ , \qquad h \in H$$

so that the numerical functions on P can indeed be identified with sections of an appropriate associated bundle ([127, Ch.II, example 5.2], also Ch VI]. These relations define the type of transformations of the scalar fields under N|H gauge transformations.

Notice that it is enough to know the functions $\mathbf{h}_{\alpha\beta}(\mathbf{x})=\mathbf{g}_{\alpha\beta}(\sigma(\mathbf{x}))$ where σ: $\mathbf{M}\to\mathbf{P}$ is a local gauge. Using G-invariance and the constraints (1), $\mathbf{h}_{\alpha\beta}$ can then be unambiguously propagated all over the bundle E.

In terms of components, we may also express the constraint (1) as follows: the scalar fields $\mathbf{h}_{\alpha\beta}(\mathbf{x})$ are subject to the linear constraint of **Ad H** invariance:

$$(2)\quad \Lambda^{\gamma}{}_{\alpha}(\mathbf{a})\, \Lambda^{\delta}{}_{\beta}(\mathbf{a})\, \mathbf{h}_{\gamma\delta}(\mathbf{x}) = \mathbf{h}_{\alpha\beta}(\mathbf{x}) \text{ for all } \mathbf{a} \in H.$$

Infinitesimally, this is expressed by the vanishing of the following Lie derivative

$$(3)\quad L_{\hat{\alpha}}\, \mathbf{h}_{\beta\gamma}(\mathbf{x}) = C_{\hat{\alpha}\beta}{}^{\delta}\, \mathbf{h}_{\delta\gamma}(\mathbf{x}) + C_{\hat{\alpha}\gamma}{}^{\delta}\, \mathbf{h}_{\beta\delta}(\mathbf{x}) = 0.$$

Counting the number of scalar fields

We have already devoted most of Ch.3 to the study of G-invariant metrics on homogeneous spaces S = H\G. In the present situation we have rather a field of metrics (from a physicists point of view, $\mathbf{h}_{\alpha\beta}(\mathbf{x})$ would be called a 'scalar field' since it does not carry space-time indices); in the language explained in Ch.6, it is a 'bundle-valued zero form'. We will give several examples in section 5.6.1.

Remark

We are not consistent when denoting these scalar fields; sometimes we write them as $\mathbf{g}_{\alpha\beta}$ (when they are considered as part of the metric $\mathbf{g}$), another time we write them as $\mathbf{h}_{\alpha\beta}$ (when we do not want to confuse them with the $\mathbf{g}_{\mu\nu}$).

5.4.4. Adapted moving frame and explicit expression for the metric.

The moving frame.

We already discussed the properties of fundamental fields in section 5.2.3 and we saw that the (n-h) fields $K_\alpha := e_\alpha$ span the vertical space at points $p \in P \subset E$ and in a neighbourhood U of P. Since we have a connection in P, we may use it to build m = dim M horizontal fields (e_μ) defined as the horizontal lifts of a basis (∂_μ) in M. Locally, E can be written as $M \times H \backslash G$ and we write

(2) $e_\mu(x,y) = \partial/\partial x_\mu - A_\mu{}^\alpha(x,y)\, K_\alpha(y)$

where

$$A_\mu{}^\alpha(x,y) = \Lambda^\alpha{}_{\hat{a}}(a)\, A_\mu{}^{\hat{a}}(x),$$

$\Lambda^\alpha{}_\beta$ being the matrix of the adjoint representation of G, $\hat{a}$ being, as usual, an index referring to Lie N(H)|H and the argument $a \in G$ of $\Lambda^\alpha{}_{\hat{a}}$ is a group element which can be any representative of the coset $y = Ha$. (In more detail the argument here goes as follows : choosing a local section $\sigma: M \rightarrow P \subset E$ we parametrize E by $x \in M$, $y \in H \backslash G$ as already discussed. Now at a point $\sigma(x)$ of P we know from Ch. 4 that the horizontal lift has the form $e_\mu(x,\sigma(x)) = \partial/\partial x_\mu - A_\mu{}^{\hat{a}}(x)\, K_{\hat{a}}(\sigma(x))$. Applying now to both sides $a \in G$ such that $y = [a]$, and making use of it at 5.3 (2) we find (2).)

The adapted moving frame is then $((K_\alpha),(e_\mu)) := (e_A)$. Notice that, although the (K_α) are orthogonal to the (e_μ), there is no reason why the (e_μ) and especially (K_α) should constitute orthonormal families : the adapted moving frame is usually not an othonormal frame and this will be important when we come to discuss the properties of spinors on E (Ch. 7 and 8). Notice also that the Killing vectors $K_i(y)$ are sometimes written $K_i{}^m(y)(\partial/\partial y^m)$, where (y^m) are coordinates on $H \backslash G$.

Expressions for the metric

In this frame, and in a neighbourhood of P, the G-invariant metrics on E can be written

$$[\hat{g}] = \begin{bmatrix} \hat{g}_{\mu\nu}(x,y) & \hat{g}_{\mu\alpha}(x,y) \\ \hat{g}_{\mu\alpha}(x,y) & \hat{g}_{\alpha\beta}(x,y) \end{bmatrix}$$

with
$$\hat{g}_{\mu\nu}(x,y) = \gamma_{\mu\nu}(x)$$
$$\hat{g}_{\mu\alpha}(x,y) = 0$$
$$\hat{g}_{\alpha\beta}(x,y) = \Lambda_\alpha{}^\gamma(a)\Lambda_\beta{}^\delta(a)h_{\gamma\delta}(x)$$

here, $a \in G$ being as above any representation of the coset $y = Ha$.

Explicitly the (contravariant) metric is

$$g^{-1} = \gamma^{\mu\nu}(x)(\partial_\mu - A_\mu{}^{\hat{a}}(x)\Lambda^\alpha{}_{\hat{a}}(a)K_\alpha(y)) \otimes (\partial_\nu - A_\nu{}^{\hat{b}}(x)\Lambda^\beta{}_{\hat{b}}(a)K_\beta(y)) + h^{\alpha\beta}(x)\,\Lambda^\gamma{}_\alpha(a)\,\Lambda^\delta{}_\beta(a)\;\; K_\gamma(y) \otimes K_\delta(y)$$

It is enlightening - although unnecessary - to express g^{-1} in P in terms of the invariant vector fields ${}^\sigma e_{\hat{a}}{}^L(g) = g^{-1}e_{\hat{a}}(g)g$ (which are only defined through the choice of a gauge σ, and satisfy $[{}^\sigma e_{\hat{a}}{}^L, K_{\hat{b}}](g) = 0$). These vector fields ${}^\sigma e_{\hat{a}}{}^L$ can be thought of as right invariant vector fields in the copy of $K = N|H$ above x. Then one obtains on P

$$g^{-1} = \gamma^{\mu\nu}(x)(\partial_\mu - A_\mu{}^{\hat{a}}(x)\;{}^\sigma e_{\hat{a}}{}^L(g)) \otimes (\partial_\nu - A_\nu{}^{\hat{b}}(x)\;{}^\sigma e_{\hat{b}}{}^L(g)) + h^{\alpha\beta}(x)\;\; {}^\sigma e_\alpha{}^L(g) \otimes {}^\sigma e_\beta{}^L(g).$$

This writing clearly exhibits the G-invariance (take the Lie derivative with respect to the fundamental fields of the G action), but destroys the explicit gauge invariance.

5.4.5 Structure functions of the adapted moving frame

The price paid for the simple form of the metric in the adapted moving frame is in the more complicated form of the commutators. In order to do any calculations (in particular, to compute the Christoffel symbols and the curvature tensor), on needs the commutators of the basis. They are given by

$$[e_\mu e_\nu](x,y) = - \Lambda^\alpha{}_{\hat{a}}(a)\, F_{\mu\nu}{}^{\hat{a}}(x)\, K_\alpha(y)$$
$$[e_\mu, K_\alpha] = 0$$
$$[K_\alpha, K_\beta](y) = f_\alpha{}^\gamma{}_\beta(a)\, K_\gamma(y)$$

where $F_{\mu\nu}{}^{\hat{a}}$ is the field strengh of the Yang-Mills field $A_\mu{}^{\hat{a}}$

$$F_{\mu\nu}{}^{\hat{a}} = \partial_\mu A_\nu{}^{\hat{a}} - \partial_\nu A_\mu{}^{\hat{a}} + C_{\hat{b}\hat{c}}{}^{\hat{a}}\, A_\mu{}^{\hat{b}} A_\mu{}^{\hat{c}}$$

and the structure functions are given by

$$f_\alpha{}^\gamma{}_\beta(a) = C_{\alpha\beta}{}^\gamma + C_{\alpha\beta}{}^{\hat{\varepsilon}} \Lambda^\delta{}_{\hat{\varepsilon}}(a) L^\gamma{}_\delta(a)$$

where $L^\gamma{}_\delta(a)$ is the inverse matrix of $\Lambda^\gamma{}_\delta(a)$. Here as everywhere in this chapter, the relation between $y \in H\backslash G$ and $a \in G$ is given by $y = y_0 a$, where $y_0 = H = [e]$ is the origin of $H\backslash G$. Everywhere, care is taken so that the formulae we use do not depend on the choice of a.

Notice also that the above formulae simplify at the points of P, since there we have

$$K_{\hat{\alpha}} = 0$$
(3) and
$$f_\alpha{}^\gamma{}_\beta = C_{\alpha\beta}{}^\gamma$$

For calculation purposes, one also needs the expression of $K_\alpha(f_\beta{}^\delta{}_\gamma)$ in terms of the structure constants $C_{ij}{}^k$ of the group G. Indeed, when one wants to compute geometrical quantities such as scalar curvature of a homogeneous space, one performs the calculation at a point $p \in P$, but what are needed are the values of the structure functions and their derivatives in the directions transversal to P. To obtain such a formula we first observe that, from (3), we have

$$f_{\beta}{}^{\delta}{}_{\gamma}(u)K_{\delta}(u) = C_{\beta\gamma}{}^{i}\, K_{i}(u).$$

Then, taking the commutator of both sides with K and using (3), we get

$$K_{\alpha}(f_{\beta}{}^{\delta}{}_{\gamma})(p) = C_{\beta\gamma}{}^{\hat{\alpha}}C_{\alpha\hat{\alpha}}{}^{\delta} \quad , \quad p \in P.$$

In particular, owing to the fact that $\mathcal{H}$ and $\mathcal{K}$ commute, we get $K_{\hat{a}}(f_{\beta}{}^{\delta}{}_{\gamma})(p) = 0$ in agreement with the results of sect. 5.3.1.

5.5 Curvature tensors for G-invariant metrics

5.5.1 Levi-Civita connection in the adapted moving frame

The structure functions of the moving frame e_A have already been given in sect. 5.4.5; from G-invariance of the metric, we get

$$e_{\mu}(g_{\alpha\beta}) = D_{\mu}g_{\alpha\beta},$$
$$e_{\alpha}{}^{R}(g_{\beta\gamma}) = C_{\alpha\beta,\gamma} + C_{\alpha\gamma,\beta}$$

where $C_{\alpha\beta,\gamma} = g_{\gamma\lambda}\, C_{\alpha\beta}{}^{\lambda}$, and $D_{\mu}g_{\alpha\beta}$ denotes the covariant derivative of $g_{\alpha\beta}$ with respect to the connection ω. When a local section $\sigma: M \rightarrow P$ is chosen, it can be explicitly written as

$$D_{\mu}g_{\alpha\beta} = \partial_{\mu}g_{\alpha\beta} + C_{\alpha\hat{a}}{}^{\lambda}\, A_{\mu}{}^{\hat{a}}\, g_{\gamma\beta} + C_{\beta\hat{a}}{}^{\gamma}\, A_{\mu}{}^{\hat{a}}\, g_{\alpha\gamma}$$

where $A = \sigma^{*}\omega$ is the Yang-Mills potential.

We now express the coefficients Γ_{ABC} of the Levi-Civita connection associated with the metric g -cf. sect.1.3.1-in terms of M-based quantities:

$$\Gamma_{\alpha\beta\gamma} = 1/2\,(f_{\alpha\gamma\beta} - f_{\gamma\beta\alpha} + f_{\beta\alpha\gamma})$$
$$\Gamma_{\mu\beta\alpha} = \Gamma_{\alpha\beta\mu} = -\Gamma_{\alpha\mu\beta} = 1/2\, D_\mu g_{\alpha\beta}$$
$$\Gamma_{\mu\alpha\nu} = -\Gamma_{\mu\nu\alpha} = -\Gamma_{\alpha\nu\mu} = -1/2\, F_{\mu\nu,\alpha}$$
$$\Gamma_{\mu\nu\rho} = \text{(the Christoffel symbols of } \gamma_{\mu\nu} \text{ on M)}.$$

Notice that in the present chapter $f_{\alpha\gamma\beta} = C_{\alpha\beta,\gamma}$ and $f_\alpha{}^\gamma{}_\beta = C_{\alpha\beta}{}^\gamma$, also $\Gamma_{ACB} = g_{CC'}\,\Gamma_A{}^{C'}{}_B$.

5.5.2 Ricci tensor and scalar curvature

We give below the formulae for Ricci and scalar curvature of P endowed with a G-invariant metric g. In the adapted moving frame (e_A) defined previously, we obtain

(1) $$R_{\alpha\beta} = R_{\alpha\beta}(H\backslash G) + 1/4\, F_{\mu\nu,\alpha}\, F_{\mu\nu,\beta} + 1/2\, h^{\gamma\lambda}\, D_\mu h_{\alpha\gamma}\, D^\mu h_{\beta\lambda} - 1/4\, D_\mu h_{\alpha\beta}.h^{\gamma\lambda} D^\mu h_{\gamma\lambda} - 1/2\, D_\mu(D^\mu h_{\alpha\beta})$$

(2) $$R_{\mu\nu} = R_{\mu\nu}(M) - 1/2\, F_{\mu\sigma,\alpha}\, F_{\nu\sigma,\alpha} - 1/4\, h^{\alpha\beta}\, h^{\gamma\lambda}\, D_\mu h_{\alpha\gamma}\, D_\nu h_{\beta\lambda} - 1/2\, D_\mu(h^{\alpha\beta}\, D_\nu h_{\alpha\beta})$$

(3) $$R_{\mu\alpha} = 1/2\, D^\sigma F_{\mu\sigma,\alpha} + 1/4\, F_{\mu\sigma,\alpha}\, h^{\beta\gamma}\, D^\sigma h_{\beta\gamma} - 1/2\, C_{\alpha\beta}{}^\gamma\, h^{\beta\lambda}\, D_\mu h_{\gamma\lambda}$$

(4) $$R = R(M) + R(H\backslash G) - 1/4\, F_{\mu\nu,\alpha}\, F_{\mu\nu,\alpha} - 1/4\, D_\mu h_{\alpha\beta} D_\mu h_{\alpha\beta} - 1/4 h_\mu h_\mu - D_\mu h_\mu$$

Comments

a) $R_{\alpha\beta}(H\backslash G)$ and $R(H\backslash G)$ are given in sect. 3.4 with understanding that $h_{\alpha\beta}$ is now a function of x.

b)The derivative D_μ in the above formulae acts both on internal and space-time indices with $A_\mu{}^\alpha$ and $\Gamma_\mu{}^\rho{}_\nu$ respectively.

c)The summation over repeated indices on the same level is performed with $h_{\alpha\beta}$ and $\gamma_{\mu\nu}$. For example the term $F_{\mu\nu,\alpha}\,F_{\mu\nu,\alpha}$ should be read as $g^{\mu\mu'}g^{\nu\nu'}g_{\alpha\alpha'}F_{\mu\nu}{}^\alpha\;F_{\mu'\nu'}{}^{\alpha'}$ (notice that we use $h_{\alpha\beta}$ and $g_{\alpha\beta}$ to denote the same object). Actually, the $F_{\mu\nu}{}^\alpha$ entering in the above expressions are non-zero only for $\alpha=\hat{a}$.

d)We call $h = \det(h_{\alpha\beta})$ and $h_\mu = h^{\alpha\beta}D_\mu h_{\alpha\beta} = h^{\alpha\beta}\partial_\mu h_{\alpha\beta} = D_\mu \ln(h)$.

e) The last term of R can be written in several ways, for example,

$$D_\mu(h^{\alpha\beta}D_\mu h_{\alpha\beta}) = -\,h^{\alpha\beta}h^{\gamma l}D_\mu h_{\alpha\gamma}D_\mu h_{\beta l} + h^{\alpha\beta}\,D_\mu D_\mu h_{\alpha\beta}$$

Later we will get field equations from an action principle on E, the quantity of interest is then $R[g]\,\gamma^{1/2}\;h^{1/2}$ where $\gamma = \det(\gamma_{\mu\nu})$. The last term of R gives

$$\begin{aligned}-D_\mu(h^{\alpha\beta}D_\mu h_{\alpha\beta})\,\gamma^{1/2}\;h^{1/2} &= -\,(D_\mu h_\mu)\,\gamma^{1/2}\;h^{1/2}\\ &= 1/2\;h_\mu h_\mu\,\gamma^{1/2}\;h^{1/2} - \partial_\mu(h_\mu\,\gamma^{1/2}\;h^{1/2})\end{aligned}$$

where appears a total divergence, and we get

$$\begin{aligned}R^E[g]\;g^{1/2} = (&R^M + R^{G/H} - 1/4\;F_{\mu\nu,\alpha}F_{\mu\nu,\alpha} - 1/4\;D_\mu h_{\alpha\beta}D_\mu h_{\alpha\beta}\\ &+ 1/4\;h_\mu h_\mu\,)h^{1/2}\gamma^{1/2} - \partial_\mu(h_\mu\,\gamma^{1/2}\;h^{1/2})\end{aligned}$$

5.6 Examples

5.6.1 Counting the number of scalar fields :

The set of all G-invariant metrics which can be defined on a given homogeneous space S=H\G is itself a (connected) manifold, which we shall call here R(G;S). The reader should be aware of the fact that a given manifold S may admit several homogeneous structures (for example, the homogeneous spaces SO(8)/SO(7), SU(4)/SU(3), Spin(7)/G_2, Sp(2)/Sp(1) are all diffeomorphic to the standard seven-sphere S^7; therefore we stress the fact that, in the following, we will consider a manifold S with a given homogeneous structure H\G, and we shall consider those metrics on S which are invariant with respect to the action of G on S. Both G and H are assumed to be compact and connected. It is well-known (see Ch. 3) that G-invariant metrics on H\G are in one-to-one correspondence with **Ad H** invariant bilinear symmetric forms on the tangent space **S** at the origin of S = H\G. Indeed, owing to the transitivity of G action, one can transport such a scalar product from the origin to any point of S, and the transport is unambiguous because of the assumed **Ad H** invariance. In order to find the dimension d of the manifold R(G,S), one has therefore to decompose the representation **Ad H** of the vector space **S** into irreducible ones :

$$(0) \qquad \mathbf{S} = \oplus_i (V_i \otimes \mathbb{R}^{r_i})$$

where the index i runs over inequivalent irreducible representations of H on V_i, r_i is the multiplicity with which V_i occurs in **S**. Since the dimension d of R(G,S) is equal to the dimension of the space of symmetric operators on **S** commuting with the representation Ad H, it follows that

$$d = \Sigma_i \, r_i(r_i+1)/2.$$

It follows that the dimension k of the gauge group **K** coincides with r_0, where i = 0 denotes the trivial representation of H

$$k = \dim \mathcal{K} = r_0.$$

It is sometimes natural to restrict the attention to G invariant metrics on S with a fixed volume element. Denoting by d_0 the dimension of the manifold of conformal equivalence classes of G invariant metrics on G\H we get $d_0 = d - 1$.

In order to find out the decomposition (0), one can use tables [171]; in practice one looks at the branching rule of **Ad** G into N = **HK**. However, one has to remember that what we need are decompositions into real-irreducible representations, while the tables (and most papers on the subject) give the branching rules in terms of complex-irreducible ones. Special care has to be taken if H or **K** is one of the groups SU(n), **Spin**(4n+2) or E_6. Indeed, these groups admit some representations (ρ) which are not self-conjugate. In such cases ρ and $\bar{\rho}$ will appear simultaneously in the reduction of the adjoint representation of G, and one has to collect together such pairs to build $\mathbb{R}$-irreducible representations.

5.6.2 A class of almost trivial examples

a) S=G/{e}, i.e., H = {e}. Now S itself is a group, and the number of scalars is the number of right invariant metrics on G. The isotropy group H is trivial and its irreducible representations are one-dimensional. We have $\mathcal{S}=\mathcal{G}=\mathcal{N}=\mathcal{K}$ and $\mathcal{H}=\mathcal{L}=0$. The number of gauge potentials (i.e. the dimension of the gauge group) is k = g = dim (G). The number of scalars is $d = g(g+1)/2$. In that case $d_0 = \dim SL(g) / SO(g)$.

b) G = G × G_1, H = **diag** G_1 = { (**a**,**a**), **a** ∈ G_1}. The homogeneous space S=H\G can be naturally identified with G_1 (as a space, not as a group !), the action of G on S being given by $\mathbf{x} \rightarrow \mathbf{a}^{-1}\mathbf{x}\mathbf{b}$, $(\mathbf{a},\mathbf{b}) \in G_1 \times G_1$. In particular H acts on S by $\mathbf{x} \rightarrow \mathbf{a}^{-1}\mathbf{x}\mathbf{a}$, so that the number of scalars is equal to the number of bi-invariant metrics on G_1. This can be determined by decomposing the adjoint representation of G into irreducible representations and applying the formula (1). The gauge group **K**=N|H is easily seen to be isomorphic to the centre of G_1 (see also sect. 2.4 or apply 5.2.5 with H=G). In particular if G_1 is simple, then d = 1 and k = 0.

c) S=H\G is an irreducible symmetric space. In that case we have a reductive decomposition $\mathcal{G} = \mathcal{H} + \mathcal{S}$ with $[\mathcal{S},\mathcal{S}] \subset \mathcal{H}$ and the adjoint representation of **Ad** H on $\mathcal{S}$ is irreducible. It follows that $\mathcal{K} = 0$ and $\mathcal{S} = \mathcal{L}$, so that the gauge group is at most discrete (k=0), and only one scalar field is present (d=1). All irreducible symmetric spaces have been thoroughly studied and classified [235]. Example: $S^7 = SO(8)/SO(7)$ is a symmetric space admitting, up to a scale, only one SO(8) invariant metric.

d) S=H\G is an isotropy-irreducible homogeneous space. The cases discussed in c) fall into this category but there are many isotropy-irreducible homogeneous spaces which are not symmetric. They are classified in [234,236]. Here again d = 1 and **N|H** are discrete. Example: S^7 = **Spin**(7)/G_2 is a simply connected isotropy-irreducible, but non-symmetric space. It admits, up to a scale, only one **Spin**(7) invariant metric (n.b. the same as the one in example c) [236]).

Another example of this type : G = **Spin**(8), H = SU(3)/Z_3, S=H\G is isotropy irreducible (not symmetric) with N $\cong$ H and a discrete gauge group **K**. The decomposition of **Ad** G into real irreducible representations of H reads :28 = 8 $\oplus$ [10+$\overline{10}$] . Notice (see the end remark of Sect.5.6.1) that 10+$\overline{10}$ is to be understood as $\mathbb{R}$-irreducible and therefore the number of scalars (i.e., the number of **Spin**(8) invariant metrics on S) is d = 1 (and not 1+1). The 8 in the above decomposition is of course $\mathcal{H}$ itself.

5.6.3 Model Building

To build a model one has to choose a global group G together with two subgroups H and K so that N = H × K is the normalizer of H in G. Then S = H\G is the internal space and K = N|H is the gauge group. We already noticed in Ch. 3 that S itself admits a principal bundle structure with basis L = G/N and structure group K = N|H.

Table 1. Simply connected irreducible symmetric spaces L=G/N, where N is not a simple group. In the last column, the real irreducible representation **Ad(N)** on L is expressed in terms of irreducible complex representations. Decompositions not appearing in this column should be computed for given values of p,q (or n).

G	N	Remarks	L
SU(4)	SO		$3\otimes 3$
SU(p+q)	$S(U_p \times U_q)$	$p\geq 2, q\geq 1$	
SO(p+q)	SO(p)×SO(q)	$p\geq 2, q\geq 2$	
SO(4+q)	SO(4)×SO(q)	$p=4, q\geq 2$	
SO(2n)	U(n)	$n\geq 3$	
USp(2n)	U(n)	$n>1$	
USp(2(p+q))	USp(2p)×USp(2q)	$p\geq 1, q\geq 1$	
E_6	SU(6)×SU(2)		$20\otimes 2$
E_6	SO(10)×U(1)		$45\otimes 1+(16\oplus\overline{16})\otimes 1+1\otimes 1$
E_7	SO(12)×SU(2)		$32\otimes 2+\overline{32}\otimes 2$
E_7	E_6×U(1)		$27\otimes 1+\overline{27}\otimes 1$
E_8	E_7×SU(2)		$56\otimes 2$
F_4	USp(6)×SU(2)		$14\otimes 2$
G_2	SO(4)	(2 cases)	$4\otimes 2$

In this table USp(2n) denotes USp(2n,ℂ) = U(n,ℍ) = Sp(2n,ℂ)∩SU(2n,ℍ) and $S(U_p \times U_q)$ = (U(p)×U(q))∩SU(p+q). Notice the following local isomorphisms: SO(4) ≅ SU(2)×SU(2), $S(U_p \times U_q) \cong$ SU(p)×SU(q)×U(1).

Table 2. Same as in Table 1 but L=G/N are simply connected irreducible and isotropy-irreducible (but not symmetric) spaces; here again we list only the cases where N is not a simple Lie group.

G	N	Remarks	$\mathcal{L}$
$SU(pq)/Z_m$	$\{SU(p)/Z_p\}\times\{SU(q)/Z_q\}$	$p\geq q\geq 2, pq\geq 4$ $m=\mathrm{l.c.m}(pq)$	
F_4	$SO(3)\times G_2$		$5\otimes 7$
F_4	$\{SU(3)\times SU(3)\}/Z_3$		$3\otimes 6+\bar{3}\otimes\bar{6}$
E_6/Z_3	$\{SU(3)/Z_3\}\times G_2$		$8\otimes 14$
E_6/Z_3	$\{SU(3)\times SU(3)\times SU(3)\}/Z_3\times Z_3$		$3\otimes 3\otimes 3+\bar{3}\otimes\bar{3}\otimes\bar{3}$
E_7/Z_2	$\{USp(6)/Z_2\}\times G_2$		$14\otimes 14$
E_7/Z_2	$SO(3)\times F_4$		$3\otimes 26$
E_7/Z_2	$\{SU(3)\times SU(6)\}/Z_6$		$\bar{3}\otimes 15+3\otimes\overline{15}$
E_8	$G_2\times F_4$		$14\otimes 52$
E_8	$\{SU(3)\times E_6\}/Z_3$		$3\otimes 27+\bar{3}\otimes\overline{27}$

Given **H** and **K** it is natural to restrict oneself to those cases where the representation of N = H.K on **G/N** is faithful and irreducible. N is then, modulo a discrete group, the normalizer of **H** in G. Symmetric irreducible spaces are listed in [235] and the non-symmetric ones can be found in [234,236]. When one goes through these lists one realizes that

N is often a simple Lie group. In such a case either H = N (in which case the gauge group **K** is trivial) or H = {e} (modulo discrete groups), so that S is a group manifold itself, S=N=K=G. The G\N simply connected with non-simple N are not common and are all listed in tables 1 and 2. Table 1 is extracted from [235] and Table 2 is extracted from [236]. We also give the reduction (over the reals) of **Ad** G with respect to N = H.K. Examples with reducible representations of N on $\mathcal{L}$ can be obtained by taking products of irreducible ones.

Let us now analyze in some detail several cases from the tables.

a) $G=E_8$, $N=(E_6\times SU(3))/Z_3$, $H = E_6$.

S = G/H is of dimension 170, L = G/N is isotropy irreducible but not symmetric. The connected component of the identity of the gauge group N|H is $SU(3)/Z_3$. The reduction of **Ad** G with respect to real irreducible representations of N=H.K is

$$\underbrace{[248]}_{G} = \underbrace{[78\otimes 1]}_{H} + \underbrace{[1\otimes 8]}_{K} + \underbrace{[27\otimes 3 + \overline{27}\otimes\overline{3}]}_{L}$$

The reduction of the subspace $\mathcal{S} = \mathcal{K}+\mathcal{L}$ with respect to H is

$$\mathcal{S} = 8[1] + 3[27+\overline{27}].$$

The dimension of the space of G invariant metrics on S is therefore

$$d = (8\times 9)/2 + (3\times 4)/2 = 42.$$

b) G=SO(10), N=SU(5)×U(1), H = U(1).

S = G/H is of dimension 20, L = G/N is is a Hermitian symmetric space. The gauge group is SU(5), and the reduction of **Ad** G with respect to N = H.K reads

$$[45] = [1\otimes 1] + [1\otimes 24] + [1\otimes 10 + \overline{1}\otimes\overline{10}].$$

The reduction of $\mathcal{S}$ with respect to H is

$$\mathcal{S} = 24[1] + 10\,[1+1].$$

(Observe that in the real domain an Abelian group may have two-dimensional irreducible representations). Therefore

$$d = (24\times 25)/2 + (10\times 11)/2 = 355.$$

c) $G=U(q+1,\mathbb{H})$, $N=U(q,\mathbb{H})\times SU(2)$, $H=U(q,\mathbb{H})$ (or $USp(2q,\mathbb{C})$)

$S=G/H$ is a quaternionic projective space ($G\backslash N$ is symmetric) $\mathbb{H}P^q$. The gauge group is SU(2). The reduction of **Ad** G with respect to H.K reads

$$[(q+1)(2q+3)] = [q(2q+1)\otimes 1] + [1\otimes 3] + [2q\otimes 2]$$

and the reduction of **S** with respect to **H** is

$$\mathbf{S} = 3[1] + 1[4q].$$

Therefore

$$d = (3\times 4)/2 + (1\times 2)/2 = 7.$$

Notice that for q=1 we obtain the SU(2) fibration of S^7 over $S^4 \simeq \mathbb{H}P^1$, i.e., the k = 1 instanton bundle [211].

d) $G=SU(4)$, $N=SU(3)\times U(1)$, $H=SU(3)$.

$S=G/H=S^7$ is an irreducible Hermitian symmetric space; the gauge group is U(1) and the reduction of **Ad** G with respect to H.K is

$$[15] = [1\otimes 8] + [1\otimes 1] + [3\otimes 1 + \bar{3}\otimes \bar{1}]$$

The H-reduction of **S** reads

$$\mathbf{S} = 1 + 1[3+\bar{3}]$$

and therefore d= 2.

e) Let us mention also that cases where the internal space is a group G but where we want to construct metrics whose isometry group is bigger than G (for instance, example 5.6.2-b) can be studied using the general results of this chapter and in particular section 5.2.5. For example, in the particular case where E is the direct product of a base space M and the three sphere S^3, we may look at the most general SU(2) invariant metrics on E. Then considering E as a principal bundle and applying results of Ch.4 and therefore building such metrics out of a triple ($\gamma_{\mu\nu}(x)$, $A_\mu(x) \in$ Lie(SU(2)), $h_{ij}(x)$ being SU(2)-invariant on the fibers $S^3{}_x$), with $x \in M$. Finally, we may also look at the most general SU(2)×U(1) invariant metrics on E, then considering E as a bundle with homogeneous fibers (SU(2)×U(1))/**diag** U(1) and therefore building such metrics out of triples ($\gamma_{\mu\nu}(x)$, $A_\mu(x) \in$ Lie(U(1)), $h_{ij}(x)$ beeing SU(2)×U(1)-invariant on the fibers $S^3{}_x$). Finally, me may look at SU(2)×SU(2) invariant metrics on E, then consider E as a bundle with homogeneous fibers (SU(2)×SU(2))/**diag** SU(2) and therefore building such metrics out of couples ($\gamma_{\mu\nu}(x)$, $h_{ij}(x)$), this time $h_{ij}(x)=f(x)h_{ij}$ being a multiple of the Killing metric on SU(2) and therefore bi-invariant; in this last case the gauge group is discrete.

5.7 <u>Action principle and the consistency requirement</u>

5.7.1. The (dimensionally reduced) scalar curvature as a Lagrangian ?

There are several possibilities of using the results of 5.5.2 to determine field equations. One possibility is to take an action $S = \int_M R\, \gamma^{1/2}\, dx$, with this choice the last term of 5.5.2(4) is already a divergence, but such a choice is quite unnatural since it does not take the "extra dimensions" seriously. Then why bring them at all? The natural choice is to take

$$S = \int_E R\, g^{1/2}\, d^{(n+s)}x = \int_E R\, \gamma^{1/2}\, h^{1/2}\, d^{(n+s)}x.$$

The scalar curvature R for G-invariant metrics being independent of $y \in G/H$, we get in this way an effective m-dimensional theory, with

$$S = \int_M R(x)\, V(x)\, \gamma^{1/2}\, dx$$

where V(x) is the volume of the fiber at x. As we shall see below, it is useful to make a conformal rescaling on the space-time metric $\gamma_{\mu\nu}$ in order to re-absorb the internal volume factor; before that we want to analyse the "consistency requirement".

5.7.2 Validity of the consistency requirement

We already explained in the summary section of Ch.4 and in sect. 4.6.2 what the consistency requirement was. Since we want our theory to admit a ground state solution of the field equations, we will consider as in Ch.4 the simple case of a Lagrangian $(R-2\Lambda)\, g^{1/2}$ with cosmological constant in d=n+s dimensions. We consider this model just for the reason that it is the simplest model (or rather class of model) which exhibits a consistent dimensional reduction. The model admits "spontaneous compactification" on **E** = **M** × G/H with **M** being, for instance a de Sitter space and S=G/H a homogeneous space endowed with a G-invariant Einstein metric (the Einstein constants have also to agree). The field equations resulting from the above Lagrangian are

$$\hat{E}_{AB} + \Lambda\, \hat{g}_{AB} = \hat{T}_{AB} \quad ,$$

where $\hat{E}_{AB} = \hat{R}_{AB} - 1/2\, \hat{R}\, \hat{g}_{AB}$ is the Einstein tensor on E and where $\hat{T}_{AB}$ describes the contribution of matter fields. In this simple model we consider $T_{AB} = 0$, and so the field equations become

(1) $\hat{R}_{\mu\nu} = 2\, \Lambda/(d-2)\, \hat{g}_{\mu\nu}$

(2) $R_{\mu\alpha} = 0$

(3) $\hat{R}_{\alpha\beta} = 2\Lambda/(d-2)\, \hat{g}_{\alpha\beta}$

i.e. they describe an Einstein space with Einstein constant $2\Lambda/(d-2)$, d=m+s. To check the consistency of the ansatz, we have to compare the above equations, which are now equations for $\mathfrak{g}_{\mu\nu}(x), A_\mu{}^{\hat a}(x), \mathfrak{h}_{\alpha\beta}(x)$, with the ones obtained from the m-dimensional action which results from integration of the original Lagrangian over the internal coordinates. Modulo a constant proportionality factor (related to the standard volume on G/H), this m-dimensional action is

$$S_{eff} = \int_M (\hat{R} - 2\Lambda)\, \mathfrak{g}^{1/2}\, \mathfrak{h}^{1/2}\, d^m x\,.$$

This action should now be varied with respect to $\mathfrak{g}_{\mu\nu}(x), A_\mu{}^{\hat a}(x)$ and $\mathfrak{h}_{\alpha\beta}(x)$ in order to obtain a set of m-dimensional equations for those fields.

Here, it is essential to take into account the constraints of Ad H invariance (sect. 5.4.3) by adding to the previous effective action the Lagrange multiplier term $\int \lambda_{\beta\gamma}{}^{\hat\alpha} L_{\hat\alpha} \mathfrak{h}^{\beta\gamma}$, where $L_{\hat\alpha}$ is the Lie derivative in the direction of the fundamental fields of H.

One gets, by explicit calculation, the following set of equations:

$$(4) \qquad \hat{R}_{\mu\nu} - 1/2\,(\hat{R} - 2\Lambda)\,\mathfrak{g}_{\mu\nu} = 0,$$

$$(5) \qquad \hat{R}_{\mu\hat a} = 0$$

$$(6) \qquad \mathfrak{g}^{1/2}\mathfrak{h}^{1/2}\,[\,\hat{R}_{\alpha\beta} - 1/2\,(\hat{R} - 2\Lambda)\,\mathfrak{h}_{\alpha\beta} + 1/2(C_{\alpha\gamma}{}^{\hat\delta}C_{\beta\hat\delta}{}^{\gamma} + C_{\beta\gamma}{}^{\hat\delta}C_{\alpha\hat\delta}{}^{\gamma}) + \lambda_{\alpha\gamma}{}^{\hat\delta}C_{\beta\hat\delta}{}^{\gamma} + \lambda_{\beta\gamma}{}^{\hat\delta}C_{\alpha\hat\delta}{}^{\gamma}\,] = 0$$

$$(7) \qquad L_{\hat\alpha}\mathfrak{h}^{\beta\gamma} = 0$$

where $\hat{R} = \hat{g}^{\mu\nu}\hat{R}_{\mu\nu} + \mathfrak{h}^{\alpha\beta}\hat{R}_{\alpha\beta}$ (notice that $\mathfrak{g}_{\mu\nu} = g_{\mu\nu}$). While eq.(4) is evidently the same as (1), eq. (5) and (6) need further discussion. Consider first eq.(6). It is seen that by choosing the Lagrange multipliers to be

$$(8) \qquad \lambda_{\alpha\hat\delta}{}^{\hat\epsilon} = -1/2\, g^{1/2}\, \mathfrak{h}^{1/2}\, C_{\alpha\hat\delta}{}^{\hat\epsilon}\,,$$

eq.(6) becomes (3). Now, one knows that the Lagrange multipliers are uniquely determined by the constraints. On the other hand, the constraints of **Ad H** invariance are compatible with (3); indeed it is easy to see that $\hat{\mathbf{R}}_{\alpha\beta}$ as given in 5.5.2 satisfies 5.4.3(2) if $\mathbf{h}_{\alpha\beta}$ does. This justifies (8) and proves that (6)+(7) is the same as (3). Eq. (5) remains to be considered. It has exactly the form of (2), except that (2) asserts that $R_{\mu\alpha} = 0$ for all α, while (5) gives this conclusion only for $\alpha = \hat{a}$ (we remind the reader that the index α runs over all basis vectors of **G/H** while $\hat{a}$ runs only over those which span the Lie algebra $\mathcal{K}$ of **N(H)|H**. Since $F_{\mu\nu,\alpha}=0$ for $\alpha \neq \hat{a}$, it remains to show that $\mathbf{X}_{\mu\alpha} := C_{\alpha\beta}{}^{\gamma}\, \mathbf{h}^{\beta\delta}\, D_\mu \mathbf{h}_{\gamma\delta} = 0$ for $\alpha \neq \hat{a}$. One can prove this as follows: first, using the Jacobi identity and the constraint equations 5.4.3(3), one can show that $\mathbf{X}_{\mu\alpha}$, with μ fixed, is a vector which is **Ad H** -invariant, and then use the fact that all **Ad H** singlets of **S** are in **K**.

This ends the proof of the consistency of the G-invariant ansatz. Observe that it is essential in this statement that every solution which is an extremum of the effective m-dimensional action is an extremum of the original d-dimensional action. The inverse statement, that is, that every constrained solution of the original field equations is a solution of the m-dimensional ones follows from the fact that the effective m-dimensional action is defined as an integral of the original one over the internal variables.

5.8 Conformal rescaling and the effective Lagrangian

The integrand of the d-dimensional action was $\mathbf{L} = \mathbf{R}^E[\mathbf{g}]\mathbf{g}^{1/2}$, and could be written, with obvious notations,

$$\mathbf{L} = \{\mathbf{R}^M[\gamma] + \mathbf{R}^{G/H}[\mathbf{h}] - 1/4\ \mathbf{F}\,\mathbf{F}\,[\gamma^{-1}][\gamma^{-1}][\mathbf{h}]$$
$$- 1/4\,[\gamma^{-1}]\,[\mathbf{h}^{-1}][\mathbf{h}^{-1}]\,[\mathbf{Dh}][\mathbf{Dh}] - 1/4\,\mathbf{h.h} - \mathbf{D}_{.}\mathbf{h}\}\ \mathbf{h}^{1/2}\,\gamma^{1/2}$$

where $h_\mu = h^{\alpha\beta} D_\mu h_{\alpha\beta}$.

The analysis is very similar to the one performed in 4.7.1: we rescale $\gamma_{\mu\nu}$ and $h_{\alpha\beta}$, i.e, we define $[\gamma'] = h^x [\gamma]$, $[h'] = h^y [h]$, with

$$x = y = 1/(m-2) \text{ with } m = \dim M.$$

A tedious but straightforward calculation leads to the following effective lagrangian:

$$L = \{R^M[\gamma'] + R^{G/H}[h'] - 1/4\, FF[\gamma'^{-1}][\gamma'^{-1}][h] - 1/4\, [h'^{-1}][h'^{-1}][\gamma'^{-1}] [Dh'][Dh'] + 1/(4(d-2))\ [h'].[h']\ \}\, \gamma'^{1/2},$$

where $d = \dim E = \dim M + \dim G/H$. We omitted the surface terms.

As in 4.7.1, it is striking to observe that the above lagrangian only differs from the first by the absence of the global $h^{1/2}$ factor and by a simple modification of the coefficient of the internal volume factor $h_\mu h^\mu$.

5.9 Normalization and units, the potential for scalar fields.

5.9.1. Behaviour of the potential for scalar fields

The potential for the scalar fields h_{ij} when there is no cosmological consttnt in the 'big' space P is $U = -R^{G/H}[h(x)]$. It has therefore a clear geometrical interpretation since its value at $x \in M$ measures the scalar curvature of the internal space (which, by G-invariance, is constant on the internal space at x). It is clear that U is in general unbounded from below, indeed if for example, we "deflate" the internal space $(G/H)_x$ by making a scaling $h_{ij} \to \rho^2 h_{ij}$, its scalar curvature may become arbitrarily big. It should be stressed here that it is probably not possible to construct a "physical" theory out of the Einstein Lagrangian of pure gravity -even in higher dimensions!-. One should add terms of the higher order in the curvature, other kind of matter fields, spinor fields for example and maybe also a cosmological constant; this may allow for further constraints like freezing the volume of the internal space and considering only for "squashing" deformations. The internal space in this chapter being a compact homogeneous space, we recover the usual Einstein-Yang-Mills theory as the lowest order of an expansion of $\mathbf{h}_{ij}(x)$ around a G-invariant Einstein metric. This indeed corresponds to a saddle point of $\mathbf{U}(\mathbf{h}(x))$ if we consider only metrics with fixed volume - remember that the critical points of the functional $[\mathbf{h}] \longrightarrow \int R[\mathbf{h}]$, when $\mathbf{h}$ varies in the space $\mathfrak{M}_0$ of metrics of fixed volume, are precisely the Einstein metrics. For a given pair (G,H), the normal metric on G/H is not always an Einstein metric (cf. remark in sect. 3.0 and 3.2.6); however, by considering squashing deformations in the direction of $\mathbf{K} = N|H$, one obtains a one-parameter family of metrics among which one usually finds one or several Einstein metrics (cf. the example of the $Sp(2)\times Sp(1)$ invariant metrics on S^7 given in sect. 3.6). The scalar curvature of these squashed metrics can be obtained directly from the dimensional reduction formula of sect. 5.5.2: One writes $S = \mathbf{H}\backslash G$ as a $\mathbf{K}$-bundle over $L = N\backslash G$; if L is isotropy irreducible, the $(1+k(k+1)/2)$ parameter family of G-invariant metrics on S is given by $\mathbf{h}_{ab} = \mu^2 \tau_{ab}$, $\mathbf{h}_{\hat{a}\hat{b}} = \psi_{\hat{a}}{}^{\hat{a}'} \psi_{\hat{b}}{}^{\hat{b}'} \tau_{\hat{a}'\hat{b}'}$, where τ_{ij} is the Killing metric on G. In particular, when $L = N\backslash G$ is symmetric, the choice $\mu^2 = 1$ and $\psi_{\hat{a}}{}^{\hat{a}'} = t\,\delta_{\hat{a}}{}^{\hat{a}'}$ gives

$$R^{G/H} = R^L + R^K - 1/4\, C_{ab}{}^{\hat{c}}\, C^{ab}{}_{\hat{c}}$$

where $R^L = \dim L/2$, $R^K = (c.\dim \mathbf{K})/(4.t^2)$ and where $C_{ab}{}^{\hat{c}}$ can be considered as the curvature of the canonical connection on S (cf. [127]); c is the index of $\mathbf{K}$ in G (i.e. $c(C_{\hat{a}i}{}^{j} C_{\hat{b}j}{}^{i}) = C_{\hat{a}\hat{c}}{}^{\hat{d}} C_{\hat{b}\hat{d}}{}^{\hat{c}}$, cf. Ch. 2). Therefore,

$$R^{G/H}(\mathbf{h}(t)) = (\dim L)/2 + (c.\dim \mathbf{K})/(4.t^2) - (1/4)(1-c)\, t^2 \dim \mathbf{K}.$$

By replacing $\mathbf{h}_{\alpha\beta}$ by $\overline{\mathbf{h}}_{\alpha\beta} = (t^2)^{-k/s}\, \mathbf{h}_{\alpha\beta}$, with $s = \dim S$, $k = \dim \mathbf{K}$ we obtain a one-parameter family of metrics with fixed volume, the scalar curvature being now

$$R^{G/H}(\bar{h}(t)) = t^{2k/s}[(\dim L)/2 + (c.\dim K)/(4.t^2) - (1/4)(1-c)\, t^2 \dim K]$$

In the case of $S^7 = Sp(2)/Sp(1)$, we get

$$R(t) = t^{6/7}(2 + 1/(2t^2) - t^2/4).$$

The rest of the discussion continues as in sect. 3.6 (one gets two Einstein metrics and the curve R(t) is analogous to the one given in sect. 4.7.2, with the replacement $(1/7,1) \rightarrow (2/5,2)$).

As we already noticed, the saddle points of $R^{G/H}(h)$ when h varies in the space of G-invariant metric (with fixed volume) coincide in many cases with G-invariant Einstein metrics on $H\backslash G$, but these saddle points are usually neither minima nor maxima. The potential $U(h) = -R^{G/H}(h)$ is clearly invariant under the whole group of diffeomorphisms of the differentiable structure of G/H ; it is in particular invariant under the group $N(H)|H \times G$. All the metrics we are considering (all the fields $h_{\alpha\beta}$) are, by assumption G-invariant but their full isometry group can of course be bigger. If $h^{\bullet}_{\alpha\beta}$ is a saddle point of $U(h)$ and if the isometry group of $h^{\bullet}$ is strictly included in $N(H)|H \times G$ (it will then be of the kind $F \times G$, with $F \subset N(H)|H$) we may say that the $N(H)|H$ gauge group is "broken" to F.

We could now discuss the problem of the cosmological constant and the way one restores physical units, but this is very similar to the discussion carried out in 4.8.2, therefore we leave it as an exercise for the reader.

5.9.2 Example of $M \times S^3$

Let $E = M \times S^3$. On E we have a global $SU(2) \times SU(2)$ action and we may in particular study metrics which are invariant under $SU(2) \times \{e\}$,

$SU(2)\times U(1)$ or even $SU(2)\times SU(2)$. In the first case, in order to apply our general techniques, we consider E as the total space a $SU(2)$-principal bundle; in the second case, we consider E as the total space of a bundle with fiber $\overline{G}/\overline{H}$ where $\overline{G} = SU(2)\times U(1)$, $\overline{H} = \mathbf{diag}(U(1))$ associated with a principal bundle with structure group $\overline{N}(\overline{H})|\overline{H} = U(1)$; in the last case we consider E as the total space of a bundle with fiber $\overline{G}/\overline{H}$ where $\overline{G} = SU(2)\times SU(2)$, $\overline{H} = \mathbf{diag}(SU(2))$ associated with a principal bundle with structure group $\overline{N}(\overline{H})|\overline{H} = Z_2$.

In the first situation, the reduction theorem gives us (besides a metric $\gamma_{\mu\nu}$ on M) a Yang-Mills field $A_\mu{}^i$ valued in Lie(SU(2)) and 6 (=3×4/2) scalar fields h_{ij}; the second situation is obtained when only one of the Yang-Mills field and two scalar fields are non-zero (we get a U(1) "Maxwell"-field); in the third situation there are no gauge fields at all and only one scalar field is present.

In the first situation and because $SO(3) = SO(\dim S^3)$ is locally isomorphic with SU(2), one can choose a special gauge where there are only three (rather than six) independent scalar fields $h_{11}(x)$, $h_{22}(x)$ and $h_{22}(x)$, $x\in M$ (cf. the discussion in sect. 2.1). In this special gauge, the metric of the fiber at $x\in M$ reads:

$$h = h_{11}\, \omega^1\otimes\omega^1 + h_{22}\, \omega^2\otimes\omega^2 + h_{33}\, \omega^3\otimes\omega^3$$

where $\{\omega^i\}$ is a base of one-forms dual with $\{e_i\}$ with $[e_i,e_j] = \epsilon_{ijk}e_k$.

Let $x\in M \to g(x)\in SU(2)$ a gauge transformation, then, as usual, Yang-Mills fields as well as scalar fields transform as $F\to gFg^{-1}$, $h\to ghg^{-1}$. In the special gauge chosen above, we compute the scalar curvature R^P of E and get the following effective Lagrangian $L = R^P\sqrt{h}\sqrt{\gamma}$:

We set $h_{ii} = \exp(2\,\varphi_i)$, $i=1,2,3$.

$$L = (R^M[\gamma] + R^G[\varphi] - 1/4\, \Sigma_i\, F_{\mu\nu}{}^i F^{\mu\nu i} \exp(2\varphi_i) - (\Sigma_i\, \partial_\mu \varphi_i)^2$$
$$-(\Sigma_i\, \partial_\mu\varphi_i \partial^\mu \varphi_i + 2\, \Sigma_{i<j,l\neq i,l\neq j}\, sh^2(\varphi_i - \varphi_j)(A_\mu{}^k)^2) - D_\mu h_\mu)\exp(\Sigma\varphi_i)\sqrt{\gamma}$$

where $R^G[\varphi] = 2(4\sigma_2 - \sigma_1{}^2)/\sigma_3$ and $\sigma_1 = 4\, \Sigma_i \exp(2\varphi_i)$

$$\sigma_2 = 16\, \Sigma_{i\neq j} \exp(2\varphi_i+2\varphi_j)$$
$$\sigma_3 = 64 \exp(2\varphi_1+2\varphi_2+2\varphi_3)$$

and $h_\mu = h^{ij} D_\mu h_{ij}$.

Under a general SU(2) gauge transformation, the Lagrangian is written as in 4.5 (with a priori 6 fields $h_{ij}(x)$).

The second case corresponds to $A^1=A^2=0$, $\varphi_1=\varphi_2$. We get a U(1)-invariant Lagrangian. Here, it is convenient to set $A=A^3$, $h_{11} = h_{22} = \rho^2$, $h_{33} = \rho^2 \exp(2\tau)$ and we find:

$$L = \{R^M[\gamma] + R^G[\rho,\tau] - 1/4\, F_{\mu\nu}F^{\mu\nu} \rho^2 \exp(2\tau) - (2\partial_\mu\rho/\rho+\partial_\mu\tau)^2$$
$$-(2\, \partial_\mu\rho\partial^\mu\rho/\rho^2 + \partial_\mu\tau\partial^\mu\tau) - D_\mu h_\mu\}\, \rho \exp(\tau) \sqrt{\gamma}$$

where $R^G[\rho,\tau] = 2/\rho^2 \cdot (4 - \exp(2\tau))$.

If we perform the (conformal) change of field variables: $\gamma_{\mu\nu}' = (\rho \exp(\tau))^{2/(\dim M -2)}\, \gamma_{\mu\nu}$ and $h_{ij}' = (\rho \exp(\tau))^{2/(\dim M -2)}\, h_{ij}$, assuming now dim $M = 4$, we get

$$L = \{R^M[\gamma] + R^G[\rho,\tau] - 1/4\, F_{\mu\nu}F^{\mu\nu} \rho^2 \exp(2\tau) + 1/5$$
$$(2\partial_\mu\rho/\rho+\partial_\mu\tau)^2$$
$$-(2\, \partial_\mu\rho\partial^\mu\rho/\rho^2 + \partial_\mu\tau\partial^\mu\tau))\} \sqrt{\gamma}$$

As in sect. 4.8.2, and before the conformal change of variables, we could have added an extra cosmological term Λ^P in the "big" space $M\times S^3$, we leave this discussion as an exercise for the reader.

The third case corresponds to the "degenerated" situation where $A=\tau=0$. Then the Lagrangian is

$$L = \{R^M[\gamma] + R^G[\rho] - (2\partial_\mu\rho/\rho)^2 - (2\, \partial_\mu\rho\partial^\mu\rho/\rho^2) - D_\mu h_\mu\}\, \rho \sqrt{\gamma}$$

with $R^G[\rho] = 6/\rho^2$

After a conformal change of variables (assuming dim $M=4$) and setting $\rho = \exp(\sigma)$, we get a typical "Liouville" interaction:

$$L = \{R^M[\gamma] + 6\exp(-2\sigma) + (4/5 - 2)\,\partial_\mu\sigma\partial^\mu\sigma\}\sqrt{\gamma}$$

5.10. Color charges, scalar charges and the particle trajectories

As in sect. 4.9 the aim here is to find out differential equations which describe projections on **M** of geodesics in **E**. A priori there is no reason at all why a projection of a geodesic should be described by a simple differential equation. However G-invariance of the metric (and consistency of dimensional reduction which follows from G-invariance) allow for a positive solution of the above problem. This solution is more complicated than the one discussed in the principal bundle case. Here, when internal spaces are homogeneous of the type G/H, rather than group manifolds, an extra charge appears which couples only to scalar fields (in fact, only to some of them). Another complication is that this charge, strictly speaking, is of a nonlinear nature. There is a simple geometrical reason for this. In the G/H case, through any point $y \in E$ there passes not one but a whole family of geodesics which all have the same projection on **M**. This is a new phenomenon comparing to the situation studied in 4.9. We will not give here a derivation of the effective equations of motion on **M**; instead we will concentrate on giving us as precise as possible formulation of the result. At first we will neglect the complication resulting from the above mentioned non-uniqueness, and only later comment upon it.

-Initial data

Given a local section $\sigma: \mathbf{x} \rightarrow \sigma(\mathbf{x}) \in P \subset E$, and corresponding to this a local parametrization $\mathbf{u} = u(\mathbf{x},y), \mathbf{x} \in M$, $y \in H\backslash G$ of $\mathbf{u}$, the following initial data completely determines a trajectory on M (which is a projection of a geodesic in E).

$(x_0{}^\mu) \in M$ - the initial point

$(\dot{x}_0{}^\mu) \in T_{x_0}M$ - the initial velocity

$(q_0{}^{\hat{a}}) \in Lie(N|H)$ - the initial "color charge"

$(\lambda_0{}^a) \in \mathcal{L}$ - the initial "Higgs charge".

Once these initial conditions are given, and once geometry determined by a G-invariant metric on E (and described by $\gamma_{\mu\nu}$, A_μ and $h_{\alpha\beta}$, as explained in sect. 5.4) is known, then the time evolution of the above data is given by the following set of differential equations :

- *Equations of motion*

$$D\dot{x}^\mu/dt = \hat{q}_{\hat{a}}\, F_{\mu\nu}{}^{\hat{a}}\, \dot{x}^\nu + 1/2\, \hat{q}^{\hat{a}}\, \hat{q}^{\hat{b}}\, D_\mu(g_{\hat{a}\hat{b}}) + 1/2 \lambda^a \lambda^b\, D_\mu(h_{ab})$$

$$D\hat{q}^{\hat{a}}/dt = C_{\hat{a}\hat{b},\hat{c}}\, \hat{q}^{\hat{b}}\hat{q}^{\hat{c}} + C_{\hat{a}b,c}\, \lambda^b \lambda^c$$

$$D\lambda^a/dt = C_{ab,c}\, \lambda^b \lambda^c$$

The symbols here have the same meaning as in 4.8.2. Here we have an extra set of equations for λ.

- *Geometrical interpretation (=gauge transformation properties) of the color and Higgs charges.*

The above equations of motion were written in a fixed gauge σ. The next question is what are the transformation properties of $q^{\hat{a}}$ and q^a under gauge transformations $\sigma(x) = \sigma'(x)[n(x)]$, $x \to [n(x)]$ being a gauge function. As these properties follow automatically from the method the equations are derived and which we do not discuss here, again we state the result only : corresponding to the gauge transformation described by the gauge function $x \to [n(x)]$, we have

$$q^{\hat{a}} \to Ad(n(x))^{\hat{a}}{}_{\hat{b}}\, q^{\hat{b}}$$

$$\lambda^a \to Ad(n(x))^a{}_b\, \lambda^b$$

This way of writing needs the following commentary :

the gauge function is $x \rightarrow [n(x)] \in N|H$ and not $x \rightarrow n(x) \in N$; what then does $Ad(n(x))$ mean ? The answer for the first of the equations is : if $n(x)$ is replaced by $n'(x)=h(x)n(x)=n(x)h'(x)$ then $Ad(n'(x))^{\dot{a}}{}_{\dot{b}}\, q^{\dot{b}} = Ad(n(x))^{\dot{a}}{}_{\dot{b}}\, q^{b}$, because of the form 3.2 (14) of the adjoint representation. The answer for the second question is the following : there is no answer unless one of the two conditions is satisfied :

i) a nonlinear nature of the Higgs charge λ^a is recognized and λ^a is considered modulo Ad H action; this interpretation happens to be a natural result of the already mentioned non-uniqueness of geodesics as it can be seen from the method of deriving the equations of motion

ii) or the bundle P admits a reduction to an N-principal bundle (observe that although N seems to be "bigger" than N|H, nevertheless we should speak about a *reduction* owing to the direction of the arrow giving a group homomorphism $N \rightarrow N|H$). As every reduction restricts the admissible gauge transformations, so, also in our case, it resolves the problem we deal with. It is to be noticed however, that a global reduction of a principal bundle is not always possible (an example would be the problem of existence of spin structures on manifolds as discussed in Ch.6).

For the derivation of the equations of motion as well as for more details concerning their interpretation and for examples see [103].

5.11 Generalized Kaluza-Klein metrics (action of a bundle of groups).

5.11.1 Motivations

The motivations are more or less the same as in sect. 4.10.1 : what is commonly referred to as "the Kaluza-Klein ansatz" in the physical articles of the 80's does not descibe a G-invariant metric. This most popular anzatz leads, after dimensional reduction performed on a space E $\cong M \times G/H$ to a gauge group which is at least G. As we shall see below the most general "popular ansatz" leads to $G \times N|H$ but this ansatz is usually inconsistent (with the definition of sect. 4.0 or sect. 4.6.2); the source of the "inconsistency" comes from the fact that, being not G-invariant, the corresponding scalar curvature is usually not constant along the fibres. Although not G-invariant, this ansatz has a precise geometrical meaning that we discuss now.

5.11.2 G×N|H -invariant metrics on the tautological bundle E × G.

Let E=E(M,H\G) being a simple G-space, therefore an associated bundle to P=P(M,N|H). The following recipe then gives the full scale ansatz (the "popular" ansatz) : (Below, a point of E is locally parametrized as $(x,y) \in M \times H\backslash G$)

1) Artificially enlarge E to $\overline{E} = E \times G$

2) Now G × G acts from the right on $\overline{E}$ by $(x,y,b)(c,d) = (x,yc,d^{-1}b)$

3) The little group of (x,y,b) is $\overline{H} = H \times \{e\}$. Therefore $\overline{E}$ can be considered a $\overline{G}/\overline{H}$ bundle over M where $\overline{G} = G \times G$, the structure group of this bundle being $\overline{N}|\overline{H} \cong N|H \times G$.

4) According to the general technique (of 5.4), we can build $\overline{G}$ = G×G invariant metrics on $\overline{E}$ (which would lead, after dimensional reductions from $\overline{E}$ to M to gauge fields valued in N|H × G). Equations, obtained from direct dimensional reduction from $\overline{E}$ to M would, of course, be consistent with equations in $\overline{E}$.

5) The above metrics on $\overline{E}$ are also G^{diag}-invariant and therefore go to the quotient $\overline{E}/G^{diag} \cong E$; the obtained metrics on E have usually no invariance left and this describes the popular (non-invariant) ansatz on E.

<u>For example</u>, supposing that E is a local product of M_4 by S^7, we write $S^7 = H\backslash G$, with G = SO(8) and H = SO(7). Then, the only gauge group emerging from a G-invariant ansatz is Z_2 (!) indeed N|H = Z_2. However the popular ansatz would lead to a gauge group SO(8) (or even SO(8)×Z_2); such a metric could be gotten from a SO(8) × SO(8) invariant metric on E×SO(8) by going to the quotient under $SO(8)^{diag}$.

Notice that in the case where the group G does not act on the bundle E, there is also another kind of non G-invariant ansatz leading to the same effective gauge group $G \times N|H$ and that will be considered in Ch.10.

5.11.3 Groups of automorphisms

The infinite dimensional gauge group **Int P** of vertical automorphisms of $P = P(M, N|H)$ acts on the space $E = E(M, H\backslash G)$ since E is associated to P (cf. sect. 4.11); we therefore have an action of a bundle of groups (remember that **Int P** can also be defined as the set of sections of the bundle of groups **Ad P**). We postpone general comments about this action to Ch.10.

5.12 Some complements on G-spaces

5.12.1 The principal orbit theorem and stratification

Let G be a compact group of transformations of a manifold E. For each $u \in E$, let $G(u)$ denote the orbit of G through u : $G(u) = \{ua; a \in G\}$; then $G(u)$ is a compact submanifold of E. When u_1 and u_2 belong to the same orbit, then their isotropy groups are conjugated ; but isotropy groups associated to points in different orbits need not be conjugate : then E decomposes into strata (a stratum is by definition made of points which have conjugated little groups). In the previous sections of this chapter we always assumed that E was a simple G-space, i.e., that there was only one stratum (i.e. one type of stabilizers). In a more general situation we get the stratification $E = \cup_i E_i$, each stratum E_i being a collection of fibres G/H_i glued together and parametrized by some manifold of orbits M_i ; moreover, each stratum E_i in turn can be written as a union of substrata P_{ij} : $E_i = \cup P_{ij}$, where the points of a given P_{ij} have the same little group (cf. the "other fibration", sect. 5.1.6); also each E_i can be considered as a bundle $E_i(M_i, G/H_i)$ associated to a principal bundle P_{ij} $(M_i, N_i/H_i)$ - here we have to extend slightly the notion of bundles to the case of "manifolds with boundaries"-. The "principal orbit theorem" [1] asserts that the stratum consisting of orbits of maximal dimension is an open dense submanifold of E; this particular stratum is usually called the "principal stratum", its orbits are "generic orbits" and its associated isotropy group have the smallest dimension in the family of all possible isotropy groups. Let us just mention a theorem by Mastow which asserts that the lattice of stabilizers is such that if H_1 and H_2 are different stabilizers (associated to two different strata), there exist always one stratum for which the stabilizer H is included in $H_1 \cap H_2$.

It is striking to observe that such an analysis can also, with proper care, be carried out in the infinite dimensional case [23].

5.12.2 Generalization of the reduction theorem for G-invariant metrics

In order to build a G-invariant metric on a stratified space, we may build a G-invariant metric by applying the reduction theorem to the principal stratum and then extending its definition to the rest of E since the principal stratum is dense in E.

In this way we get "almost everywhere" a Yang-Mills field N(H)|H, but also, in some places, Yang-Mills fields valued in $N(H_i)|H_i$, where H_i is a "bigger" subgroup than H.

However, notice that we cannot assert, in general, that the extension of the metric to the whole of E will not generate degeneracies or singularities at the boundary of the principal stratum.

5.13 Pointers to the literature

Simple G-spaces and stratified spaces: 23, 41, 20, 152, 153, 166, 167

Kaluza-Klein theory on simple G-spaces: 4, 21, 35, 41, 168, 189, 193, 207,213,230

Consistency requirement: 45, 59, 61

VI

GEOMETRY OF MATTER FIELDS

6.0 Summary section

Geometrically, matter fields are described by sections of associated vector bundles (we do not consider non-linear σ-models). In components, matter fields are described as carrying two kinds of indices:

tensorial covariant and contravariant indices i.e. $\varphi_\mu{}^\nu{}_\sigma$ and also, possibly, "internal" indices i.e. $\varphi_\mu{}^i$ or $\varphi^i{}_j$. For the internal indices, it is necessary to specify a (matrix) representation say $\rho = (\rho^i{}_j)$ of the structure group of the principal bundle which acts on these indices as an effect of gauge transformations. Transormation properties of tensors are uniquely defined (apart of a possible weight) by the covariance or contravariance of the tensor indices. The conventional spinor fields make sense only on a background of a given Riemannian structure (and even a so called "spin structure") and spinorial indices transform under a two-valued representation of the orthogonal group or, better, under a representation of its two-fold covering.

We consider a principal bundle P with structure group G and base manifold **M**. **P** is endowed with a principal connection ω, which can be considered either as a Lie G valued form $\omega = \omega^i T_i$, $T_i \in$ Lie G, on P or, via a local section σ: $M \rightarrow P$, as a Lie G-valued 1-form $A_\mu = A_\mu{}^i T_i$ on M. A matter field of type ρ under gauge transformations is typically a k-form (on P, or locally, on M) with values in a vector space F carrying a representation ρ of G (and thus, of Lie G).

The most important covariant operation on such matter fields is the exterior covariant derivative $\varphi \rightarrow D\varphi$ which makes a k-form into a (k+1)-form of the same type. Its explicit form is given by the formula 6.2(7"). The iteration of this operation produces the curvature Ω of the connection as we have

$$(D^2\vec{\varphi})_{\mu_1 \dots \mu_{k+2}} = 1/(2!\,k!) \quad \delta_{\mu_1 \dots \mu_{k+2}}{}^{\nu_1 \dots \nu_{k+2}} \; \Omega_{\nu_1 \nu_2} \; \vec{\varphi}_{\nu_3 \dots \nu_{k+2}}$$

where Ω-the curvature of ω acts on the values of $\vec{\varphi}$ via the (matrix) representation of Lie G on F (cf. Formula 6.2(8')).

When $\vec{\varphi}$ is a zero-form, i.e. when it does not carry tensorial indices, one usually denotes the 1-form $D\vec{\varphi}$ as

$$(D\vec{\varphi})_\mu = \nabla_\mu \vec{\varphi} = \partial_\mu \vec{\varphi} + A_\mu \vec{\varphi}$$

To perform covariant differential operations on tensors which are not exterior forms it is necessary to have M endowed with an affine connection (with or without torsion). An affine connection is a particular example of a connection in a principal bundle : in this case the bundle of linear frames or, if it preserves some metric, in a bundle of orthonormal frames; the structure group is GL(m) or O(m) respectively. Such an affine connection allows us to define a covariant derivative $\nabla : \mathbf{S} \to \nabla \mathbf{S}$ which makes k-covariant l-contravariant F-valued tensors of type ρ into k+1 - covariant l-contravariant $\nabla \mathbf{S}$ (with the notation $(\nabla \mathbf{S}_{\sigma..}{}^{\nu..})_{\mu} = \nabla_{\mu} \mathbf{S}_{\sigma..}{}^{\nu..}$) of the same type

$$\nabla_{\mu} \mathbf{S}_{\sigma..}{}^{\nu..} = \partial_{\mu} \mathbf{S}_{\sigma..}{}^{\nu..} + \Gamma_{\mu}{}^{\nu}{}_{\lambda} \mathbf{S}_{\sigma..}{}^{\lambda..} - \Gamma_{\mu}{}^{\lambda}{}_{\sigma} \mathbf{S}_{\lambda..}{}^{\nu..} + \mathbf{A}_{\mu} \mathbf{S}_{\sigma..}{}^{\nu..}$$

where $\Gamma_{\mu}{}^{\nu}{}_{\lambda}$ are the components of the affine connection in a local moving frame $\mathbf{e}_{\mu} = \partial/\partial x_{\mu}$

$$\nabla_{\mu} \mathbf{e}_{\nu} = \Gamma_{\mu}{}^{\sigma}{}_{\nu}\, \mathbf{e}_{\sigma}.$$

If there is no torsion, then the exterior covariant derivative D can be expressed in a simple way by ∇_{μ} ; for a tensor $\mathbf{S}_{\sigma..}{}^{\nu..}$ which is antisymmetric in some of its covariant indices, say $\sigma_1 \sigma_2 ... \sigma_k$, we may decide to consider it as a (tensor$\otimes$F) -valued k-form, then D produces (tensor$\otimes$F)-valued k+1 form (the word "tensor" refers to the tensor indices other than $\sigma_1 \sigma_2 ... \sigma_k$). In a holonomic frame (i.e. assuming $[\partial\mu, \partial\nu] = 0$) we have then

$$(D\mathbf{S})^{\nu}{}_{\sigma 0..\sigma 1.\sigma 2..\sigma \mathbf{k}} = 1/k!\ \delta_{\sigma 0 ....\sigma \mathbf{k}}{}^{\mu 0 ...\mu \mathbf{k}}\ \ \nabla_{\mu 0} \mathbf{S}^{\nu ..}{}_{...\mu 1 ...\mu 2 ...\mu \mathbf{k}}.$$

When the affine connection has torsion $T_{\mu}{}^{\sigma}{}_{\nu}$, it appears on the right side as if exemplified by the unit tensor $I = \delta_{\mu}{}^{\nu}$ considered as vector valued 1-form, in which case we have

$$(DI)_{\mu}{}^{\sigma}{}_{\nu} = T_{\mu}{}^{\sigma}{}_{\nu} \neq \nabla_{\mu} \delta_{\nu}{}^{\sigma} - \nabla_{\nu} \delta_{\mu}{}^{\sigma} = 0.$$

If **M** is endowed with a metric then it is possible to build a second-order invariant differential operator, the best known example being the Laplace-Beltrami operator acting on F-valued k-forms. Its expression in terms of the "rough" Laplacian and the curvature tensors is given by the Weitzenbock formula (cf.6.4). Spinors do not always exist on a Riemannian manifold. If a manifold does not admit neutral spinors, it may admit spinors carrying a U(1), or some color-charge. This is related to the topological problems concerning an existence of spin structures and spin-G structures. These problems do not arise in the Einstein-Cartan formulation of gravity and spinor fields. In this theory the physical fields : connection form, soldering form and spinors are, at the beginning, not related to geometry of **M**. An interpretation of the connection as affine connection, soldering form as a moving frame, and spinors as space-time spinors, is possible only for certain solutions of the Einstein - Cartan theory - those for which the soldering form is of a maximal rank.

In 6.1 we discuss matter fields as sections of associated bundles, and different possible ways of global and local description of these fields. Also bundle-valued forms are discussed there.

In 6.2 we provide a general formalism of covariant and exterior covariant derivatives, and discuss connection and curvature forms.

In 6.3 we consider affine connection as a particular example of a connection in a principal bundle and in 6.4 we discuss covariant derivative of bundle-valued tensors, given an affine connection and a Yang-Mills potential.

In 6.5 and 6.6 we discuss spin structures and spin-G-structures as well as the Dirac operator for charged spinors.

Finally in 6.7 we discuss geometry of the Einstein-Cartan theory, where we stress the fact that this theory admits degenerate (and even vanishing) metric tensor.

6.1 Description of matter fields

Each matter field is characterized by the property that "it transforms under a certain representation of a certain group". To give a precise meaning to this description, consider first tensorial properties of matter fields. According to the old-fashioned (but full of meaning) definition, a tensor (at a given point x) is a "something" characterized by the fact that it has "components" (which are numbers) in every frame (at x), and that these components transform linearly when one frame is replaced by another. The law of transformation tells us then the kind of the tensor (covariant, contravariant), and its weight (when we deal with a tensor density). To take the simplest example, a vector v is a something that has components v^a in a frame e_a, and when e_a is replaced by $e_{a'} = \Lambda^a{}_{a'} e_a$ then the components of v become $v^{a'} = \Lambda^{-1a'}{}_a v^a$. It is therefore suggestive to write $v = v^a e_a = v^{a'} e_{a'}$ (or even better $v = e_a v^a$), which notation explains the above transformation law. Now, the physical matter fields usually have not only tensorial properties, but also carry charges of different kind, and, sometimes, they take even values in a manifold rather than a vector space. In a precise mathematical language they are described by *sections of associated bundles*. That means the following: first of all, the bundle of frames of a manifold E (which we will denote FE) is replaced by a more general principal bundle, say U, over E with a structure group, let us call it T. U is a union of fibers T_y, ($y \in E$) an element $u \in T_y$ plays the role of a *generalized frame* at y. If u and u' are two such frames, they are related by a transformation $u' = u\, r$, $r \in T$, which is an analogue of $e_{a'} = e_a \Lambda^a{}_{a'}$ (often, u is seen as a- non abelian in general-"phase factor" rather than a frame). Thus the group T knows how to act on U from the right, its action is assumed to be free and transitive on the fibers. Now, let F be a manifold in which our field takes its values. For instance, F can be a vector space, a group, or a homogeneous space. The field character is then fixed by the specification of a representation, say ρ, of T on F, and by replacing the rule $v^{a'} = \Lambda^{-1a'}{}_a v^a$ by its generalization $\vec{v}' = \rho(r)\, \vec{v}$, $r \in T$ (often we will use $\vec{\ }$ to distinguish objects taking values in a typical fiber F). Since F does not need to be a vector space, it is better to call ρ, in such a case a "realization" or better, an *action* of T on F (from the left). It is then suggestive to write this "something" with the above transformation properties, symbolically, as $v = u.\vec{v} = u'.\vec{v}'$, which generalizes the vector notation $v = e_a v^a = e_{a'} v^{a'}$. According to this idea one constructs the associated bundle $\mathfrak{F} = U\times_\rho F$ by dividing the *Cartesian* product U×F through the equivalence relation which identifies $(u,\vec{v})$ with $(ur,\rho(r^{-1})\vec{v})$, $r\in T$ and denoting $v = u.\vec{v}$ the equivalence class of the pair $(u,\vec{v})$. Notice that the group T acts on U from the right, on F from the left, but does not act any more on the associated bundle (as we have taken quotient through it).

A matter field Φ is now a *section* of $\mathfrak{F}$, i.e. a function Φ which to each $y\in E$ ascribes an object $\Phi(y) \in F_y$. Choosing a "frame" $u\in U$ over y we can write then

$$\Phi(y) = u.\vec{\Phi}(u)\ ^1 ,$$

where $\vec{\Phi}(u) \in F$ is uniquely determined by u and is an equivariant function on the fiber

$$\vec{\Phi}(ur) = \rho(r^{-1})\,\vec{\Phi}(u)\,,\ r\in T.$$

Locally Φ can be described as an F-valued function on E: if σ: E⟶U is a local section of U then $\Phi(y)$ is determined by its "coordinates" ${}^\sigma\vec{\Phi}(y) \in F$ in the "frame" $\sigma(y)$:

$$\Phi(y) = \sigma(y).{}^\sigma\vec{\Phi}(y) = \vec{\Phi}(\sigma(y))\,.$$

We have therefore three equivalent descriptions of matter fields:

---

[1] Our notation convention here is that F-valued objects are denoted $\vec{\Phi}$ while $\mathfrak{F}$ -valued objects are denoted Φ; observe that even if F is a vector space and ρ is a linear representation, $\mathfrak{F}$ is not a vector space as addition of elements from different fibers of $\mathfrak{F}$ is not defined (unless in very special circumstances as e.g. when $\mathfrak{F}$ is endowed with a flat connection).

1) As sections Φ of an associated bundle $\mathcal{F}$ -we write $\Phi \in \Gamma\mathcal{F} = \Lambda^0(E,\mathcal{F})$,

2) As equivariant F-valued functions $\vec{\Phi}$ on the principal bundle U - we write $\vec{\Phi} \in \Lambda_{eq}{}^0(E,F)$,

3) Locally, as section-dependent (gauge dependent) F-valued functions on the base E, we write ${}^{\sigma}\vec{\Phi} \in \Lambda^0(E,F)$.

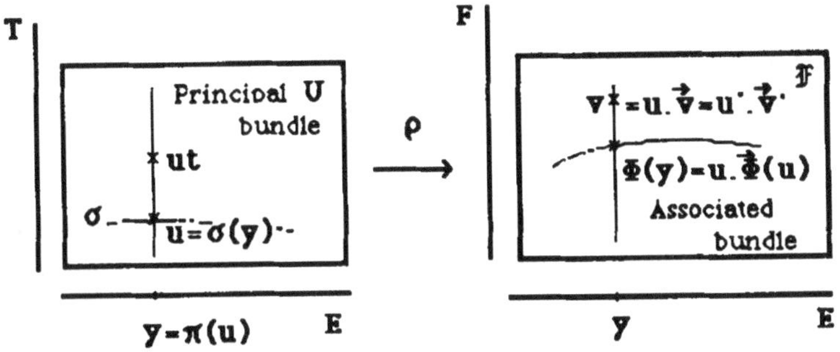

σ= local section = moving frame

matter field: (3 descriptions)
$\Phi : E \to \mathcal{F}$
$\vec{\Phi} : U \to F$
${}^{\sigma}\vec{\Phi} : E \to F$

Sometimes matter fields are *forms* (i.e. differential forms) on E with values in $\mathcal{F}$ (this makes sense only if F is a vector space) and we write $\Phi \in \Lambda^k(E,\mathcal{F})$. However, a k-form on E with values in $\mathcal{F}$ can be considered as an element of $\Lambda^0(E,\Lambda^kE \otimes F)$ i.e. the value of Φ at x can be considered as an element of the vector space $\Lambda_x{}^kE \otimes F_x$ where $\Lambda_x{}^kE$ is the space of antisymmetric k-covariant tensors at x. It is well known [Kobayashi, Trautman!] that there is a one-to-one correspondence between $\mathcal{F}$ valued k-forms Φ on E and F-valued equivariant horizontal k-forms $\vec{\Phi}$ on U. (An F-valued k-form $\vec{\Phi}$ on U is called *horizontal* provided it vanishes as soon as one of its arguments is vertical; it is called *equivariant* if $\vec{\Phi}(X_1\mathbf{a}, \ldots, X_k\mathbf{a}) = \rho(\mathbf{a}^{-1})\,\vec{\Phi}(X_1, \ldots, X_k)$).

The correspondence is given by the relation

$$\Phi(\pi(X(u))) = u.\hat{\Phi}(X(u))$$

where X(u) stands for a vector (or vectors) tangent to U at u. *Locally* (i.e. with respect to σ: M→P) such $\mathcal{F}$-valued k-forms are, of course, described as F-valued k-forms *on* E.

6.2 Covariant derivative and curvature

The purpose of this paragraph is to justify and discuss in some detail the geometrical meaning of the formulae given in the summary section (in this section we replace U,E,T by P,M,G of the summary section). We first recall some well known constructions of differential geometry.

Let ω be a connection form on a principal bundle **P** of base **M** and structure group G; ω is then a 1-form on P, valued in **Lie(G)** and **Ad**-equivariant i.e.

(1) $$\omega_{pa}(\xi_p a) = a^{-1}\,\omega_p(\xi_p)a$$

where ξ_p is any vector tangent to **P** at **p**, and $a \in G$.

Moreover, denoting $e_i := e_i^R$ the fundamental fields on P corresponding to a basis T_i of **Lie(G)** (cf. Ch.IV), the connection form has preassigned values on e_i-s:

(2) $$\omega_p(e_i(p)) = T_i$$

for every $p \in P$. A vector ξ tangent to P is vertical if its projection $\pi_*(\xi)$ vanishes, π: P→M, being the bundle projection. Equivalently, ξ is vertical if and only if it is a linear combination of the fundamental fields e_i. The concept of verticality does not require any connection form.

However, it is the connection form that allows us to define *horizontal* vectors: ξ is called horizontal if $\omega(\xi) = 0$. Denoting H_p as the subspace of T_pP consisting of all horizontal vectors at p of the connection ω, the projection π sets up a 1-1 correspondence between vectors tangent to M at $x = \pi(p)$ and vectors in H_p. If $\xi \in H_p$ and if $\pi_* \xi$ is its projection on T_xM, then ξ is said to be the horizontal lift of $\pi_* \xi$. We can now split any vector $\xi \in T_pP$ into its vertical part $v\xi$ and horizontal part $h\xi$. Using both the exterior derivative and the horizontality defined via the connection ω, we can define the *covariant exterior derivative* on P, denoted by D acting on F-valued k-forms on P (F beeing a vector space) by a simple formula

(3) $(D\vec{\phi})(\xi_1,\ldots,\xi_n) = (d\vec{\phi})(h\xi_1,\ldots,h\xi_n).$

For this definition to work, $\vec{\phi}$ does not need to be either horizontal or equivariant. However notice that

(4) i) $D\vec{\phi}$ is always horizontal

ii) if $\vec{\phi}$ is equivariant then $D\vec{\phi}$ is equivariant too.

In particular we can apply D to the connection form itself, we get then the curvature Ω of ω (which is a Lie(G) valued, horizontal, Ad-equivariant, 2-form on P)

(5) $\Omega := D\omega$ [2]

The result of acting with D on ω can be calculated explicitly, leading to the Cartan structure equation

(6) $D\omega = d\omega + 1/2\ [\omega \wedge \omega]$.

On the other hand, supposing $\vec{\phi}$ horizontal and equivariant of type ρ, one finds the following explicit formula

(7) $D\vec{\phi} = d\vec{\phi} + \rho(\omega) \wedge \vec{\phi}$.

---

[2] According to our convention we should write $\vec{\Omega}$ and $\vec{\omega}$, since both are vector-valued; we shall not do this in order to simplify the notation.

The formulae (6) and (7) imply the Bianchi identity

(8) $$D\Omega = D^2\omega = 0.$$

Observe that there is no contradiction (the factor 1/2 in (6)) between the formulae (6) and (7), since (7) is valid only for Φ horizontal, which is certainly not the case with ω as shown explicitly by the property (2).

For a form Φ which is equivariant and horizontal we find that a repeated application of D reproduces the curvature form:

(9) $$D^2\Phi = \rho(\Omega)\wedge\Phi .$$

Therefore (9) could be even taken as defining the curvature in a representation ρ.

There are two warnings necessary concerning our notation. The first is that $\rho(\omega)$ and $\rho(\Omega)$ should be in fact replaced by a derived representation ρ' of the Lie algebra induced by the representation ρ of the group. The second is that expressions like $[\omega\wedge\omega]$ or $\rho(\omega)\wedge\Phi$ should be understood as: one takes the ordinary formula for the exterior product of forms with the understanding that instead of the usual product of numbers one takes commutator in the first case or action of an operator (or a matrix) on a vector in the second one.

The explicit formulae (6), (7), (9) have the advantage over the original defining formula (3) because the operator d and $\wedge$ go without change through pull-back, with the result that these latter formulae look exactly the same in a local gauge-dependent description. Let σ M→P be a local section; we can pull back then the objects ω,Ω and Φ on M and define in this way the Yang-Mills field ${}^{\sigma}A := \sigma^*\omega$, its curvature ${}^{\sigma}F := \sigma^*\Omega$ and matter fields ${}^{\sigma}\Phi := \sigma^*\Phi$ i.e.

(6) $$\begin{aligned} {}^{\sigma}A_x(v_x) &= \omega_{\sigma(x)}(\sigma_* v_x) \\ {}^{\sigma}F_x(v_x{}^1, v_x{}^2) &= \Omega_{\sigma(x)}(\sigma_* v_x{}^1, \sigma_* v_x{}^2) \\ {}^{\sigma}\Phi_x(v_x{}^1, \ldots, v_x{}^n) &= \Phi_{\sigma(x)}(\sigma_* v_x{}^1, \ldots, \sigma_* v_x{}^n) \end{aligned}$$

The formulae (6),(7) and (8) read then

$$(6') \qquad {}^{\sigma}F = d\,{}^{\sigma}A + 1/2\ [\,{}^{\sigma}A \wedge {}^{\sigma}A\,]$$
$$(7') \qquad D\,{}^{\sigma}\Phi = d\,{}^{\sigma}\Phi + \rho({}^{\sigma}A) \wedge {}^{\sigma}\Phi$$
$$(8') \qquad D^2\,{}^{\sigma}\Phi = \rho({}^{\sigma}F) \wedge {}^{\sigma}\Phi$$

We choose a (local) moving frame (ϵ_μ) on **M** (in general non-holonomic i.e. $\epsilon_\nu = \epsilon_\nu{}^\mu\, \partial_\mu$, (∂_μ) being a local coordinate frame (hence holonomic): $[\partial_\mu,\partial_\nu] = 0$ but $[\epsilon_\mu,\epsilon_\nu] = f_{\mu\nu}{}^\rho\, \epsilon_\rho$), we will call ϵ^μ the dual basis (and therefore $d\epsilon^\rho = -1/2\ f_\mu{}^\rho{}_\nu \epsilon^\mu \wedge \epsilon^\nu$). Notice that ϵ_μ (as well as ∂_μ) acts as a (first order) differential operator on $C^\infty(\mathbf{M})$ [Attention: in other chapters like Ch. 4,5,..., we use the notation ∂_μ for a non holonomic frame]. We choose also a basis $\mathbf{e}_\alpha$ (α=1,2,...,dim **F**) in **F**. We may then write (abusing the notation and omitting the symbol σ), if Φ is a zero form (i.e. a section of $\mathcal{F}$),

$$\Phi = \Phi^\alpha\, \mathbf{e}_\alpha,$$

$$D\Phi = \mathbf{e}_\alpha\ (D\Phi)^\alpha{}_\mu\ \ \epsilon^\mu = \ (D\Phi)^\alpha{}_\mu\ \ \mathbf{e}_\alpha\ \epsilon^\mu$$

and, if Φ is a p-form:

$$D\Phi = 1/p!\ \ D\Phi^\alpha{}_{\mu 1 \ldots \mu p}\ \ \mathbf{e}_\alpha\ \ \epsilon^{\mu 1} \wedge \epsilon^{\mu 2} \wedge \ldots \wedge \epsilon^{\mu p}.$$

Often instead of $(D\Phi)^\alpha{}_\mu$ one writes $D_\mu\Phi^\alpha$, etc. The latter should, however, be avoided as it may cause confusion. For reasons explained in sect. 6.5, we will also introduce the symbol ∇; $\nabla_\mu\Phi^\alpha = (D_\mu\Phi)^\alpha$ in the particular case of Φ being a zero form.

Let $T_i = ((\,T_i)_\beta{}^\alpha)$ (i=1,...,dim G) be the matrix representation of the Lie algebra of G on the vector **F** derived from the representation ρ of the group G (Attention: in other chapters we denoted by the same letters matrices representing these generators). We also introduce the standard notations:

$$\mathbf{A} = \mathbf{A}_\mu \, \epsilon^\mu, \qquad \mathbf{A}_\mu = A_\mu{}^i \, T_i$$

for the matrix representation of the connection $\omega = (\omega_\beta{}^\alpha)$ whose elements are the one-forms

$${}^\sigma\omega^\alpha{}_\beta = (\mathbf{A}_\mu)^\alpha{}_\beta \, \epsilon^\mu = A_\mu{}^\alpha{}_\beta \, \epsilon^\mu.$$

Similarly for the curvature:

$$\mathbf{F} = 1/2 \; \mathbf{F}_{\mu\nu} \, \epsilon^{*\mu} \wedge \epsilon^{*\nu}, \qquad \mathbf{F}_{\mu\nu} = F_{\mu\nu}{}^i \, T_i$$

we denote $F^\alpha{}_\beta = (1/2)\,(\mathbf{F}_{\mu\nu})^\alpha{}_\beta \, \epsilon^\mu \wedge \epsilon^\nu$ the matrix of the curvature two-form in the representation ρ: [2]

$$F_{\mu\nu}{}^\alpha{}_\beta := (\mathbf{F}_{\mu\nu})^\alpha{}_\beta = F_{\mu\nu}{}^i \, (T_i)^\alpha{}_\beta.$$

Omitting the symbols σ for the local section ("moving frame") and ρ for the representation, the formulae (6'),(7') can be written as

(6") $$\mathbf{F}_{\mu\nu} = \partial_\mu \mathbf{A}_\nu - \partial_\nu \mathbf{A}_\mu - f_{\mu\nu}{}^\sigma \mathbf{A}_\sigma + [\, \mathbf{A}_\mu, \mathbf{A}_\nu \,]$$

or

(6"a) $$F_{\mu\nu}{}^i = \partial_\mu A_\nu{}^i - \partial_\nu A_\mu{}^i - f_\mu{}^\sigma{}_\nu A_\sigma{}^i + C_{kl}{}^i A_\mu{}^k A_\nu{}^l$$

(7") $$\begin{aligned}(D\mathbf{\Phi})^\alpha{}_{\mu 0 \ldots \mu k} = {} & \Sigma_{i=0\ldots k} \, (-1)^i \; \partial_{\mu i} \mathbf{\Phi}^\alpha{}_{\mu 0..\hat{\mu} i..\mu k} \\ & - \Sigma_{i<j} \, (-1)^{i+j} \, f_{\mu i}{}^{\mu 0}{}_{\mu j} \, \mathbf{\Phi}^\alpha{}_{\mu 0..\hat{\mu} i..\hat{\mu} j.\mu k} \\ & + \Sigma_{i=0\ldots k} \, (-1)^i \, \mathbf{A}_{\mu i}{}^\alpha{}_\beta \, \mathbf{\Phi}^\beta{}_{\mu 0..\hat{\mu} i..\mu k}\end{aligned}$$

where $\mathbf{A}_\mu = A_\mu{}^j T_j$ (Remember also that $[T_i,T_j]=C_{ij}{}^k \, T_k$, and that often we write $f_i{}^k{}_j = C_{ij}{}^k$ - cf. footnote in Ch.1 -).

---

[2] Sometimes, discussing connections in vector bundles one writes $\Gamma_\mu{}^\alpha{}_\beta$ for $A_\mu{}^\alpha{}_\beta$ and $R_{\mu\nu}{}^\alpha{}_\beta$ for $F_{\mu\nu}{}^\alpha{}_\beta$.

In particular for Φ a zero form i.e. a section of the bundle $\mathfrak{F} = P\times_\rho F$ we get the well-known simple expression

$$(10) \qquad D_\mu \Phi^\alpha := (D\Phi)^\alpha{}_\mu = \partial_\mu \Phi^\alpha + \mathbf{A}_\mu{}^\alpha{}_\beta \Phi^\beta .$$

In this case we denote $D_\mu\Phi$ by $\nabla_\mu\Phi$ and call it the covariant derivative of the section Φ. Locally we may choose sections $\vec{e}_\alpha$ such that $^\sigma\vec{e}_\alpha = e_\alpha$ as the basic e_α ($\alpha=1..\dim F$); for Φ, more precisely, we take local sections of $\mathfrak{F}$ such that $(^\sigma\vec{e}_\alpha)^\beta = \delta_\alpha{}^\beta$. We now apply the derivative ∇_μ to each of these sections. According to the formulae (6) we get:

$$(11) \qquad \nabla_\mu \vec{e}_\alpha = \mathbf{A}_\mu{}^\beta{}_\alpha \, \vec{e}_\beta$$

where we have taken into account the fact that $\partial_\mu\vec{e} = \partial_\mu\delta_\beta{}^\alpha = 0$.

Note In the rest of this chapter as well as in chapters 7 and 8, we shall no longer pedantically distinguish between $\mathfrak{F}$-valued and F-valued objects (in particular we will omit the arrows $\vec{\ }$), thus allowing for a possible confusion of the reader. We will also not distinguish between A_μ and $\mathbf{A}_\mu$, the meaning of the symbols to be read from the context.

6.3 The case of M endowed with affine connection and / or metric.

6.3.1 Covariant derivative of vector fields on M

Consider now the case of P being not some abstract bundle, externally related to M, but the bundle of frames FM (F stands for "frames") of the manifold M itself . An element p of P is denoted now as e and is a linear frame (basis) in the vector space T_xM tangent to M at x. Instead of p we shall therefore write $(e_\mu, \mu=1, ...,m)$ with $e_1, ..., e_m \in T_xM$. The structure group of FM is $GL(m,\mathbb{R})$, its elements are denoted Λ, and they act on frames from the right by $(e\Lambda)_\mu = e_\nu\Lambda^\nu{}_\mu$. A connection form is now matrix-valued $\omega=(\omega^{\mu}{}_{\nu})$. We choose now a local section $\sigma = x \rightarrow (e_{\mu}(x))$ of the frame bundle (in fact, often we denote such a section e). The formulae 6.2(1-8) of course hold, and (11) reads

$$(1) \qquad \nabla_{\mu} e_{\nu} = \Gamma_{\mu}{}^{\sigma}{}_{\nu} \; e_{\sigma} \; ,$$

where we used the notation $\Gamma_{\mu}{}^{\sigma}{}_{\nu}$ instead of $A_{\mu}{}^{\sigma}{}_{\nu}$.

Observe that in sect. 6.2 we have choosen two moving frames: e_{μ} in the tangent bundle and e_{α} in the vector bundle. Here the two bundles coincide so that there is no need to choose them differently. Sometimes, however, it is convenient to write the above formula as $\nabla_{\mu} e_{a} = \Gamma_{\mu}{}^{b}{}_{a} \, e_{b}$ with the understanding that the index μ refers to one frame, usually holonomic, while the indices a,b refer to another one, for example orthonormal). It is important to observe that the tangent bundle **TM** can be considered as a bundle associated to the frame bundle **FM** via the natural action of $GL(m,\mathbb{R})$ on $\mathbb{R}^m$.

We can also consider the identity tensor $I = (\delta_{\mu}{}^{\nu})$ as a one-form on **M** with values in the tangent bundle. Its exterior covariant derivative is called the torsion of the affine connection ω. Using the formula (7") we find that the torsion $T_{\mu}{}^{\sigma}{}_{\nu}$ is given by

$$(2) \qquad T_{\mu}{}^{\sigma}{}_{\nu} = (DI)_{\mu}{}^{\sigma}{}_{\nu} = - f_{\mu}{}^{\sigma}{}_{\nu} + \Gamma_{\mu}{}^{\sigma}{}_{\nu} - \Gamma_{\nu}{}^{\sigma}{}_{\mu} \; ,$$

where $f_{\mu}{}^{\sigma}{}_{\nu}$ are the structure functions of the moving frame $e_{\mu}(x)$.

Suppose now that a metric is given on **M**. Then instead of the $GL(m)$ bundle **FM** we can use a sub-bundle $O_{+}FM$ of **FM** consisting of oriented orthonormal frames for **g**. Let us distinguish here the orthonormal frames by the subscript e_a. Thus we have $g(e_{\mu},e_{\nu})=g_{\mu\nu}$, but $g(e_a,e_b) = \eta_{ab}$, $(\eta_{ab}) = \mathrm{diag}(-1,\dots,-1,+1,\dots,+1)$, $p+q = m$.

To assume that connection ω leaves the metric tensor invariant is the same as to assume that ω is a connection in the bundle $O_{+}FM$. Then ω takes values in $\mathrm{Lie}(SO(\eta))$, $SO(\eta)$ being the special orthogonal group of η : $\Lambda \in SO(\eta)$ iff $\Lambda^T \eta \Lambda = \eta$ and $\det \Lambda = 1$. We thus have in particular $\Gamma_\mu{}^{ab} = -\Gamma_\nu{}^{ba}$, where $\Gamma_\mu{}^{ab} = \eta^{bc} \Gamma_\mu{}^a{}_c$ and where $\Gamma_\mu{}^a{}_b = (\omega_\mu)^a{}_b$ is given by the formula

$$(3) \qquad \nabla_\mu e_a = \Gamma_\mu{}^b{}_a \, e_b.$$

This time $\nabla_\mu e_a = (De_a)(\partial_\mu)$, ∂_μ being any frame $(\partial_\mu)(x) \in T_x$ for M, for instance a coordinate basis $\partial_\mu = \partial/\partial x_\mu$. The curvature 2-form $\Omega_{\mu\nu}$ of the connection ω is also Lie(SO(η))-valued ; we write $R_{\mu\nu}{}^a{}_b$ for $(\Omega_{\mu\nu})^a{}_b$. The expression for $R_{\mu\nu}{}^a{}_b$ in terms of $\Gamma_\mu{}^a{}_b$ which follows from (6") has already been given in Chap.1.

Assuming that **M** is endowed with a metric, one can construct several natural differential operators. The most important differential operators are δ (the adjoint of **d**), the Laplace-Beltrami operator $d\delta+\delta d$, both acting on differential forms, and the Lichnerowicz operator acting (for example) on symmetric tensors. They will be discussed in 6.4, in a more general context.

6.3.2 Covariant derivative of arbitrary tensor fields (tensor-valued 0-forms) on M

Let S be an arbitrary tensor field on **M**, for instance it may be p-covariant and q-contravariant ; we interpret it therefore as a section of the vector bundle $(TM)^{\otimes q} \otimes (TM^*)^{\otimes p}$; this vector bundle is associated to the frame bundle **FM** via the natural action of GL(m,$\mathbb{R}$) on $\mathbb{R}^{mq} \otimes \mathbb{R}^{*mp}$ - S being a section of **TM(p,q)** - or a zero-form valued in **TM(p,q)** -, we know how to define its covariant derivative **DS** (according to the general formalism of 6.2) which is a 1-form on **M** valued in **TM(p,q)** (it can therefore be also considered as tensor of type (p+1,q)) - when this one-form is evaluated on a vector $\mathbf{u} \in$ **TM**, we get the covariant derivative $\nabla_u S$ (which is now a zero-form valued in **TM(p,q)**). To *remember* that S is considered as a zero form valued in **TM(p,q)**, we will write explicitly the fiber indices of S "under" the **D** symbol, for instance if S $\in$ **TM(2,1)**, we will write

(1) $$DS_{\nu\rho}{}^{\mu} = dS_{\nu\rho}{}^{\mu} + \omega^{\mu}{}_{\tau}S_{\nu\rho}{}^{\tau} - \omega^{\tau}{}_{\nu}\, S_{\tau\rho}{}^{\mu} - \omega^{\tau}{}_{\rho}\, S_{\nu\tau}{}^{\mu}$$

hence

(2) $$\nabla_{\sigma}S_{\nu\rho}{}^{\mu} = \partial_{\sigma}S_{\nu\rho}{}^{\mu} + \Gamma_{\sigma}{}^{\mu}{}_{\tau}\, S_{\nu\rho}{}^{\tau} - \Gamma_{\sigma}{}^{\tau}{}_{\nu}\, S_{\tau\rho}{}^{\mu} - \Gamma_{\sigma}{}^{\tau}{}_{\rho}S_{\nu\tau}{}^{\mu}.$$

Notice that indices $(\nu\rho\mu)$ in the above play the role of the index α in 6.2(10). Finally, if (e^{μ}) is the moving co-frame dual to the basis (e_{μ}), we get (for fixed ρ) the relation

$$\nabla_{\mu}e^{\rho} = -\Gamma_{\mu}{}^{\rho}{}_{\nu}\, e^{\nu}.$$

Here, we apply 6.3.2(2) to one-forms $e^1,\ldots,e^m$, one by one, and take into account the defining relation of a coframe: $(e^{\mu})_{\nu}=\delta^{\mu}{}_{\nu}$.

6.3.3. Covariant derivative of tensor-valued r-forms on M

The general formulae of 6.2 also tell us how to compute the exterior covariant differential of $\mathcal{F}$-valued r-forms on M, in particular $\mathcal{F}$ can be the tensor bundle TM(p,q). What is sometimes confusing is that TM(p,q)-*valued r-forms on* M *can also be considered as zero-form valued in* TM(p+r,q) ; but then, the general formalism of section 6.2 will give a different answer for its covariant differential. We will get a (r+1)-form valued in TM(p,q) in the first case (it will be antisymmetric in r+1 indices) but a 1-form valued in T(p+r,q) in the second (it will only be antisymmetric in r-indices). In order to *remember* how a given object is considered, *we will write explicitly the fiber indices (not the form indices)* under the D symbol, for instance if S is a 2-form valued in TM we will write

$$S^{\mu} = 1/2\; S_{\nu\rho}{}^{\mu}\, \varepsilon^{\nu} \wedge \varepsilon^{\rho}$$

and

$$DS^{\mu} = dS^{\mu} + \omega^{\mu}{}_{\tau} \wedge S^{\tau}$$

which is a 3-form valued in TM (As a rule, we will never use the symbol

∇ when acting on something other than a zero form, while for a zero-form both symbols D and ∇ may be used). The reader can then either consider DS as a 3-form valued in **TM** and compute the exterior differential **DDS** ($= \Omega \wedge S$), which is a 4-form valued in **TM**, or he can consider DS as a zero-form valued in TM(3,1) and compute ∇DS which is a 1-form valued in TM(3,1) !

Let us show in an example how to generalize the formulae of section 6.3.2. Take S as a 2-form valued in TM(2,1) i.e.

$$S_{\nu\rho}{}^{\mu} = 1/2! \; S_{\nu\rho}{}^{\mu}{}_{\sigma\tau} \; \epsilon^{\sigma} \wedge \epsilon^{\tau}$$

then

$$DS_{\nu\rho}{}^{\mu} = dS_{\nu\rho}{}^{\mu} + \omega^{\mu}{}_{\tau} \wedge S_{\nu\rho}{}^{\tau} + \omega^{\tau}{}_{\nu} \wedge S_{\tau\rho}{}^{\mu} + \omega^{\tau}{}_{\rho} \wedge S_{\nu\tau}{}^{\mu}.$$

6.4 The case of a bundle P = P(M,G) endowed with a principal connection ω and whose base is endowed with affine connection

Let us assume now that **M** is endowed with an affine connection $\Gamma_{\mu}{}^{\sigma}{}_{\nu}$, and consider *in addition* a principal bundle **P** over **M** with a connection form **A** valued in Lie(G), G being the structure group. Let $\mathfrak{F} = P \times_{\rho} F$ be a bundle associated to P via a (linear) representation ρ on F as discussed in sect. 6.1.

A matter field can now be any $\mathfrak{F}$-valued tensor on **M**. It may be considered as a zero form valued in $TM(p,q) \otimes \mathfrak{F}$ which is a vector bundle of base **M** and fibers $TM_x(p,q) \times F_x$ above x on which we have a connection Γ + **A**. The general formalism of sect. 6-2 then applies. We have then the following formula

$$(1) \quad \nabla_{\mu} \vec{\Phi}_{\nu ..}{}^{\sigma ..} = \partial_{\mu} \vec{\Phi}_{\nu ..}{}^{\sigma ..} + A_{\mu} \vec{\Phi}_{\nu ..}{}^{\sigma ..} + \Gamma_{\mu}{}^{\sigma}{}_{\lambda} \vec{\Phi}_{\nu ..}{}^{\lambda ..} + \Gamma_{\mu}{}^{\lambda}{}_{\nu} \vec{\Phi}_{\lambda ..}{}^{\sigma ..} + \ldots$$

If the affine connection has a non-vanishing torsion, then one has to be careful; for $\vec{\Phi}$ being e.g. a section of the bundle $\mathfrak{F}$, we find (assuming for simplicity $f_{\mu}{}^{\sigma}{}_{\nu} = 0$)

$$(2) \qquad \Omega_{\mu\nu}\vec{\varphi} = (D^2\vec{\varphi})_{\mu\nu} = (\nabla_\mu \nabla_\nu - \nabla_\nu \nabla_\mu)\vec{\varphi} + T_\mu{}^\sigma{}_\nu \nabla_\sigma \vec{\varphi},$$

where $\Omega_{\mu\nu}\vec{\varphi} = \Omega_{\mu\nu}{}_i{}^j \varphi_j$.

This last formula may serve as a good example of how confusing a tensor notation can be: the reader may remember a simple formula for a curvature which reads

$$(3) \qquad \Omega(X,Y) = \nabla_X \nabla_Y - \nabla_Y \nabla_X - \nabla_{[X,Y]} ;$$

applying this formula to a section $\vec{\varphi}$, and calculating the result for $X = \partial_\mu$, $Y = \partial_\nu$, (so that $[X,Y] = 0$) we do not find any trace of torsion in (3) which clearly contradicts (2). The answer to this puzzle consists in the following: in (2) we apply ∇_μ to $\nabla_\nu \vec{\varphi}$; according to our conventions $\vec{\varphi}$ was a section of the bundle $\mathcal{F}$ but $\nabla_\nu \vec{\varphi}$, which has an extra index ν, is a section of the bundle $T^*M \otimes \mathcal{F}$. So, when evaluating ∇_μ on $\nabla_\nu \vec{\varphi}$ both $\Gamma_{\mu\nu}{}^\sigma$ and A_μ are used (they may coincide). On the other hand, in (3) one applies ∇_Y to a zero-form $\vec{\varphi}$, the result being another zero-form, denoted $\nabla_Y \vec{\varphi}$; then, to this zero-form ∇_X is applied. In other words, in evaluating the r.h.s. of (4) only A_μ (which may coincide with Γ_μ, but is applied only to the internal index of $\vec{\varphi}$) is used. This is also the case for $X = \partial_\mu$, $Y = \partial_\nu$. Or, put in yet another way, in evaluating $\nabla_\mu \nabla_\nu \varphi$ according to (3), the index ν is supposed to have a fixed numerical value, while in (2) it is a tensorial index.

From now untill the end of this paragraph we *shall assume vanishing torsion*. Notice the following explicit formula relating the exterior covariant derivative D and the covariant derivative ∇_μ :

$$(D\vec{\varphi})_{\mu_0 \cdots \mu_q} = \Sigma_{i=0..q} (-1)^q \, \nabla_{\mu_i} \vec{\varphi}_{\mu_0 \cdot \hat{\mu}_i \cdot \mu_q} + \Sigma_{i<j} (-1)^{i+j} f_{\mu_i}{}^\sigma{}_{\mu_j} \vec{\varphi}_{\sigma \mu_0 \cdot \hat{\mu}_i \cdot \cdot \hat{\mu}_j \cdot \cdot \mu_q}$$

where $f_\mu{}^\sigma{}_\nu$ are the structure constants of the moving frame e_μ in which the components $\varphi_{\mu_0 \cdots \mu_q}$ of φ are taken.

Exercise. The study of an even more general case: r-forms or M valued in $TM(p,q) \otimes \mathcal{F}$ is left to the reader.

Given a metric $g = (g_{\mu\nu})$ on **M** we can introduce the Hodge *-operator mapping p-forms into (m-p)-forms according to the formula

$$({}^*\varphi)_{\mu_1 \cdots \mu_q} = 1/p! \; g^{1/2} \, \epsilon_{\nu_1 \cdots \nu_p \mu_1 \cdots \mu_q} \, g^{\nu_1 \sigma_1} \ldots g^{\nu_p \sigma_p} \, \varphi_{\sigma_1 \cdots \sigma_q}$$

where $g = |\det g_{\mu\nu}|$. Observe the formula

$$\varphi \wedge {}^*\psi = (-1)^{p(m-p)} \, \psi \wedge {}^*\varphi \, , \quad \text{when } \psi \text{ and } \varphi \text{ are p-forms.}$$

Assuming **M** compact and oriented, and assuming also that a G-invariant, possibly x- dependent ($x \in M$), scalar product k_{ij} is given in **F**, we can introduce a scalar product in the space of $\mathcal{F}$-valued forms on **M** :

$$(\varphi, \psi) = \int_M \; k_{ij} \, \vec{\varphi}^i \wedge {}^* \vec{\psi}^j .$$

(This scalar product is, of course, indefinite if the metric $g_{\mu\nu}$ is pseudo-Riemannian).

Notice: If **M** is not oriented then the * operator can also be defined, but then it maps differential forms into the "De Rham currents" [De Rham-Variétés Différentiables, Dieudonné Vol 3]. Anyhow the product (φ, ψ) is independent of any orientation of **M**; it can also be written

$$(\varphi, \psi) = 1/p! \int_M k_{ij} \, \varphi^i{}_{\mu_1 \cdots \mu_p} \, \psi^{j \, \mu_1 \cdots \mu_p} \, |\det g_{\mu\nu}|^{1/2} \, dx$$

Also notice that if **M** is even dimensional then the * operator applied to m/2 - forms makes use only of the conformal structure and not of the full metric.

The operator D^* adjoint to D with respect to the above scalar product (it maps p-forms into (p-1)-forms) is then given by (when acting on p-forms)

$$D^* = (-1)^p *^{-1} D * .$$

Explicitly, we have

$$(D^*\omega)_{\mu_1\cdots\mu_{p-1}} = -\nabla^\mu\omega_{\mu\mu_1\cdots\mu_{p-1}}.$$

The Laplace-Beltrami operator Δ (on $\mathcal{F}$ valued p-forms) is defined to be the operator $DD^* + D^*D$. It is related to the "rough" Laplacian $\bar{\Delta} = -\nabla^\mu\nabla_\mu$ by the formula (sometimes known as a Weitzenböck formula)[3] .

$$(\Delta\vec{\varphi})_{\mu_1\cdots\mu_p} =$$

$$(DD^* + D^*D)\vec{\varphi}_{\mu_1\cdots\mu_p} = \bar{\Delta}\vec{\varphi}_{\mu_1\cdots\mu_p} + \Sigma_{i=1..p} R_{\mu_i}{}^{\mu}\vec{\varphi}_{\mu_1\cdots\mu_i\cdots\mu_p}$$

$$+ \Sigma_{i,j=1..p} R_{\mu_i}{}^{\nu\sigma}{}_{\mu_j}\vec{\varphi}_{\mu_1\cdots\cdots\underset{(i)}{\nu}\cdots\cdots\underset{(j)}{\sigma}\cdots\cdots\mu_p}$$

$$- \Sigma_{i=1..p} F_{\mu_i}{}^{\mu}\vec{\varphi}_{\mu_1\cdots\underset{(i)}{\mu}\cdots\mu_p}.$$

The action of the Yang-Mills strength F on $\vec{\varphi}$ in the last term is of course given via the representation ρ of Lie(G) on F. Observe that on zero-forms the Laplace-Beltrami Δ and the "rough" Laplacian $\bar{\Delta}$ coincide.

Remark: Often one can find the following notation : instead of ω one uses the symbol ∇ to denote a connection (=covariant derivative) in a vector bundle, instead of D (respect. D^*) one then writes d^∇ (respect.

[3] In some of the literature, Bianchi identities are used to write the result with a factor 1/2 in front of the curvature term and with a summation over different indices of the curvature tensor than in our formula.

δ^∇). The symbol for the Laplace-Beltrami operator then becomes $\Delta = d^\nabla \delta^\nabla + \delta^\nabla d^\nabla$.

The Laplace-Beltrami operator can also be applied to other tensors as well. For instance we may consider a 2-covariant tensor $h_{\mu\nu}$ as a 1-form valued in 1-covariant tensors, but also as a zero form valued in 2-covariant tensors. Moreover, when considered as a 1-form, we can take the first or the second index as the form index. Applying the Weitzenböck formula in each of these cases we get respectively the following expressions (observe that the **F**-term now becomes e.g. $+R_\mu{}^\lambda{}_\nu{}^\nu\, h_{\lambda\sigma}$)

1-form	$\bar{\Delta}\, h_{\mu\nu} \ + R_\mu{}^\lambda\, h_{\lambda\nu} + R_\mu{}^\lambda{}_\nu{}^\sigma\, h_{\lambda\sigma}$
1-form	$\bar{\Delta}\, h_{\mu\nu} \ + R_\nu{}^\lambda\, h_{\mu\lambda} + R_\nu{}^\lambda{}_\mu{}^\sigma\, h_{\sigma\lambda}$
0-form	$\bar{\Delta}\, h_{\mu\nu}$

Taking the sum of these expressions with factors $(-1)^k$ where $k=0$ for 0-forms and $k=1$ for 1-forms, we get the so-called *Lichnerowicz* operator, which, on symmetric tensor reads

$$\Delta_L\, h_{\mu\nu} = \bar{\Delta}\, h_{\mu\nu} \ + R_\mu{}^\lambda\, h_{\lambda\nu} + R_\nu{}^\lambda\, h_{\mu\lambda} + \ 2\, R_\mu{}^\lambda{}_\nu{}^\sigma\, h_{\lambda\sigma}.$$

Observe that if $h_{\mu\nu}$ is traceless (i.e. if $g^{\mu\nu}\, h_{\mu\nu} = 0$) then $\Delta_L\, h_{\mu\nu}$ is also traceless.

6.5. <u>Spin structures and spinors.</u>

6.5.1. Spin structures on an orientable manifold E.

The usual definition of a spin structure goes as follows : assume **E** is endowed with a Riemannian metric **g** (whose canonical diagonal form is $(\eta_{ab})=\eta$) and an orientation. Denote O_+FE the $SO(\eta)$ - principal bundle of oriented orthonormal frames of **E**. Then a spin structure of the pair (**E**, **g**) is a pair (SE, λ), where SE is a principal **Spin**(η)-bundle over E, and λ : SE $\rightarrow$ O_+FE is a bundle homomorphism satisfying

$$\lambda(pA) = \lambda(p)\ \Lambda(A),\ p \in P,\ A \in \mathbf{Spin}(\eta)$$

where $A \rightarrow \Lambda(A)$ is the 2:1 covering group homomorphism from **Spin**(η) to SO(η).

Suppose that the metric g is not Riemannian but pseudo Riemannian. Then, usually, by a spin structure, one understands a spin↑-structure, where **Spin↑** is the two-fold covering of SO↑(η) (this last group being the connected component of identity of SO(η)[5]. Actually, corresponding to the inclusion SO↑(η) $\subset$ SO(η) $\subset$ O(η) we have the inclusion of two-fold spin coverings **Spin↑**(η) $\subset$ **Spin**(η) $\subset$ **Pin**(η). Existence of a spin↑-structure on a manifold E implies that the (pseudo) Riemannian manifold E is oriented and "time"-oriented (the "time" may have more than one dimension).

A manifold must obey certain topological requirement in order that there exists any spin structure at all; on the other hand it may have different (inequivalent) spin structures. For more details see [147].

6.5.2. The bundle of spinors

Let us start with a manifold E endowed with metric g and spin structure (SE,λ). The structure group of SE is **Spin**(η) which is a two-fold covering of SO(η). Let now **W** be a vector space real or complex (or even quaternionic !) on which a representation of the Clifford algebra $\mathbb{C}(\eta)$ of η is given. We denote by Γ_a (a = 1,2, ..., dim E) the linear operators on **W** representing the generators of $\mathbb{C}(\eta)$. We have

(1) $$[\Gamma_a, \Gamma_b]_+ = 2\,\eta_{ab}.$$

---

[5] SO↑(η) can be defined directly in terms of matrices by requiring that both the determinants of the submatrices associated with "space" and "time" are positive.

Now[6], since **Spin**($\mathfrak{y}$) is a subset of $\mathfrak{C}(\mathfrak{y})$, we have automatically a representaion S of **Spin**($\mathfrak{y}$) on **W**. For **A** $\in$ **Spin**($\mathfrak{y}$) let S(**A**) denote the operator on **W** representing **A**. Because of the definition of **Spin**($\mathfrak{y}$)

(2) $$S(\mathbf{A})\Gamma_a S(\mathbf{A})^{-1} = \Gamma_b \Lambda(\mathbf{A})^b{}_a \ .$$

We do not need an explicit form of the representation S ; it is enough to know the derived representation $\mathbf{s}$ of **Lie(SO(**$\mathfrak{y}$**))** = **Lie(Spin(**$\mathfrak{y}$**))** in terms of Γ -s. The Lie algebra of SO($\mathfrak{y}$) consists of matrices $\mathbb{M}=(M_{ab})$ satisfying $M^{ab} + M^{ba} = 0$, where $M^{ab} = \eta^{bc} M_{ac}$. Using the properties (1) it is then straightforward to verify that for

(3) $$\mathbf{s}(\mathbb{M}) := 1/8\, M^{ab}[\,\Gamma_a\,, \Gamma_b\,] = 1/2\, M^{ab}\,\Sigma_{ab}$$

where $\Sigma_{ab} = 1/4\,[\,\Gamma_a\,, \Gamma_b\,]$
we have:

(4) $$[\mathbf{s}(\mathbb{M}), \mathbf{s}(\mathbf{N})] = \mathbf{s}([\mathbb{M}, \mathbf{N}])$$

and

(5) $$[\mathbf{s}(\mathbb{M}), \Gamma_a\,] = \Gamma_b\, M^b{}_a .$$

These equations tell us that (3) gives the realization of the SO($\mathfrak{y}$) = **Spin**($\mathfrak{y}$) - Lie algebra on **W**, as defined by the representation S.

The principal bundle SE will be called the spin bundle. Using the representation S of **Spin**($\mathfrak{y}$) on **W**, we build the (associated) bundle $\mathfrak{W}$ = $SE\times_S\mathbf{W}$: the bundle of spinors. Its sections will be called spinor fields.

---

[6] Cf. for example [45'].

6.5.3 The spin-connection

There exists a one to one correspondence between connections $\hat{\omega}$ on the orthonormal frame bundle O_+FE[7] and connections ω on the spin bundle SE. Indeed let $\hat{t}_{\dot{e}}$ be a tangent vector at $\dot{e} \in SE$, then we define $\hat{\omega}(\tau_{\dot{e}}) := \omega(\lambda_*(\tau_e))$ where $\lambda = SE \to OFE$ is the bundle homomorphism defining the spin structure (cf.5.1) and where we also made use of the isomorphism between Lie(SO(η)) and Lie (Spin(η)).

The affine connection ω was written $\omega = (\omega^a{}_b)$; therefore using (3), we find the spin connection $\hat{\omega} = s(\omega) = 1/8\ \omega^{ab}\ [\Gamma_a, \Gamma_b]$ and using $\Gamma^a := \eta^{ab}\Gamma_b$, we get

$$\hat{\omega} = 1/8\ \omega_{ab}\ [\Gamma^a, \Gamma^b] = 1/4\ \omega_{ab}\ \Gamma^a\Gamma^b$$

where (a,b, ...) indices refer to a local orthonormal frame (e_a).

If (∂_μ) is a coordinate frame and (e^{*c}) the dual basis of (e_c), we can of course write

$$\hat{\omega}_{ab} = \hat{\omega}_{cab}\ e^c = \omega_{\mu ab}\ dx^\mu.$$

The curvature operator $R = D\hat{\omega}$ can be gotten in the same way: we had $\mathbb{R} = (R^a{}_b)$ for the curvature of ω, using (3) we find

$$\hat{R} = 1/4\ \ R_{ab}\,\Gamma^a\,\Gamma^b\ , \text{with} \quad R_{ab} = 1/2\ R_{\mu\nu ab}\ dx^\mu \wedge dx^\nu.$$

If we use only orthonormal indices, then $R_{ab} = 1/2\ R_{cdab}\ e^c \wedge e^d$ and it is easy to show using the relations (1) that

$$R_{cdab}\,\Gamma^d\,\Gamma^a\,\Gamma^b = 2\,\rho_{ce}\ \Gamma^e \ , \text{(}\rho\text{ here being the Ricci tensor)}$$

and

[7]Remember that ω defines an affine connection on E which is compatible with the given metric g used in the construction of O_+FE. Notice that ω may carry a non-zero torsion.

$R_{cdab}\,\Gamma^c\Gamma^d\,\Gamma^a\,\Gamma^b = -2\,\tau$, (τ here being the scalar curvature)

It is maybe useful to remind the reader what our conventions are for the curvature tensors:

Riemann:[8] $R_{\mu\nu} = \partial_\mu\Gamma_\nu - \partial_\nu\Gamma_\mu + [\Gamma_\mu,\Gamma_\nu]$

$R_{\mu\nu}{}^a{}_b = \partial_\mu\Gamma_\nu{}^a{}_b - \partial_\nu\Gamma_\mu{}^a{}_b + \Gamma_\mu{}^a{}_c\Gamma_\nu{}^c{}_b$

Ricci: $\rho_{\mu\nu} = R_{\sigma\mu}{}^\sigma{}_\nu$

6.5.4. Covariant derivative on spinors. The Dirac operator.

We follow the general formalism of 6.2. Let ψ be a field of spinors (i.e. a section of $\mathfrak{W}$), i.e., locally, a zero-form valued in the vector space $\mathbf{W}$. Then $D\psi$ is locally a one-form valued in $\mathbf{W}$ and is given by

$$D\psi = d\psi + \hat{\omega}\psi = d\psi + 1/4\,\omega_{ab}\,\Gamma^a\,\Gamma^b\,\psi.$$

Its value taken on a tangent vector v is denoted $\nabla_v\psi := D\psi(v)$. Spinor fields satisfying $D\psi = 0$ are called "parallel spinors", or sometimes "constant spinors".

Let (e_a) be a local orthonormal moving frame on E, then we may construct the covariant derivatives $\nabla_a\psi := \nabla_{e_a}\psi$ (which locally are $\mathbf{W}$ valued zero-forms) and we define the Dirac operator $\mathbf{P}$ by

$$P\psi := \Gamma^a\,\nabla_a\psi.$$

Notice that P is also $\mathfrak{W}$-valued zero-form on E. Spinor fields satisfying $P\psi = 0$ are called "harmonic spinors" .

---

[8] Warning : $R_{\mu\nu}$ denotes here the curvature 2-form whereas in previous chapters we used this symbol to denote the Ricci tensor.

It is sometimes convenient to introduce y-dependent Γ-matrices ($y \in E$) associated to an arbitrary (not orthonormal) frame. For instance let (∂_μ) be a coordinated frame : we have $\partial_\mu = (e_\mu{}^a) e_a$ where $(e_\mu{}^a)$ is of course a y-dependent matrix. We then locally define $\Gamma_\mu(y) = (e_\mu{}^a)(y) \Gamma_a$ and $\Gamma^\mu = g^{\mu\nu} \Gamma_\nu$. . Then, the Dirac operator reads

$$P = \Gamma^a \nabla_a = \Gamma^\mu \nabla_\mu.$$

The geometrical meaning of $\Gamma_\mu(y)$ is the following : one first builds the Clifford algebra bundle $SE \times_{Spin} \mathfrak{C}(\eta) = \mathfrak{C}(E)$ by using the adjoint action of **Spin** (η) on $\mathfrak{C}(\eta)$ (alternatively, one can define $\mathfrak{C}(E)$ as the Clifford algebra of the tangent bundle). The tangent bundle is canonically embedded in $\mathfrak{C}(E)$: we have a 'fundamental' form $\Gamma : v \in TE \rightarrow \Gamma(v) \in \mathfrak{C}(E)$. For instance if we take $v = \partial_\mu$, we get $\Gamma_\mu := \Gamma(\partial_\mu)$. These objects can be considered as zero forms valued in $TE^* \otimes W \otimes W^*$ for which we have a covariant derivative ∇_μ; one can check that $\nabla_\mu \Gamma_\nu = 0$ and that $[\Gamma_\mu, \Gamma_\nu]_+ = 2 g_{\mu\nu}$.

The reader may convince himself that the subbundle of $\mathfrak{C}(E)$, fibers of which are the spin groups of the tangent space $T_x E$ is a bundle of groups (not principal) which coincides with the **Ad**(SE) bundle (its sections are the vertical automorphism of the spin-bundle SE). cf. also sect. 4.11.

Notice that we used a spin structure to construct $\mathfrak{C}(E)$; since, however, $\mathfrak{C}(E)$ happens to coincide with the bundle of the Clifford algebras of the tangent bundle, it follows that for the construction of $\mathfrak{C}(E)$ no spin structure is necessary. This explains why sometimes, in the mathematical literature, the Dirac operator is studied on $\mathfrak{C}(E)$-modules, of which $\mathfrak{W}$ is an example.

The square $\Delta = P^2$ of the Dirac operator is called the spinor Laplacian; its relation with the "rough" Laplacian $(-\eta^{ab} \nabla_a \nabla_b)$ is

$$\lambda = -\eta^{ab}\nabla_a\nabla_b + 1/4\,\tau$$

where τ is the scalar curvature and we assumed that ω was the Levi-Civita connection (no torsion).

6.5.5 Spinors before metric ?

In sect. 6.7 we will discuss in some details the Einstein-Cartan geometry. This geometry involves a definition of spinors which is not the usual one; although we will develop the theory from scratch in sect. 6.7, we want here make a few comments about an alternative definition of the concept of spin structure. Assume that there is given a principal bundle U over E with structure group **Spin**(η). An Einstein-Cartan geometry will be defined as consisting of pairs (ω,θ), ω being a **Spin**(η)-connection in U, and θ is a one-form on E with values in the (fibers of the) associated vector bundle $U\times_{Spin(\eta)}\mathbb{R}^d$ (d = dim E). We can define then a spin structure for E as consisting of a pair (U,θ), with U as above and θ nondegenerate. This is not a usual definition ! (The usual definition of a spin structure was given in sect. 6.5.1). We can now see that the two definitions of the spin structure are essentially equivalent. Indeed, given a pair (U,θ) with θ nondegenerate, we can use θ to identify the vector bundle $U\times_{Spin(\eta)}\mathbb{R}^d$ with **TM**; this identification induces a metric **g** and an orientation on **M**, and also allows us to identify the orthonormal frame bundle O_+FE of (E,**g**) with the quotient U/Z_2 of the "spin frame" U by Z_2 = {-1, +1}, i.e. by the 2π rotation. The map λ: $U \rightarrow O_+FE = U/Z_2$ is canonical in this case. Conversely, given (E,**g**) and a spin structure (U,λ), we can identify the $U\times_{Spin(\eta)}\mathbb{R}^m$ bundle with **TM** and take for θ the identity which is certainly nondegenerate. We omit the details of the above mentioned canonical identifications as they are more or less evident.

6.6 Generalized spin structures

There is a generalization of the concept of a spin structure which is important for physical applications. It is the concept of $spin_G$-structure. Suppose G is a compact Lie group with a distinguished central Z_2 subgroup. Let $Spin_G(\eta)$ be the group $(Spin(\eta)\times G)/(diag\ Z_2)$ where Z_2 of $Spin(\eta)$ is the subgroup of $Spin(\eta)$ generated by -1 (i.e. by the 2π rotation). We have, of course, again a group homomorphism $Spin_G(\eta) \to SO(\eta)$. In fact we have exact sequences

$$1 \to G \to Spin_G(\eta) \to SO(\eta) \to 1\ ,$$
$$1 \to Spin(\eta) \to Spin_G(\eta) \to G \to 1.$$

We define then a $spin_G$-structure exactly as we did for spin structures. We may also call $P := U/Spin(\eta)$. The most known example is the $spin_{U(1)}$-structure, also called a $spin_c$-structure, in which case $G = U(1)$ and the Z_2-generator of $U(1)$ is $\exp(i\pi)$.

Physically $spin_G$-structures are introduced to describe spinors which carry a (color) charge. It should be stressed that (owing to taking quotient through Z_2) a charged spinor is not "a spinor which has an extra charge", in fact, it is known that a spin-structure may exist on a manifold which does admit an ordinary spin structure (i.e. which does not admit neutral spinors). Notice that it follows from the second exact sequence that a $spin_G$-structure reduces to a spin structure iff the principal bundle with structure group G is trivial. Such "charged" spinor fields arise naturally in the process of dimensional reduction as discussed in Ch.8.

Since it costs us no extra effort we will assume our spinors to carry a color charge i.e. they will be described by sections of a bundle associated to a $Spin_G(\eta)$-principal bundle U. Whenever necessary G may be put equal to Z_2 itself, then we come back to the ordinary (neutral) spinors. So, U is a principal bundle over E with structure group $Spin_G(\eta)$. Let W be a vector space in which spinor fields take values. We demand from $\mathbf{W}$ two things : first, it should carry representation of the Clifford algebra $\mathfrak{C}(\eta)$ of η :

$$\{\Gamma_a, \Gamma_b\} = 2\,\eta_{ab} \quad , a,b=1,....,d = \dim E,$$

and second, it should carry a representation, say S, of the group $Spin_G(\eta)$ with the property

$$S(A)\Gamma_a S(A)^{-1} = \Lambda(A)^b{}_a \Gamma_b$$

Λ being the homomorphism $Spin_G(\eta) \rightarrow SO(\eta)$. For the Lie algebras we have $Lie(Spin_G(\eta)) = Lie(G) + Lie(Spin(\eta))$, so that we have, in fact, two representations of the Lie algebra $Lie(Spin(\eta))$: one obtained from the representation S of $Spin_G$ and the oher obtained from the representation of the Clifford algebra whose generator are $1/4\,[\Gamma_a,\Gamma_b]$. *We shall assume that the two representations coincide*. (In Ch.8 we will see that this property is automatically guaranteed when charged spinor fields come as a result of dimensional reduction.)

A spinor field ψ is then a section of the associated bundle $\mathfrak{W} = U \times_S \mathbf{W}$. Locally (i.e. respect to a local section $\sigma : E \rightarrow U$ of the bundle U of "charged spin" frames) it is given by a function $\vec{\psi}: E \rightarrow \mathbf{W}$. Supose now a principal connection is given in U. It has values in $Lie(Spin_G) = Lie(G) + Lie(Spin(\eta))$ and therefore the connection form splits into two pieces : one, denoted $\mathbf{A}_\mu$, with values in $Lie(G)$, and the other, denoted Γ_μ with values in $Lie(Spin(\eta))$. Taking into account the fact that the representation of $Lie\ (Spin(\eta))$ is given by the commutator $[\Gamma_a,\Gamma_b]$, we can write the covariant derivative $\nabla_\mu\psi$ of a spinor field as

$$\nabla_\mu \vec{\psi} = \partial_\mu \vec{\psi} + \mathbf{A}_\mu \vec{\psi} + 1/8\ \Gamma_\mu{}^{ab}\,[\Gamma_a,\Gamma_b]\,\vec{\psi}.$$

Notice that till now we did not "solder" the bundle U to the frame bundle FE of E, now, let us suppose that a soldering 1-form θ is given (not necessarily nondegenerate - cf. sect. 6.7). θ is a 1-form on E with values in the associated bundle $F = U \times_{Spin_G(\eta)} \mathbb{R}^d$, where a representation of $Spin_G(\eta)$ on $\mathbb{R}^d$ is given via the first exact sequence and $SO(\eta)$ acts naturally on $\mathbb{R}^d$. Locally θ is given by a matrix $\theta_\mu{}^a$ - index "μ" referring to a local frame ϵ_μ of E, and index "a" to a local basis e_a in F determined by σ. We can then introduce a first-order operator denoted $\mathbb{D}$ defined as

$$\mathbb{D}\vec{\psi} = 1/(d-1)!\ \epsilon_{a_1 \cdots a_{d-1} a}\ \theta^{a_1} \wedge \ldots \wedge \theta^{a_{d-1}} \wedge \Gamma^a D\vec{\psi}$$

where $D\vec{\psi}$ is the 1-form $(D\vec{\psi})_\mu = \nabla_\mu \vec{\psi}$. $\mathbb{D}\vec{\psi}$ is a d-form, so it has essentially only one component $(D\vec{\psi})_{1 \ldots m}$. Suppose now $\theta_\mu{}^a$ is inversible; then, by an explicit calculation we find this component (in the ϵ_μ basis) to be equal to $\det(\theta_\mu{}^a)\, P\psi$, P being the Dirac operator $P\psi = \Gamma^\mu D_\mu \psi$, with $\Gamma^\mu = \theta^\mu{}_a \Gamma^a$, where $\theta^\mu{}_a$ is the inverse matrix of $\theta_\mu{}^a$.

6.7 An example: Einstein-Cartan theory with spinor fields

As an illustration which makes use of many concepts discussed so far, we consider the Einstein-Cartan theory of gravitation with a spinorial matter (spin 1/2 or 3/2). Since a precise geometrical formulation of this theory is not that easy to find, and as we take here a somewhat unorthodox point of view (which is, in our opinion, the only consequent one), we shall elaborate the scheme in some more detail than needed only from the illustration point of view. In the formulation that we present below, the gravitational field (metric) is not a primary field, and need not be nondegenerate. However, this possible degeneracy does not cause any problem, at least at the classical level. What happens at the quantum level ? To our knowledge, nobody considered this question seriously, although some of the Sakharov papers speculate on the possibility of changing space-time signature "inside" elementary particles, which phenomenon must involve degeneracy of the metric at the boundary of the region (theories involving phenomena of this kind should be however based upon a group other than $SO(\eta)$ (for example $SL(m,\mathbb{R})$ or $GL(m,\mathbb{R})$).)

We start with a manifold **M** and a principal bundle P over **M**. At the beginning, **M** is endowed with topology and differentiable structure but *no metric* and *no connection*. P is a principal bundle like in any Yang-Mills theory. It can be trivial or not. Eventually, path integral would anyhow involve sum over toplogies of **M** and of P, so we assume M and P are given. For the structure group of P, we shall take SO($\mathfrak{y}$) - the special orthogonal group of a scalar product $\mathfrak{y}$ = diag(-1,...,-1,+1,....,+1), with signature (p,q), and p+q = m = dim **M**. In fact, since we are going to discuss spinorial matter, we take the structure group of P to be the two-fold covering **Spin**($\mathfrak{y}$) of SO($\mathfrak{y}$). The points of P are called spin-frames. We shall consider two associated bundles: one corresponding to the representation of **Spin**($\mathfrak{y}$) on $\mathbb{R}^m$ via the 2:1 homomorphism **Spin**($\mathfrak{y}$)$\rightarrow$SO($\mathfrak{y}$) $\subset$ GL(m,$\mathbb{R}$). We call this bundle $\mathfrak{V}$, and the other bundle $\mathfrak{W}$ corresponding to a representation of **Spin**($\mathfrak{y}$) on a (real or complex) vector space **W** induced by a representation of the Clifford algebra of $\mathfrak{y}$. Elements of $\mathfrak{V}$ are called vectors, elements of $\mathfrak{W}$ are called *spinors* . However these vectors have (so far) nothing to do with any metric structure of **M**; in fact *there is yet no metric* on **M**. It is important to make this last statement clear. A metric (a usual Riemannian metric on a manifold) allows one to compute scalar products of *tangent* vectors to the manifold (a tangent vector is defined either as an equivalence class of tangent curves, or as a differential operator on functions on the manifold, or as an element of the tangent bundle of the manifold). What we have here (the elements of the vector bundle $\mathfrak{V}$) may be called vectors but they are not *tangent* vectors since we do not know yet how to interpret them as elements of the tangent bundle of **M** (the principal bundle P we start with is not the orthonormal frame bundle of a given metric on **M** but just some SO(p,q) principal bundle exrenal to **M**). The data of the principal bundle P allow us to compute scalar products of these "vectors" of $\mathfrak{V}$, but, let us stress again, one should not mix these "vectors" with tangent vectors and their scalar product with metric. Once a field θ, wich we shall now introduce, is fixed, *only then* does the previous scalar product (hidden in the definition of P) become a possibly degenerated metric on M. But the field θ, which plays the role of a "vielbein", or a "soldering form" is a *dynamical* field to take variations over. So, *we take now, as our matter fields, two, fields: a one-form with values in* $\mathfrak{V}$ *denoted* θ, *and a section (resp. one-form) with values in* $\mathfrak{W}$ *denoted* ψ. There are two possible cases of ψ being either a zero-form or a one-form corresponding to spin 1/2 and 3/2[9] respectively. We shall write now a Lagrangian describing interaction of this matter, represented by θ and ψ, with *geometry*. *Geometry is represented by a connection form* ω in P. A configuration of the theory consists therefore of a triple (θ, ψ, ω); observe that all three are completely *independent* of each other, making no trouble at all when taking independent variations. This last property distinguishes vividly the above formulation from others in which it is usually a pain to answer the question of how to take variations over the metrics keeping spinors (and spin structures) fixed. Observe also that, in the case of spin 3/2, all three objects are one-forms indicating a possibility of a common origin.

Now we will write the Lagrangian m-form describing the interaction of the fields θ, ψ and ω. It will have two parts: one describing interaction of θ and ω, and one describing interaction of ψ, ω and θ. To write down the formulae we will choose a local section σ of the bundle P; observe however that, since the ϵ - symbol is $SO(\eta)$-invariant, the Lagrangian m-forms below do not depend on a choice. In the formulae below we shall skip the subscript σ as it was already done in (6") and (7") of the previous section. For the Lagrangian m-form of the theory we take $L = L_g + L_\psi$ with

$$L_g = \epsilon_{a_1 \dots a_m} (\lambda_0\, \theta^{a_1} \wedge \dots\dots \wedge \theta^{a_m} + \lambda_1\, \Omega^{a_1 a_2} \wedge \theta^{a_3} \dots\dots \wedge \theta^{a_m}$$
$$+ \lambda_2\, \Omega^{a_1 a_2} \wedge \Omega^{a_3 a_4} \wedge \theta^{a_5} \wedge \dots \theta^{a_m} + \dots\dots + \lambda_{m/2}\, \Omega^{a_1 a_2} \wedge \dots\dots \wedge \Omega^{a_{m-1} a_m})$$

and L_ψ is defined by

---

[9]Actually a pure spin 3/2 field is a section of a virtual bundle.

$$L_\psi = \lambda\, \epsilon_{a_1 \ldots a_m}\, v^{a_1} \wedge \theta^{a_2} \wedge \ldots\ldots\ldots \wedge \theta^{a_m}$$

for spin 1/2, while

$$L_\psi = \lambda\, \epsilon_{a_1 \ldots a_m}\, v^{a_1 a_2} \wedge \theta^{a_3} \wedge \ldots\ldots\ldots \wedge \theta^{a_m}$$

for the Rarita-Schwinger field. In these expressions Ω denotes the curvature two form of ω on **M** with values in **Lie(Spin(η))** = **Lie(SO(η))** $\subset$ **Lie(GL(m))**, so that Ω is just a matrix $\Omega^a{}_b$. The 1-form v^a appearing in L_ψ is defined as follows: we assume that there is a sesquilinear form, written $\overline{\Psi}\psi$ on **F** which is **Spin(η)** invariant; then $v^a := \overline{\Psi}\gamma^a D\psi$ where $\gamma^a := \eta^{ab}\gamma_b$ and γ_b are the generators representing the Clifford algebra of η on **F**. Similarly, $v^{ab} := \overline{\psi}\gamma^a\gamma^b D\psi$ for the Rarita-Schwinger field. Finally, $\epsilon_{a_1 \ldots a_m} = 0, \pm 1$ is the totally antisymmetric Kronecker symbol: $\epsilon_{a_1 \ldots a_m} = \delta_{a_1 \ldots a_m}{}^{1 \ldots m}$.

Comments

1) Given a configuration (ω, θ, ψ) we can construct a symmetric tensor $g_{\mu\nu} := \eta_{ab}\, \theta_\mu{}^a \theta_\nu{}^b$. It needs not, however be nondegenerate. Indeed it was assumed or required that $\det \theta_\mu{}^a \neq 0$ (here $\det \theta_\mu{}^a$ is the determinant of the matrix $\theta_\mu{}^a$ of θ , the matrix elements being calculated for a local coordinate system x^μ on **M** and a local σ section of **P**). Observe that both the lagrangian and the field equations which follow from the Lagrangian make perfect sense for a degenerate $\theta_\mu{}^a$ and $g_{\mu\nu}$ (for instance the lagrangian *vanishes* for vanishing θ).

2) The λ_0 term is essentially the cosmological constant term; for a nondegenerate θ, it is equal to $\lambda_0\, \mathrm{sgn}(\det \theta_\mu{}^a)\, |\det g_{\mu\nu}|^{1/2}\, d^m x$.

3) The λ_1 term is essentially the Einstein-Hilbert term including torsion. Indeed, provided θ is nondegenerate, we can introduce an affine connection $\Gamma_{\mu\nu}{}^\sigma$ on **M** as a unique affine connection on **M** for which $\theta_\mu{}^a$

considered as a $T^*M \otimes \mathcal{F}$ -valued zero-form on **M** is parallel; explicitly $\Gamma_{\mu\nu}{}^{\sigma}$ is determined by the equation

$$D_\mu \theta_\nu{}^a := \partial_\mu \theta_\nu{}^a - \Gamma_\mu{}^\sigma{}_\nu \theta_\sigma{}^a + \omega_\mu{}^a{}_b\, \theta_\nu{}^b = 0.$$

This connection $\Gamma_\mu{}^\sigma{}_\nu$ is automatically metric (i.e. $D_\mu g_{\nu\sigma} = 0$) and, in general it has torsion. Its torsion $T_\mu{}^\sigma{}_\nu$ is determined by the equation (again assuming $\theta_\mu{}^a$ nondegenerate)

$$(D\theta)_{\mu\nu}{}^a = T_{\mu\nu}{}^\sigma\, \theta_\sigma{}^a.$$

Now the λ_1 term is equal to λ_1 sgn(**det** $\theta_\mu{}^a$) R | **det** $g_{\mu\nu}$ | $^{1/2}$ $d^m x$, where R is the scalar curvature of the connection $\Gamma_\mu{}^\sigma{}_\nu$. Observe that neither scalar curvature (since it involves inverse metric) nor sgn(**det** $\theta_\mu{}^a$) make sense for **det** $\theta_\mu{}^a \rightarrow 0$. However, as we see now, the *product* of the two things is well defined in the limit and defines the λ_1 - term of the Lagrangian.

4) The λ_2-term is non-trivial only for m>4, as for m=4 it is a boundary term. It naturally arises from string theory. The higher terms may appear in still higher dimensions.

5) The L_ψ term written explicitly is again nothing but the Dirac Lagrangian. Again assuming θ nondegenerate we may write it down as

$$L_\psi = \lambda\ \mathrm{sgn}(\mathbf{det}\ \theta_\mu{}^a)\ |\ \mathbf{det}\ g_{\mu\nu}\ |^{1/2}\ \bar{\Psi}\gamma^\mu \nabla_\mu \psi\ \ d^m x$$

where $\gamma^\mu := g^{\mu\nu}\theta_\nu{}^a\gamma_a$. Observe that this latter expression makes sense only when $g_{\mu\nu}$ is non-degenerate and blows up for **det** $\theta_\mu{}^a \rightarrow 0$, whereas the original form of L_ψ is completely regular in the limit $\theta_\mu{}^a \rightarrow 0$; instead of blowing up, it just vanishes! The covariant derivative $D_\mu\psi$ of the spinor field can be written explicitly as

$$D_\mu \psi = \partial_\mu \psi + 1/2\ \omega_\mu{}^{ab}\, \Sigma_{ab}\psi$$

where

$$\Sigma_{ab} = 1/4\,[\,\gamma_a, \gamma_b\,]\ .$$

6) θ corresponds to what is sometimes called "vielbein", and sometimes "soldering form". Indeed, θ serves as an identification of the bundle $\mathfrak{F}$ with the tangent bundle of **M**. Once this identification is acccomplished (provided it is topologically possible), then θ plays the role of the canonical form on the frame bundle of **M**.

7) We remind the reader that all these formulae are written with respect to a local section of the bundle **P** they have however an invariant meaning.

6.8 Miscellaneous

We have discussed "genuine" spinors in sect. 6.2, 6.3, 6.4, however, one could consider also p-forms on **E** valued in the bundle of spinors ; the whole machineary described in sect. 6-2 applies here. Since we have both a Levi-Civita connection and a (spin) connection in **W**, we may even consider arbitrary spinor valued tensor fields on **E** (not necessarily p-forms) and define on them covariant derivatives etc. We have this as an exercise to the reader.

6.9 Pointers to the literature

Spin structures 1, 4, 14, 73, 82, 84, 91, 97, 135, 147
Covariant derivatives 76, 127, 231
Vector bundles 96, 127, 210
Dirac operator 55, 78, 91, 135, 170

VII

HARMONIC ANALYSIS AND DIMENSIONAL REDUCTION

7.0 Summary Section

It is well known that, when a group G acts on a manifold E, the "matter fields" defined on this manifold can be analysed in terms of irreducible representations of the group G. For example, periodic functions on the real line (i.e. functions on the circle S^1) can be analysed via Fourier analysis (in this well-known situation, $G = U(1)$ and any irreducible representations is characterized by an integer $\gamma \in Z$); to each function f and to each integer γ we associate a Fourier component :

$$\hat{f}(\gamma) = \int f(\theta)\, e^{-i\gamma\theta}\, d\theta/2\pi$$

and we can recover f from the knowledge of its Fourier components via the relation:

$$f(\varphi) \cong \Sigma_\gamma \hat{f}(\gamma)\, e^{i\gamma\varphi} \quad .$$

This Fourier series does not need to be finite (when it is, $f(\varphi)$ is called a trigonometric polynomial) but in any case, the space of such finite sums is dense in the space of L^2 periodic functions. These well-known properties can be generalized in two ways: first by replacing U(1) by a compact group G (or by a homogeneous space G/H), then, by replacing the functions on G (or on G/H) by "generalized" functions (matter fields) on M×G (or M×G/H), M being some manifold (space-time for instance).

The first generalisation, non-abelian harmonic analysis [125] is recalled in more details in sect. 7.1 and goes as follows : to each function f on a compact group G, and to each irreducible representation α of G, we associate its Fourier component

(1) $$\hat{f}(\alpha) = \int_G f(g)\alpha^*(g^{-1})\, dg$$

where α^* is the representation contragredient to α :

$$\alpha^*(g) = \alpha(g^{-1})^*$$

and "dg" stands for the (normalized) Haar measure on G.

Notice that $\hat{f}(\alpha)$ is a linear operator in the representation space $W(\alpha)$ - more precisely in its dual $W^*(\alpha)$. We can then recover f from the knowledge of its Fourier components $\hat{f}(\alpha)$ via the inverse Fourier transform relation

(2) $$f(k) \cong \sum_{\alpha\in\hat{G}} d(\alpha)\, \mathrm{Tr}\, (\hat{f}(\alpha)\alpha^*(g)),$$

where α runs in the set $\hat{G}$ of irreducible representations of G (up to equivalence) and where $d(\alpha) = \dim W(\alpha)$ is the dimension of the representation α; this last equality can be easily recast in a form which is more adapted to further generalisations, namely

$$(3) \qquad f(g) \cong \sum_{\alpha \in \hat{G}} d(\alpha) \int_G [R(g)f](k)\, \chi_\alpha(g^{-1})\, dg$$

where χ_α is the character G of α and where $R(g)$ denotes the (left) regular representation defined by

$$(4) \qquad [R(g)f](k) = f(g^{-1}k) \quad .$$

The next generalisation involves the replacement of G by a space $E = M \times G$ (we write here a direct product but it may be "twisted"). There are essentially no changes in the discussion : when analysing functions $f(x,g)$ on E, one can think of $x \in M$ as a parameter in eq. (3), playing essentially no role.

When the group G is replaced by the homogeneous space G/H, the only essential change in the previous formalism is the replacement of $f(g)$ or $f(x,g)$ by $f([g])$ or $f(x,[y])$, $[g] = gH$.

Matter fields are usually not just real or complex functions on $E = M \times G/H$ but maps from E to some vector space F which carries also a representation ρ of some other group T ; for instance T is the (covering of) the orthogonal group of E and (ρ,F) is the spinor representation (we study then spinor fields on E). The harmonics appearing in the analysis of such matter fields will, in a sense, be harmonics of type (ρ,α), α referring to the global action of G and ρ to the "local" action of T (as already stressed many times, there is no (global) action of T on such fields -in the same sense that there is no (global) action of GL(n) on vectors at a point of a manifold: one has first to choose an (arbitrary) frame). The main result explained in more detail in sect. 7.2 is then, in plain terms, the following :

Let E be a manifold which can be written locally as $M \times G/H$, endowed with a (global) action of a compact group G. Let T be some other group and $\Gamma\mathcal{F}$ the space of matter fields on E which transform (locally) under T according to the representation ρ of T on the vector space F. Then the space of generalized trigonometric polynomial is dense in $\Gamma\mathcal{F}$, in other words, any matter field (written in some gauge as $y \in E \rightarrow \Phi(y) \in F$) can be decomposed as follows

$$\Phi(y) \doteq \sum_{\alpha \in \hat{G}} [\Pi_\alpha \Phi](y) \quad , \tag{5}$$

where

$$[\Pi_\alpha \Phi](y) = d(\alpha) \int_G [R(g)\Phi](y)\, \chi_\alpha(g^{-1})\, dg. \tag{6}$$

In this formula, R denotes the (usually reducible and infinite-dimensional) representation of G on $\Gamma\mathcal{F}$ "induced" by the representation ρ of T on F.

This is a generalisation of the Peter-Weyl theorem.

The next step in our analysis is then to construct the space $\Gamma_\alpha\mathcal{F} = \Pi_\alpha(\Gamma\mathcal{F})$ of harmonics of type α, and for that we can be guided by the theory of induced representation recalled in sect. 7.3. There, we restrict M to a single point and choose $T=H$; more precisely we consider the group G as a collection of subgroups H parametrised by G/H, choose a representation ρ of H on a vector space F and study the space $\Gamma\mathcal{F}$ of "matter fields of type ρ" on G/H (we refer to chapt.6 for a discussion of the several ways of describing these fields locally or globally). This infinite-dimensional space $\Gamma\mathfrak{F}$ carries a natural representation R(g) of G inherited from the G action on G/H. For every irreducible representation α of G, the representation R restricted to $\Gamma_\alpha\mathfrak{F}$ is a multiple of α. Intuitively, vectors of the subspace $\Gamma_\alpha\mathfrak{F}$ of $\Gamma\mathfrak{F}$ can be gotten from W_α by the action of operators intertwining the representation R and α of G on the vector spaces $\Gamma\mathfrak{F}$ and W_α respectively. The space of these operators, in turn, can be considered as the subspace (call it $\Gamma_{inv}\mathfrak{C}_\alpha$) of those elements of $\Gamma\mathfrak{C}_\alpha := \Gamma\mathfrak{F}\otimes W_\alpha{}^*$ which are invariant under the representation $R\otimes\alpha^*$ of G; the precise statement is then to say that $\Gamma_\alpha\mathfrak{F}$ and $W_\alpha\otimes\Gamma_{inv}\mathfrak{C}_\alpha$ are isomorphic. We therefore restrict our attention to the subspace $\Gamma_{inv}\mathfrak{C}_\alpha$; the last observation (sometimes referred to as the "Frobenius reciprocity theorem") consists in showing that there is a one to one correspondence between elements of $\Gamma_{inv}\mathfrak{C}_\alpha$ and vectors of $F\otimes W_\alpha{}^*$ which are invariant under the representation $\rho\otimes\alpha^*$ of H (we call this space $F\otimes_H W_\alpha{}^*$). In practice, one has to look at the singlets appearing into the decomposition of the representation $\rho\otimes\alpha^*$ into irreducible representations of H (notice that α^* is an irreducible representation of G but is usually reducible with respect the subgroup H).

In sect. 7.4, we analyse the situation when M is no longer a single point and T is not equal to H: we have then to generalise the above. The effect of the replacement of H by T does not greatly affects the above. Indeed, there exists a homomorphism λ from H to T which characterises the G action. This map tells us how the T -frames are "rotated" at the point y when we perform a transformation of $E = M\times G/H$ which does not affect the point y. Therefore H still acts on F via the composition of maps $h \rightarrow \lambda(h) \rightarrow \rho(\lambda(h))$ and the vector space $F \otimes_H W_\alpha{}^*$ is well determined. The effect of M is then to "add a parameter x" in other words to replace the above vector space by the space of maps from M to $F \otimes_H W_\alpha{}^*$.

In this summary section we have mostly used a "local" description of matter fields but this should not precisely cloud the fact that there is a hidden (by the vocabulary) local gauge group. We have to answer the following question: what is the local gauge group "acting on" the generalized matter fields valued in $F \otimes_H W_\alpha^*$? This question is studied in sect. 7.5 and we get the following: the gauge group is $N(\overline{H})|\overline{H}$ where $\overline{H}$= $diag(H,\lambda(H)) \subset G\times T$ and $N(\overline{H})$ is the normalizer of $\overline{H}$ in $G\times T$. (This answer should not be surprising since the problem is essentially the same as the one already discussed in Ch.5 (sect. 5.2.5), cf. also Ch.9). Moreover $N(\overline{H})|\overline{H}$ is locally isomorphic with $N(H)|H \times Z$ where Z is the centralizer of $\lambda(H)$ in T .

Let us summarise the discussion (a more precise theorem may be found at the end of sect. 7.5) :

1) The space of all matter fields $\Gamma\mathfrak{F}$ on $E \cong M\times G/H$, transforming locally as the representation (ρ,F) of the group T, carries an infinite-dimensional representation $R(g)$ of G.

2) Every matter field on E can be approximated by linear combination of harmonics labelled by the irreducible representations α of G.

3) The space $\Gamma_\alpha\mathfrak{F}$ of field harmonics of type α can be reconstructed from the space $\Gamma\mathfrak{F} \otimes_G W_\alpha^*$ of intertwining maps; indeed $\Gamma_\alpha\mathfrak{F} = (\Gamma\mathfrak{F} \otimes_G W_\alpha^*) \otimes W_\alpha$. Calling $\mathfrak{C}_\alpha = \mathfrak{F} \otimes W_\alpha^*$ and $\Gamma_{inv}\mathfrak{C}_\alpha = \Gamma\mathfrak{F} \otimes_G W_\alpha^*$, we may consider these intertwining maps themselves (the space $\Gamma_{inv}\mathfrak{C}_\alpha$) as equivariant generalized matter fields on E (they carry an extra index associated to the representation α^*).

4) The "interesting" space $\Gamma_{inv}\,\mathfrak{E}_\alpha$ can itself be considered as the space of <u>all</u> matter fields on **M** (dimensional reduction) valued in the vector space $F \otimes_H W_\alpha{}^*$, i.e. as the space of all sections of a vector bundle $\overline{\mathfrak{E}}_\alpha$ of base **M** and typical fiber $\overline{F}_\alpha = F\otimes_H W_\alpha{}^*$.

5) The local gauge group of these effective matter fields is $N(\overline{H})|\overline{H}$ where $N(\overline{H})$ is the normalizer of $\overline{H} = diag(H,\lambda(H)) \subset G\times T$ in $G\times T$, λ being an homomorphism from H to T characterizing the G action on the space of T -frames.

The following pictures illustrate the constructions previously mentioned and should help the reader to remember the notations used in this chapter.

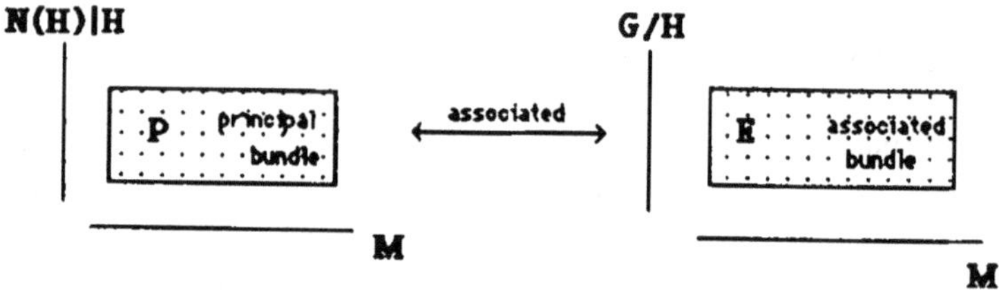

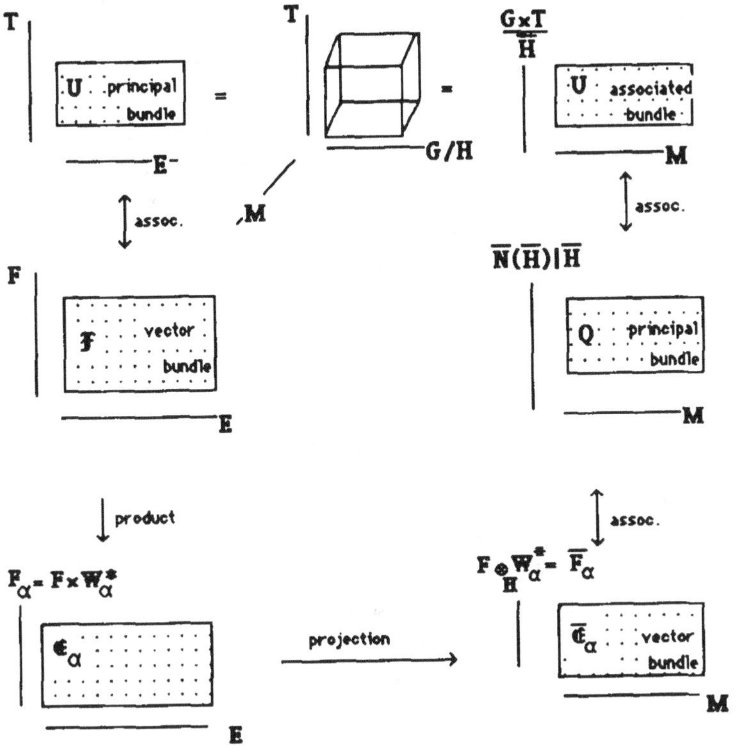

The following definitions and notations should be kept in mind:

$$\Gamma\mathfrak{F} = \oplus_{\alpha\in\hat{G}} \; \Gamma_\alpha \mathfrak{F}$$

$$\Gamma_\alpha \mathfrak{F} = \Gamma_{inv}\, \mathfrak{E}_\alpha \otimes W_\alpha$$

$$\Gamma_{inv}\, \mathfrak{E}_\alpha = \Gamma\mathfrak{F} \otimes_G W_\alpha \cong \Gamma\bar{\mathfrak{E}}_\alpha .$$

Sect. 7.6 contains a brief discussion of dimensional reduction of differential linear operators acting on the matter fields along with a discussion of the physical "consistency problem" when matter fields are added to the Lagrangian on E. It is argued that the theory is inconsistent if one truncates the theory by coupling gravity on E (with a G-invariant ansatz) to matter fields which are not invariant. There is of course no surprise here: the situation is analoguous to what happens in usual field theory when one wants to couple electrons and photons but is willing to forget at the same time the influence of the electrons on the electromagnetic field. The theory is of course inconsistent but nothing prevents us from studying an "external field problem", i.e., in our context, a metric on E being <u>given</u>, we study the behaviour in it of matter fields of type α.

7.1 <u>A particular case of 7.2: non-abelian harmonic analysis</u>

In order to become more familiar with the general formalism, let us recall what happens when the space **M** is a single point, $E = G$, a compact Lie group and $F = \mathbb{C}$, the complex numbers.

"Matter fields" can now be described either as smooth sections of the trivial bundle $\mathfrak{F} = G \times \mathbb{C}$ i.e. as maps $k \in G \rightarrow \Phi(k) = (k, \tilde{\Phi}(k)) \in \mathfrak{F}$ or, more simply by $\mathbb{C}$-valued smooth maps $\tilde{\Phi}(k)$ on G. The action of G on the space $\Gamma\mathfrak{F}$ of these sections of $\mathfrak{F}$ is given by

$$
\begin{aligned}
[R(g)\Phi](k) &:= g\Phi(g^{-1}k) = g(g^{-1}k, \tilde{\Phi}(g^{-1}k)) \\
&= (k, \tilde{\Phi}(g^{-1}k))
\end{aligned}
$$

or by

$$[R(g)\tilde{\Phi}](k) = \tilde{\Phi}(g^{-1}k).$$

This is the usual (left) regular representation of the group G.

The Fourier transform $\hat{\Phi}$ of the C-valued function $\tilde{\Phi}$ associates to each representation $\alpha \in G$ an endomorphism $\hat{\Phi}(\alpha)$ of the vector space $W_\alpha{}^*$: $\hat{\Phi}(\alpha)$ is the "Fourier coefficient" of $\tilde{\Phi}$ in the representation α. $\tilde{\Phi}$ is recovered from the knowledge of its Fourier coefficients via the inverse Fourier transform relation :

$$\Phi(k) = \sum_{\alpha \in \hat{G}} d(\alpha) \operatorname{Tr} [\hat{\Phi}(\alpha)\, \alpha^*(k)].$$

This non-commutative Fourier analysis on compact groups holds usually in $L^2(G)$, a space much bigger than $\Gamma\mathcal{F}$. We will now modify the previous relation in order to introduce the projection operators Π_α and the harmonics $\Pi_\alpha \Phi$ of type α :

$$\Phi(k) = \sum_{\alpha\in\hat{G}} d(\alpha) \operatorname{Tr} \int_G \Phi(g)\, \alpha^*(g^{-1})\, \alpha^*(k)\, dg \quad ,$$

from the definition of $\hat{\Phi}$,

$$= \sum_{\alpha\in\hat{G}} d(\alpha) \operatorname{Tr} \int_G \Phi(g)\, \alpha^*(g^{-1}k)\, dg \quad ,$$

since α^* is a contragradient representation,

$$= \sum_{\alpha\in\hat{G}} d(\alpha) \operatorname{Tr} \int_G \Phi(kg)\, \alpha^*(g^{-1})\, dg \quad ,$$

by the change of variable $g \rightarrow g^{-1}k$

$$= \sum_{\alpha\in\hat{G}} d(\alpha) \operatorname{Tr} \int_G [R(g)\Phi](k)\, \alpha^*(g)\, dg \quad ,$$

by definition of R,

$$= \sum_{\alpha\in\hat{G}} d(\alpha) \int_G [R(g)\Phi](k)\, \chi_\alpha(g^{-1})\, dg,$$

by definition of the character χ_α :

$$\chi_\alpha(y) = \operatorname{Tr} \alpha(g) = \operatorname{Tr} \alpha(g)^* = \operatorname{Tr}(\alpha^*(g^{-1})).$$

The previous formula defines projection operators Π_α and the inverse Fourier transform relation, written $\Phi(k) = \Sigma_{\alpha\in\hat{G}} [\Pi_\alpha \Phi](k)$ will justify our starting point in the next section, namely eq. 7.2(4).

7.2 Harmonic expansion (generalized Peter-Weyl theorem)

Let E be a manifold with a given action of a compact group G. For instance, E can be a multidimensional universe with G its global symmetry group [Ch.5]. On E consider linear matter fields of some fixed type. In general these matter fields will be described by sections of a certain vector bundle $\mathfrak{F} = \mathfrak{F}(E,F)$, not necessarily trivial, with base E and fiber F; we suppose that F is a finite dimensional vector space (real or complex). Such a situation will in particular arise if we start from a principal bundle $U = U(E,T)$ with base E and structure group T and from a representation ρ of T on the vector space F; $\mathfrak{F}$ is then defined as the associated bundle $\mathfrak{F} = U \times_\rho F$. We also assume that $\mathfrak{F}$ is "equivariant" i.e. that the symmetry group G "knows" not only how to act on E, the base space, (by diffeomorphisms) but also, by bundle automorphisms, on $\mathfrak{F}$, the bundle space (usually this action will come from an action of G on the principal bundle U -in this chapter we assume that the G action is a left action-)[1] ; for any $g \in G$, we have

a map: $y \in E \rightarrow gy \in E$,

a map: $u \in U \rightarrow gu \in U$,

with $\pi(gu) = g\pi(u)$, π being the projection map $\pi: U \rightarrow E$

and $\forall k \in T$, $g(uk) = (gu)k$.

Finally,

a map: $\Phi = u.\Phi_u \in \mathfrak{F} \rightarrow g\Phi = (gu).\Phi_u \in \mathfrak{F}$, $\Phi_u \in F$.

[1] The reader will have noticed that this assumption is opposite to the one we made in the previous chapters. Actually, from this left G action and from the right T action, we will build in sect. 7.4 a right $\overline{G} = G \times T$ action on the total space of U, and it will then be possible to apply directly the results of chapter 5.

The situation is depicted by the following figure:

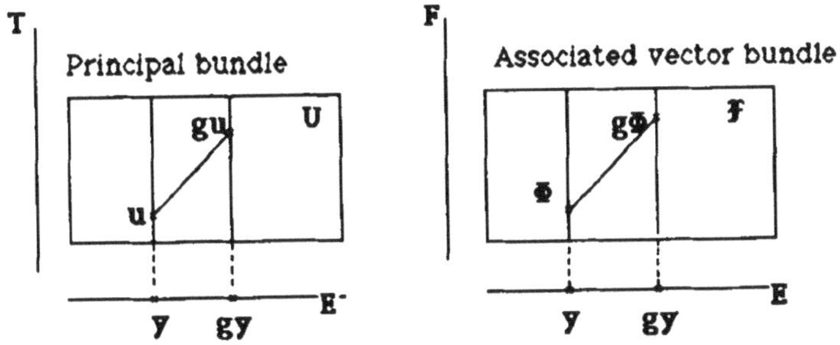

Notice that the G action on E, U, $\mathfrak{F}$ is written as a left action because we want to forget about parenthesis (a better justification for it is that we have indeed a homomorphism from G to the group of automorphisms of the bundle U, and the action of automorphisms on a right principal bundle is usually written as a left action, cf. sect 4.10). Let $\Gamma\mathfrak{F}$ denote the space of all sections Φ of $\mathfrak{F}$, $\Gamma\mathfrak{F}$ is an infinite-dimensional vector space and its elements are matter fields. Given the action of G on E and on $\mathfrak{F}$, there is a natural representation of G on $\Gamma\mathfrak{F}$:

(1) $$(R(g)\Phi)(y) = g\Phi(g^{-1}y).$$

We say that R is the representation of G on $\Gamma\mathfrak{F}$ <u>induced</u> by the representation ρ of T on the vector space F. Notice that if we describe matter fields as ρ - equivariant maps $\tilde{\Phi}$ from U to the vector space F (sec. 6.1),

$$\tilde{\Phi}(uk) = \rho(k^{-1})\tilde{\Phi}(u),$$

then the induced representation R is written differently :

(2) $$(R(g)\tilde{\Phi})(u) = \tilde{\Phi}(g^{-1}u).$$

The vector space of fixed points of the representation R consists of particular matter fields (called invariant) which satisfy the property

$$(3) \qquad \forall g \in G,\ (R(g)\Phi)(y) = \Phi(y) \quad \text{i.e.}\quad \Phi(gy) = g\Phi(y) \quad \text{or} \quad \tilde{\Phi}(gu) = \tilde{\Phi}(u).$$

Their space $\Gamma_{inv}\mathcal{F}$ will be called space of invariant sections. The situation can be described by the following picture :

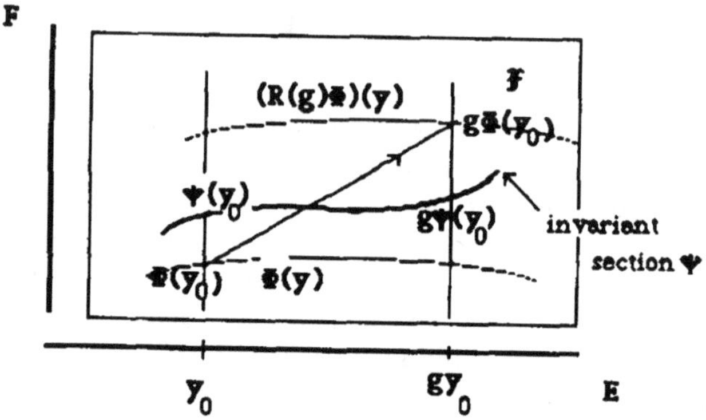

It is clear that the (infinite-dimensional) representation R(g) of G on the vector space $\Gamma\mathcal{F}$ is in general reducible (for example the subspace $\Gamma_{inv}\mathcal{F}$ of invariant sections carries the trivial representation of G). We now want to decompose $\Gamma\mathcal{F}$ into subspaces $\Gamma_\alpha\mathcal{F}$, each consisting of field configurations transforming according to a given irreducible representation α. For each irreducible representation α let W_α be the representation space of α (W_α is finite-dimensional since G is assumed to be compact), $d(\alpha) = \dim W_\alpha$ and χ_α be the corresponding character (unless stated otherwise, F and W are assumed to be complex vector spaces).

It is also convenient to introduce $\hat{G}$ - the dual of G - which is the (discrete) space of (equivalence classes of) irreducible representations of G; notice that $\hat{G}$ does not carry any group structure when G is not abelian.

We then define, for every $\alpha \in \hat{G}$, a projection operator $\Pi_\alpha : \Gamma\mathcal{F} \to \Gamma\mathcal{F}$ by the formula

$$(4) \qquad (\Pi_\alpha(\Phi))(y) = d(\alpha) \int_G [R(g)\Phi](y)\, \chi_\alpha(g^{-1})\, dg$$

where dg is the normalized Haar measure on G.

The range of Π_α will be denoted by $\Gamma_\alpha\mathcal{F}$ and its elements will be called *harmonics of type* α. It is intuitively clear (and the proof can be adapted from [198] [199] [224]) that the algebraic direct sum $\oplus_\alpha(\Gamma_\alpha\mathcal{F})$ is dense in $\Gamma\mathcal{F}$. More precisely, every field configuration $\Phi \in \Gamma\mathcal{F}$ can be approximated with an arbitrary accuracy by a finite superposition of harmonics (a generalized trigonometric polynomial) in the following sense : for every $\epsilon > 0$ and for every compact $A \subset E$ there is a finite subset $\Sigma \subset \hat{G}$ and $\Phi_\Sigma \in \bigoplus_{\alpha\in\Sigma} \Gamma_\alpha\mathcal{F}$ such that $\sup_{y\in A} \|\Phi(y) - \Phi_\Sigma(y)\| < \epsilon$; the norm here is any continuous norm on the fibers of $\mathcal{F}$. This approximation theorem is sufficient for our purpose and we do not need to investigate in what sense the "inverse Fourier transform" formula 7.0-5 holds. In particular we are not interested in making $\Gamma\mathcal{F}$ into a Hilbert space - This is because E is thought of as a model for an (extended) space time of hyperbolic signature, and in field theory it is the space of all *solutions* of a hyperbolic differential equation rather than the space of *all fields configurations* which carries a useful Hilbert space structure.

The physical conclusion of the above is that, in order to study matter fields of some given type on the extended space time E (for example spinor fields $\Psi(y)$), it is enough to study independently the harmonics labelled by $\alpha \in \hat{G}$ (i.e. matter fields $\Psi_\alpha(y)$). We will see in section 7.4 and 7.5 in which sense these harmonics can be thought of as matter fields on the (usual) space time M.

The general formalism will of course also apply if we take E = G. Consider the left action of G on itself, choose U as the frame bundle of G, then T =GL(dim G), F=$\mathbb{R}^{\dim G}$ and $\mathfrak{F}$ is the tangent bundle of G; sections of $\mathfrak{F}$ are just vector fields on G. It is interesting to notice that harmonics of type α = identity (i.e., R(g) = Id.) satisfy $\Phi(k) = g^{-1}\Phi(gk)$, i.e. are left invariant vector fields whereas harmonics of type Ad, as we will see later are right invariant vector fields.

7.3 A particular case of 7.4: induced representations

The purpose of this subsection is not to study in detail this important subject (cf. for instance [223]) but to show how it fits in our general framework and to introduce sect.7.4 (the Frobenius reciprocity theorem will be our guideline in the next subsection). We therefore choose U = G, a compact Lie group and write it as a T = H bundle over B = G/H.

Now, since G is a principal bundle with base G/H and structure group H, we can study " fields on G/H ", which are sections of bundles associated to G. H being the structure group of G, we choose therefore a representation ρ of H on some vector space F and construct the associated bundle $\mathfrak{F} = G\times_\rho F$; sections of this vector bundle may be called "fields of type ρ". We refer to Ch.6. for a discussion of different possible ways of local and global description of these fields. Here however we have a particular situation : the principal bundle itself has a group structure! In fact, as already mentioned in Ch.4, the left action of G on itself maps cosets of G/H onto cosets and commutes with the right action of H, therefore producing automorphisms of the principal bundle G $\rightarrow$ G/H. Hence (cf. the previous subsection) G knows how to act on elements of $\mathfrak{F}$ and on sections of $\mathfrak{F}$[2]: if $\Phi \in \Gamma\mathfrak{F}$ is given by an equivariant function $\hat{\Phi}: G \rightarrow F$ of type ρ

$$(1) \qquad \hat{\Phi}(gh)=\rho(h^{-1})\hat{\Phi}(g),\ g\in G,\ h\in H$$

[2] The reader will notice that in this particular situation G is a finite dimensional subgroup of the infinite dimensional group of automorphisms (non vertical) of the bundle G = G(G/H,H).

(cf. sect. 6.1) then the "induced" representation $R(g)$ of G on $\Gamma\mathfrak{F}$ takes the form

(2) $$(R(g)\Phi)(g') = \Phi(g^{-1}g')$$

which is a particular case of the formula (2) of the previous subsection[3]. The representation R will be in general reducible (even if ρ is irreducible). The representation space $\Gamma\mathfrak{F}$ will decompose into the direct sum $\Gamma\mathfrak{F} = \oplus_{\alpha\in\hat{G}} \Gamma_\alpha\mathfrak{F}$ so that, for every irreducible representation α of G, the representation R restricted to $\Gamma_\alpha\mathfrak{F}$ is a multiple of α. What we want now is to construct $\Gamma_\alpha\mathfrak{F}$ explicitly. Let us proceed to this end in several clearly marked steps.

<u>Step 1.</u> Let W_α denote a (complex) vector space carrying an irreducible representation α of G (for instance, for G=SU(2) irreducible representations are labelled by $\alpha=(j,m)$ with $j=0, \pm 1/2, \pm 1, \ldots$ and $m=-j, -j+1, \ldots, j-1, j$, with $W_{j,m} = \mathbb{C}^{2j+1}$). Denote by W_α^* the dual of W_α. It carries then the representation α^* contragradient to α. In components, denoting (w^i) the components of vectors of W_α and (w_i) those of W_α^*, let $w^i \to A(g)^i{}_j w^j$ denote the matrix form of α and $w_i \to A(g^{-1})^j{}_i w_j$ the matrix form of α^*.

<u>Step 2.</u> We have started with a vector bundle $\mathfrak{F}$. Its sections are described by functions $\Phi: G \to F$ which are H-equivariant (eq.1). Now we extend this bundle to $\mathfrak{E}_\alpha := \mathfrak{F} \otimes W_\alpha^*$. $\mathfrak{E}_\alpha$ is a vector bundle whose fibers are obtained by taking tensor products of the fibers of $\mathfrak{F}$ with a fixed vector space W_α^*. The space $\Gamma\mathfrak{E}_\alpha$ of sections of $\mathfrak{E}_\alpha$ is then the tensor product

---

[3] In term of the section Φ, (i.e. locally, maps from G/H to F), the induced action reads (cf. also 7.2(1), 7.2(2)):

$$((R(g)\Phi)(y) = g\Phi(g^{-1}y) \text{ with } y \in G/H.$$

(3) $$\Gamma\mathfrak{E}_\alpha = \Gamma\mathcal{F} \otimes W_\alpha^*$$

Sections ϕ of $\mathfrak{E}_\alpha$ can therefore be considered as maps $\vec{\phi}: G \to F \otimes W_\alpha^*$ which are ρ equivariant i.e.

(4) $$\vec{\phi}(gh) = [\rho(h^{-1}) \otimes id]\vec{\phi}(g),$$

where by "id" we denote the trivial action of H on W_α^*. In components sections ϕ of $\mathfrak{E}_\alpha$ are given by functions $\phi_i^\alpha(g)$ satisfying

(5) $$\phi_i^\beta(gh) = (\rho(h^{-1}))^\beta{}_{\beta'}\, \phi_i^{\beta'}(g)$$

<u>Step 3.</u> We introduce now the representation $R_\alpha := R \otimes \alpha^*$ of G on $\Gamma\mathfrak{E}_\alpha$ [4]. Taking into account the explicit form (2) of R as well as (3) we find

(6) $$(R_\alpha(g)\vec{\phi}(g') = id \otimes \alpha^*(g)\phi(g^{-1}g')$$

This time "id" denoting the trivial action of G on F.

In components

(7) $$(R_\alpha(g)\phi)^\beta{}_i\,(g) = A(g^{-1})^j{}_i\,\phi^\beta{}_j(g^{-1}g')$$

<u>Step 4</u> We have extended F to $F \otimes W_\alpha^*$, $\mathcal{F}$ to $\mathfrak{E}_\alpha = \mathcal{F} \otimes W_\alpha^*$ and $\Gamma\mathcal{F}$ to $\Gamma\mathfrak{E}_\alpha = \Gamma\mathcal{F} \otimes W_\alpha^*$. Now we shall *restrict* our attention to the subspace

---

[4] Notice that this is not a *canonical* action (i.e. is not the induced representation) of G on $\mathfrak{E}_\alpha$ considered as a bundle associated to G via the representation $\rho \otimes id$ of H on $F \otimes W_\alpha^*$.

$\Gamma_{inv}\mathfrak{E}_\alpha$ consisting of those sections $\psi \in \Gamma\mathfrak{E}_\alpha$ which are invariant[5] under R_α. Thus $\psi \in \Gamma_{inv}\mathfrak{E}_\alpha$ if and only if $R_\alpha \psi = \psi$, i.e.

$$(8) \qquad \vec{\psi}(gg') = id \otimes \alpha^*(g)\, \vec{\psi}(g')$$

In components:

$$(9) \qquad \psi \in \Gamma_{inv}\mathfrak{E}_\alpha \iff \psi^{\beta}{}_i(g^{-1}g') = A(g)^j{}_i\, \psi^{\beta}{}_j(g')$$

<u>Step 5</u> We have now the following theorem (cf. [223]) (sometimes called The Frobenius Reciprocity Theorem although the result gotten in step 6 is also called by the same name): the space $\Gamma_\alpha\mathfrak{F}$ of sections of the original bundle $\mathfrak{F}$ which, under the action of G, transform according to a given representation α of G -the space we are interested in- is naturally isomorphic to the tensor product of W_α and $\Gamma_{inv}\mathfrak{E}_\alpha$[6]

$$\Gamma_\alpha\mathfrak{F} \cong W_\alpha \otimes \Gamma_{inv}\mathfrak{E}_\alpha$$

It remains to know explicitly $\Gamma_{inv}\mathfrak{E}_\alpha$ and an explicit form of the isomorphism.

<u>Step 6</u> Let us first identify $\Gamma_{inv}\mathfrak{E}_\alpha$. It follows from the condition (8) which characterizes elements of $\Gamma_{inv}\mathfrak{E}_\alpha$ that $\vec{\psi}(g)$ is known for all g provided $\vec{\psi}(e) \in F \otimes W_\alpha^*$ is known. However $\vec{\psi}(e)$ can not be arbitrary as $\vec{\psi}$ has to satisfy the condition (4) characterizing all

---

[5] The space $\Gamma_{inv}\mathfrak{E}_\alpha$ can also be denoted $\Gamma \otimes_G W_\alpha^*$ and is isomorphic with the space $Hom_G(\Gamma\mathfrak{F}, W_\alpha)$ of operators which intertwine the representations R and α on the spaces $\Gamma\mathfrak{F}$ and W_α.

[6] Intuitively, this corresponds to the fact that elements of the vectorial subspace $\Gamma_\alpha\mathfrak{F}$ of $\Gamma\mathfrak{F}$ have, under $R(g)$ the same behaviour as those of W_α under α and can be gotten from W_α by the action of intertwining operators (cf. previous footnote).

elements of $\Gamma\mathcal{E}_\alpha$. Putting $g=e$ in (4) , and making use of (8) for $g=h$, we find

$$\rho(h)\alpha^*(h)\vec{\psi}(e)=\vec{\psi}(e); \qquad h\in H.$$

Thus, there is a one to one correspondance between elements of $\Gamma_{inv}\mathcal{E}_\alpha$ and vectors of $F\otimes_H W_\alpha^*$, i.e. vectors of $F\otimes W_\alpha^*$ which are invariant under the action of the representation $\rho\otimes\alpha^*$ of H[7]. Observe that originally α^* was a representation of G but, since H is a subgroup of G, α^* restricts to a representation (in general reducible) of H. In components : there is a one to one correspondence between the functions $\psi^{\beta}{}_i(g)$ sastifying (5) and (7), and elements $v^{\beta}{}_i \in F\otimes W_\alpha^*$ sastifying

$$R(h)^{\beta}{}_{\beta'}\, v^{\beta'}{}_i = A(h^{-1})^j{}_i\, v^{\beta}{}_j$$

Knowing ρ and α , the decomposition of $\rho\otimes\alpha^*$ into irreducible representations of H and, in particular, the singlets of $F\otimes W_\alpha^*$ can be now found from the tables of group representations. Thus we may assume that $\Gamma_{inv}\mathcal{E}_\alpha$ is known.

<u>Step7</u> It remains to give explicitly the isomorphism $W_\alpha\otimes\Gamma_{inv}\mathcal{E}_\alpha \rightarrow \Gamma\mathcal{E}_\alpha$. The simplest way is to use components : if $w^i \in W_\alpha$ and $\vec{\psi}$: $g \rightarrow \psi^{\beta}{}_i(g)$ is in $\Gamma_{inv}\mathcal{E}_\alpha$ then $w\otimes\psi$: $g\rightarrow w^i\, \psi^{\beta}{}_i(g)$ is in $\Gamma_\alpha\mathcal{F}$. What is very easy to see is that $w^i\, \psi^{\beta}{}_i(g)$ is in $\Gamma\mathcal{F}$. It is also easy to check, using the explicit form of the projection operator π_α given in sect. 7.2, that $w\otimes\psi$ is in $\Gamma_\alpha\mathcal{F}$. It is also staighforward to check that this map $W_\alpha\otimes\Gamma_{inv}\mathcal{E}_\alpha \rightarrow \Gamma_\alpha\mathcal{F}$ is onto. So that the essence of the theorem given in step 5 consists of the statement that the map is one to one .

[7] This result can also be written as follows:

$$Hom_G(W_\alpha, \Gamma\mathcal{F}) = Hom_H(W_{\alpha|\text{Restricted to } H}, F)$$

This property is also referred to as "Frobenius reciprocity".

Remark An alternative construction is also possible. One constructs the bundle $\mathfrak{C}_\alpha{}^{new} := G\times_{\rho\otimes\alpha^*}(F\otimes W_\alpha{}^*)$, with a *canonical* action of G given as in (2) (i.e., in the present case, induced by the representation $\rho\otimes\alpha^*$ of H), and considers the space $\Gamma_{inv}\mathfrak{C}_\alpha{}^{new}$ of invariant sections of this bundle. One also constructs the bundle $\mathfrak{W}_\alpha = G\times_\alpha W_\alpha$. Then we have an isomorphism $\Gamma_\alpha\mathfrak{F} = \Gamma\mathfrak{W}_\alpha\otimes(\Gamma_{inv}\mathfrak{C}_\alpha{}^{new})$, where the tensor product on the right is taken fiberwise. Comparing to the construction given in details above, this alternative construction may be considered as geometrically more natural. However, it introduces two opposite "twists" which effectively cancel together, so it is of no use for our purposes.

7.4 Harmonic expansion -2- (generalized Frobenius theorem)

We now want to reconstruct $\Gamma_\alpha\mathfrak{F}$ in the general situation discussed in sect. 7.2 (the picture given in sect. 7.0 illustrates the present discussion). E is now a (local) product $M\times G/H$ (in Ch V we assumed right action of G on E, while here it is more convenient to use a left action; this is why we take G/H instead of as H\G). U is a principal bundle over E with structure group T, and G acts on U by automorphisms. $\mathfrak{F}$ is a vector bundle associated to U via a representation ρ of T on a vector space F. The case previously discussed is a particular case of this more general situation, indeed, in sect. 7.3 we had

i) the action of G on E was transitive (E=G/H)

ii) each orbit of G acting on E had a distinguished point; indeed there was only one orbit and the distinguished point was [e]

iii) the groups T and H coincided

Now, in the general case, we have many orbits, and the space of orbits, called M, may have a nontrivial topology. Such a nontrivial topology may prevent the fibration $E\rightarrow M$ to have a smooth section, so that a smooth marking of a point on each orbit may be impossible. Finally, we must carefully deal with the groups T and H as they are different here .

Let us start with analyzing this last question first. We have assumed G acts on U by bundle automorphisms that is

$$a(ur) = (au)r$$

for all $a \in G$, $u \in U$, $r \in T$. The full group acting on U is therefore

$\overline{G} := G \times T$, G acting from the left and T from the right. To apply the formalism of Ch.5 to the new situation we will make the action of $\overline{G}$ on U into a right action by taking the convention

$$u(a,r) := a^{-1}ur.$$

We suppose that this action of $\overline{G}$ on U has only one orbit type so that all the stability groups of this action are mutually conjugate. Let us therefore investigate one such stability group. We choose a point $y_0 \in E$ with stability group H , and let u_0 be a point in the fibre of U over y_0 . Denote by $\overline{H}$ the stability group of y_0. We have then $(a,r) \in \overline{H}$ iff $u_0(a,r) = u_0$ iff $u_0r = au_0$. Taking the projection $\pi: U \to E$ of both sides we find $ay_0 = y_0$ i.e. $a \in H$. Now, since the element $r \in T$ is uniquely determined by $a \in H$ and the equation $u_0r = au_0$, we can write $r = \lambda(a)$. Then it is immediate that the map $\lambda: H \to T$ defined by the above condition i.e. by the equation

$$hu_0 = u_0\lambda(h) \, , \; h \in H$$

is a group homomorphism. It will play a crucial role in the following.

Summarizing this part of our discussion : we have found that the action of G on U determines a group homomorphism $\lambda : H \to T$, and that the typical stability group $\overline{H}$ of the action of $\overline{G} := G \times T$ on U is $\overline{H} = \{(h,\lambda(h)) : h \in H\}$. We shall also use the notation $\overline{H} = \mathrm{diag}(H \times \lambda(H))$. Thus we see, comparing with the particular case dealt with in sect. 7.3, that although the groups H and T are different in the present case, nevertheless H is homomorphically mapped into T . In most applications the map λ will be an embedding so that, in these cases, H can be identified with a subgroup of T . There are however possible situations (with a non-simple H), where λ may have a nontrivial kernel. The following picture illustrates the above remarks.

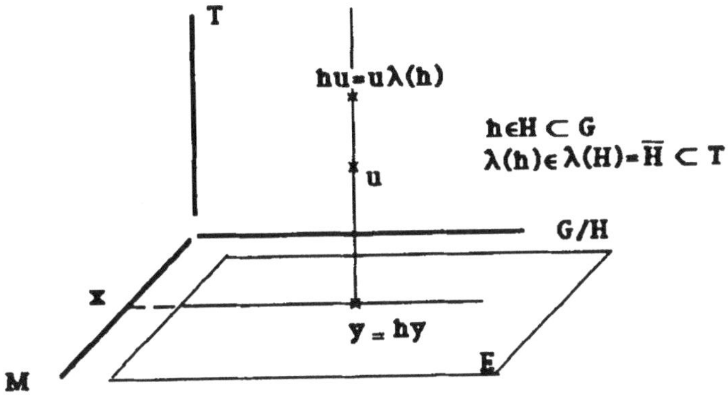

Let us now come back to the original problem of characterizing the subspace $\Gamma_\alpha \mathcal{F}$. The beginning follows exactly the steps of sect.7.3: we construct the vector bundle $\mathcal{C}_\alpha := \mathcal{F} \otimes W_\alpha^*$ with fibre $F \otimes W_\alpha^*$, associated to U via the representation $\rho \otimes id$ of T on $F \otimes W_\alpha^*$, and we introduce the action $R_\alpha := R \otimes \alpha^*$ of G on $\mathcal{C}_\alpha$ by

$$(R_\alpha(a)\vec{\psi})(u) = id \otimes \alpha^*(a)\vec{\psi}(a^{-1}u) .$$

In components, and using the notation of sect.7.3, we have

$$(R_\alpha(a)\psi)^\beta{}_i(u) = A(a^{-1})^j{}_i \psi^\beta{}_j(a^{-1}u).$$

Then we consider the space $\Gamma_{inv}\mathfrak{C}_\alpha$ of sections of $\mathfrak{C}_\alpha$ invariant under R_α.

$\Gamma_{inv}\mathfrak{C}_\alpha$ consists of those $\phi \in \Gamma\mathfrak{C}_\alpha$ which satisfy the equation

$$\phi^\beta{}_i(a^{-1}u) = A(a^{-1})^j{}_i \ \phi^\beta{}_j(u).$$

It can be then shown, similarly as in sect.7.3 (cf. also [224]) that $\Gamma_\alpha\mathfrak{F}$ is naturally isomorphic to $W_\alpha \otimes \Gamma_{inv}\mathfrak{C}_\alpha$. The isomorphism is given as before by $(w^i)\otimes(\phi^\beta{}_i) \mapsto w^i\phi^\beta{}_i$. What remains to do is to characterize the space $\Gamma_{inv}\mathfrak{C}_\alpha$. In sect.7.3, when $E=G/H$ and $M=E/G$ was just a point, $\Gamma_{inv}\mathfrak{C}_\alpha$ was isomorphic to the space $F\otimes_H W_\alpha{}^*$ of H-singlets for the representatioon $\rho\otimes\alpha^*$ of H on $F\otimes W_\alpha{}^*$. In the present general situation we have the following natural generalization of the previous result.

The space $\Gamma_{inv}\mathfrak{C}_\alpha$ is naturally isomorphic to the space $\Gamma\overline{\mathfrak{C}}_\alpha$ of sections of a certain vector bundle $\overline{\mathfrak{C}}_\alpha$ over M with typical fibre $F \otimes_H W_\alpha{}^*$ consisting of all singlets of the representation $(\rho\circ\lambda)\otimes\alpha^*$ of H on $F\otimes_H W_\alpha{}^*$.

In order to become more familiar with this result let us add a few comments: F was defined as a representation space for T but it is also a representation space for H, via the representation $\rho\circ\lambda$, : $h \in H \rightarrow \rho(\lambda(h)) \in \mathrm{End}\, F$. It is then meaningfull to consider the space $F \otimes_H W_\alpha{}^*$ of invariant vectors in $F \otimes W_\alpha{}^*$, since H acts on F as above and on $W_\alpha{}^*$ via the restriction to H of the representation α^* of G. The whole analysis of the last section then goes through, the role of M itself is to add an index to the right hand side of the usual Frobenius reciprocity theorem and one also gets $\Gamma\mathfrak{F} \otimes_G W_\alpha{}^* = \Gamma\overline{\mathfrak{C}}_\alpha$ where $\Gamma\overline{\mathfrak{C}}_\alpha$ is the space of sections of a vector bundle $\overline{\mathfrak{C}}_\alpha$with base M and typical fiber $F \otimes_H W_\alpha{}^*$. Indeed, $\mathfrak{F} \otimes_G W_\alpha{}^*$ admits an obvious projection π to M, the fiber of which, $\pi^{-1}(x)$ is isomorphic with $F \otimes_H W_\alpha{}^*$.This is obvious locally (we add a "continuous index" $x \in M$ to the right hand side of the Frobenius theorem) but we still do not know how to construct $\overline{\mathfrak{C}}_\alpha$ (for example we should know what is the "effective" structure group of this new bundle); this will be taken care of in the next subsection.

7.5 Harmonic expansion and dimensional reduction

We have seen that the natural building blocks out of which every field configuration can be reconstructed are harmonics, i.e. G-invariant sections of vector bundles $\mathfrak{C}_\alpha = \mathfrak{F} \otimes W_\alpha{}^*$, $\alpha \in \hat{G}$. These sections are still "matter fields" on E; we will now show that the space $\Gamma_{inv}\mathfrak{C}_\alpha$ is isomorphic with the space $\Gamma\bar{\mathfrak{C}}_\alpha$ of all sections of a certain vector bundle $\bar{\mathfrak{C}}_\alpha$ over $M = E/G$.

Let us remember (Ch.4) that if E is a simple G space, the submanifold P of E consisting of all points with the stability group H is a principal bundle over M with structure group N(H)|H and that $E = P \times_{N(H)|H} G/H$ is associated with P. Now U = U(E,T) is a principal bundle over E with structure group T and G acts also on U by bundle automorphisms (such a situation will also occur, for example when considering symmetries of Yang-Mills fields, cf Ch.9). The action of G on U is characterized by a certain homomorphism $\lambda : H \to T$.

By applying the same method which was used to define P but now replacing E by U, we see that the total space of U can also be considered as a $(G \times T)/\bar{H}$ bundle over M, where $\bar{H} = \mathrm{diag}\,(H, \lambda(H)) \subset G \times T$. We define Q as a submanifold of U which is a principal bundle over M with structure group $N(\bar{H})|\bar{H}$, where $N(\bar{H})$ is the normalizer of $\bar{H}$ in $G \times T$; U can be also fibrated as the associated bundle $Q \times_{N(\bar{H})|\bar{H}} (G \times T)/\bar{H}$. One should refer to the Fig. given in sect.7.0 for a picture of the situation.

One also proves (sect.5.0 and 5.2.5) that $\bar{N}|\bar{H} = N(\bar{H})|\bar{H}$ is locally the product of $N|H = N(H)|H$ and of the centralizer $Z = Z(\lambda(H))$ of $\lambda(H)$ in T. With the above information in mind, let us consider again our vector bundle $\mathfrak{C}_\alpha$; we already noticed that the subspace $F \otimes_H W_\alpha{}^*$ of $F \otimes W_\alpha{}^*$ consists of vectors invariant under the subgroup H acting by $\rho \circ \lambda$ on F and by α^* on $W_\alpha{}^*$ i.e. of vectors invariant under the subgroup $\bar{H} =$

$\mathrm{diag}(\lambda(H),H) \subset T \times G$. Now, the group $T \times G$ acts on the set $F \otimes W_\alpha{}^*$ by $\rho \otimes \alpha^*$, and the subgroup $N(\overline{H}) \subset T \times G$ leaves the subset $F \otimes_H W_\alpha{}^*$ invariant. Therefore $\rho \otimes \alpha^*$ restricts to a representation of $N(\overline{H})|\overline{H}$ on $F \otimes_H W_\alpha{}^*$. We then construct the associated bundle $Q \times_{N(\overline{H})|\overline{H}} (F \otimes_H W_\alpha{}^*) = \mathfrak{C}_\alpha$ which has base M.

Let us summarise the whole discussion by the following theorem:

Let E be a manifold with a simple action of a compact Lie group G, call M the space of orbits and G/H the typical orbit; let U be a principal bundle with base E and structure group T, let ρ be a representation of T on a vector space F and $\mathfrak{F} = U \times_\rho F$ the associated vector bundle. Assuming that the action of G lifts to U (and thus to $\mathfrak{F}$), the smooth sections of $\mathfrak{F}$ can be approximated by finite linear combinations of generalized harmonics Φ_α, $\alpha \in \hat{G}$; the space $\Gamma_\alpha \mathfrak{F}$ of the harmonics of type α can itself be reconstructed from the study of the space $\Gamma_{inv} \mathfrak{C}_\alpha = \Gamma\mathfrak{F} \otimes_G W_\alpha{}^*$. This space, in turn, is isomorphic with the space of all sections of a vector bundle $\mathfrak{C}_\alpha$ over M whose typical fiber is $\overline{F}_\alpha = F \otimes_H W_\alpha{}^*$ and whose structure group is $N(\overline{H})|\overline{H}$, $\overline{H} = \mathrm{diag}(H,\lambda(H)) \subset G \times T$, λ being a homomorphism from H into T characterizing the G action on U, and $N(\overline{H})$ being the normalizer of $\overline{H}$ in $G \times T$. Moreover this structure group is isomorphic locally with $N(H)|H \times Z$, $N(H)$ beeing the normalizer of H in G and Z the centralizer of the image of H in T. The space $\Gamma_\alpha \mathfrak{F}$ of harmonics of type α is the image of the homomorphism $i_\alpha : W_\alpha \otimes \Gamma_{inv}\mathfrak{C}_\alpha \longrightarrow \Gamma\mathfrak{F}$ defined by $i_\alpha(w \otimes \Phi) = \Phi(w)$. When W_α is a real vector space and α is a real irreducible representation, then $\Gamma_\alpha \mathfrak{F}$ is isomorphic with $W_\alpha \otimes' \Gamma_{inv}\mathfrak{C}_\alpha$, where $\otimes'$ is taken over the commuting ring of α which is $\mathbb{R}$, $\mathbb{C}$ or $\mathbb{H}$ (the quaternion field); in other words, the above homomorphism becomes an isomorphism if we replace $\otimes$ by $\otimes'$.

In the very simple case where $E = G$, $F = \mathbb{C}$, $\Gamma_\alpha \mathfrak{F}$ carries the regular representation of G and we have a decomposition $\Gamma_\alpha \mathfrak{F} = \oplus_\alpha W_\alpha$ where W_α

appears with a multiplicity equal to $\dim W_\alpha$; the above results tell us that $\Gamma_\alpha \mathcal{F}$ is isomorphic with $W_\alpha \otimes \mathrm{Hom}_G(W_\alpha, \Gamma\mathcal{F})$ -cf. sect.7.3 (step 6 and footnote). This is indeed true since $\dim \mathrm{Hom}_G(W_\alpha, \Gamma\mathcal{F}) = \dim W_\alpha$.

7.6 Generalised homogeneous differential operators and the consistency problem for matter fields

We call $D : \Gamma\mathcal{F} \rightarrow \Gamma\mathcal{F}$ a generalized homogeneous differential operator on the simple G space E if D commutes with the induced action of G on $\Gamma\mathcal{F}$. It is clear that such a D induces, for each $\alpha \in G$, a linear differential $D_\alpha : \Gamma_\alpha\mathcal{F} \rightarrow \Gamma_\alpha\mathcal{F}$; also, D induces a linear differential operator $\tilde{D}_\alpha$ on the space of section $\Gamma\bar{\mathcal{F}}_\alpha$ of the bundle $\bar{\mathcal{F}}_\alpha$ of base M and typical fiber $\bar{F}_\alpha = F \otimes_H W_\alpha^*$. Indeed one first extends D to $D \times Id$ on $\Gamma\mathcal{F} \otimes W_\alpha^*$ and observes that $\Gamma\mathcal{F} \otimes_G W_\alpha^*$ is invariant under $D \times Id$; thus $\tilde{D}_\alpha$ is defined as the image of $D \times Id$ restricted to $\Gamma\mathcal{F} \otimes_G W_\alpha^*$ under the isomorphism $\Gamma\mathcal{F} \otimes W_\alpha^* \rightarrow \Gamma\bar{\mathcal{F}}_\alpha$. It remains to prove that $\tilde{D}_\alpha$ so defined is again a <u>differential</u> operator and the proof can be adapted from [29]. Studying D_α on $\Gamma_\alpha\mathcal{F} \cong W_\alpha \otimes \Gamma\bar{\mathcal{F}}_\alpha$ clearly amounts to study $\tilde{D}_\alpha$ on $\Gamma\bar{\mathcal{F}}_\alpha$ (the W_α factor will only modify, for example, the multiplicity of the eigenvalues).

Notice that D, D_α and $\tilde{D}_\alpha$ are not necessarily elliptic operators since they can be defined via some (non necessarily euclidean) metric on M; however, we always assume that the fibers $(G/H)_x$ of E are endowed with an Euclidean G-invariant metric $h_{\alpha\beta}(x)$; $x \in M$. In all cases of interest the operator D will be constructed out of a <u>given</u> G-invariant metric on E (i.e. we study the behaviour of matter fields in a background gravitational field). We will be typically interested in a (hyperbolic) equation of the form $D\Phi - a\Phi = o$ where a is a real given parameter; because of the G-invariance of the problem it is enough to study this problem in each space $\Gamma_\alpha\mathcal{F}$. We are therefore led to an infinite number of differential equations of the type $\tilde{D}_\alpha\Phi_\alpha + a\Phi_\alpha = o$, α labelling an irreducible representation of G. Now, in each particular case (cf. examples in the next section) it will be possible to split $D_{eff} = \mathfrak{D}_\alpha$ into three pieces $D_{eff} = D_M + D_{G/H} + \Delta$ where D_M is an hyperbolic differential operator involving covariant derivatives for the geometry of **M** and gauge derivatives with respect to the gauge group $N(\overline{H})|(\overline{H})$, $D_{G/H}$ is an elliptic operator -actually a family of elliptic operators-with parameter $x \in M$ associated with the metric $h_{\alpha\beta}(x)$ of $(G/H)_x$ above $x \in M$ and Δ is a non minimal interaction term.

Because of the restriction to $\Gamma\overline{\mathfrak{E}}_\alpha$, it is natural to study more particularly those matter fields (of type α) which are eigenfunctions of the internal differential operator $D_{G/H}$ i.e. which satisfy $D_{G/H}\,\Phi = \lambda\,\Phi$; of course the eigenvalue λ is a function of $x \in M$ since what we have here is a collection of operators $D_{G/H}$ labelled by x. Interesting candidates for describing (for example) elementary particles would then be matter fields satisfying an equation of the type :

$$(1) \qquad \{D_M + \Delta + (a-\lambda(x))\}\,\Phi = o.$$

Notice that when the metric on the fibers of E is constant with x (for instance if we choose a fixed Killing metric) the $\lambda(x)$ are constant and directly related to the Casimir of G in the representation α; provided one then introduces dimensionful parameter in the problem, (a,λ) is interpreted as a mass (or a $(\text{mass})^2$) term [19]. From the point of vue of physical interpretation, it is clear that one could get equation (7.6(1)) by adding an appropriate piece to the Lagrangian of sect. 5.7, but this would also in principle modify the right hand side of Einstein equations in E by adding a term proportional to the stress energy tensor of Φ. One would also get another set of equations by using a G-invariant ansatz, performing dimensional reduction and integrating over G/H ; there is no reason why these two sets of equations should be compatible (unless Φ is invariant), indeed one expects a "reaction" of Φ on the given external field and there is no reason for this reaction to be G-invariant if Φ is not. However nothing prevents us to study an "external field problem" i.e. to build a Hilbert space out of the space of solutions of equation (7.1) and to implement usual quantum mechanics (or even quantum field theory) as we do with usual Klein Gordon equation (for that we take **M** as our usual 4 dimensional space time, possibly curved).

7.7 Pointers to the literature

Induced representations 8, 44, 125, 190, 198, 199, 224, 229
Harmonic analysis and dimensional reduction 16, 44
Homogeneous differential operators 16, 22, 29, 86, 224, 229
Group action on vector bundles 133, 191, 198

VIII

DIMENSIONAL REDUCTION OF THE ORTHOGONAL BUNDLE AND OF THE SPIN BUNDLE

8.0 Summary

Here we apply the general formalism of chapters 6 and 8 to important particular cases : so far we considered quite a general situation of an abstract space of matter fields over $E = M \times G/H$ (locally), in particular we had to assume that some action of G on these fields was somehow given with the proviso that it had to agree with the action of G on E but did not follow from the later. In the following we shall focus our discussion on the particular cases of tensor or spinor fields on E; for this we have to apply the reduction procedure to the particular cases of orthonormal or spin frames on E. The general situation described in the previous section was depicted by the following

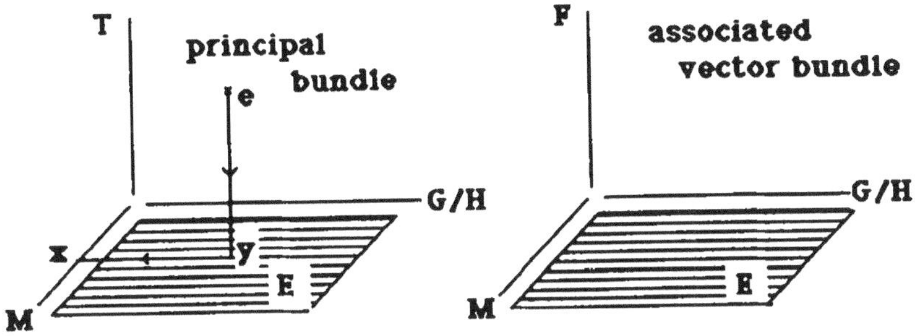

with a given representation ρ of T on the vector space F.

The particular situation(s) that we study now is the case where U is the bundle of orthonormal frames (or spin frames) over E, where T is related to the orthogonal group O(dim E) or to its covering and where (ρ, F) is some tensor (or spinor) representation.

Let us assume that E is equipped with a G-invariant metric g_E ; point transformations under G induce transformations of tangent vectors and of frames and, being isometric, they map orthonormal frames into orhtonormal ones. In sect.8.1 we study the space OE of orthonormal frames and realize that this space is unecessarily large for our purpose. It is enough to consider only "adapted frames" (whose space is called AOE); roughly speaking, an adapted orthonormal frame is made up of three kinds of vectors: those which are tangent to M, those which are tangent to N(H)|H and those which are tangent to

G/N(H). The reader should remember that, as usual, we consider G/H as a collection of (group) spaces N(H)|H labelled by G/N(H), N(H) being the normalizer of H in G, or, at the infinitesimal level we write the tangent space at the origin of G/H as $\mathfrak{S}$=Lie(N(H)|H)+$\mathfrak{L}$, $\mathfrak{L}$ being the tangent space at the origin of G/N(H). The group acting on adapted orthonormal frames is no longer O(m+s) with m=dim(M), s=dim(G/H) but T = O(m) × O(k) × O(s-k) with k = dim(N(H)|H), s-k=dim(G/N(H)).

In sect.8.1 we also apply the dimensional reduction mechanism discussed in part 7 with U = AOE and T as above; for this we have first to identify the homomorphism λ: H ⟶ T discussed in sect.7.4 and sect.7.5. Intuitively, H -transformations leave the origin fixed but "rotate" points (and therefore vectors) around it and there is therefore a map from H to the rotation group T. It is intuitively clear that λ is related to the adjoint representation of H in $\mathfrak{S}$, moreover N(H)|H does not "feel" these H-rotations : more precisely we will show that $\lambda(H) \subset$ O(s-k). Following the conclusions of ch. 7, we see that every tensor field (of some given type (ρ,F)) on E can be approximated by linear combination of harmonics labelled by irreducible representations (α, W_α) of G; the space of harmonics of type α on E can be reconstructed from the space of all matter fields on M valued in the vector space $F \otimes_H W_\alpha$, the local gauge group of these matter fields on M is $N_{G\times T}(\overline{H})|\overline{H}$ where $\overline{H}$=diag(H,λ(H)) and $N_{G\times T}(\overline{H})$ is the normalizer of $\overline{H}$ in G × T; in our case the local gauge group is locally isomomorphic to the product N(H)|H × O(m) × O(k) × Z, where Z is the centraliser of Ad(H) in O(s-k). For example, take G = U(2, $\mathbb{H}$) = Sp(2), H = U(1, $\mathbb{H}$) = Sp(1), E=M×S^7, then $N_{G\times T}(\overline{H})|\overline{H}$ = SU(2) × O(m) × O(3) × SU(2); also in the case when H is trivial we get a local gauge group G × O(m) × O(n), with n = dim G.

In order to find the dimensionally reduced Laplace (and also Dirac) operator from those defined in E, we need explicit expressions for the affine connection ω_{ABC}. This expression will in general depend on the choice of an adapted orthonormal moving frame e =($e_{\underline{\mu}}$; $e_{\underline{\alpha}}$)=($e_{\underline{\mu}}$; $e_{\hat{\underline{a}}}$, $e_{\underline{a}}$) in E; we already know the expression of the reduced connection in a frame $(e_\mu; \epsilon_{\hat{a}}, \epsilon_a)$ made up of invariant horizontal fields e_μ and of fundamental fields $(\epsilon_{\hat{a}}, \epsilon_a)$, at least in the neighbourhood of points of P $\subset$ E, (locally, P = M×N(H)|H, cf. ch.5). To find its expression in an adapted orthonormal frame (we underline the indices whenever the frame is orthogonal) it is then necessary to introduce an "internal Vielbein" (change of basis) : $e_{\underline{\alpha}} = \epsilon_\beta \varphi^\beta{}_{\underline{\alpha}}$. The ω_{ABC} are given in sect.8.2 and can be gotten easily from sect.5.5.1 via the above change of basis and keeping in mind that ω_{ABC} does not "transform" like a tensor (but like a connection).

Notice that $|| \varphi^\beta{}_{\underline{\alpha}} || = || \varphi^{\underline{\alpha}}{}_\beta ||^{-1}$ and that if $h_{\alpha\beta}$ denotes the internal metric, we get:

$$h_{\alpha\beta} = \varphi^{\underline{\alpha}}{}_\alpha \varphi^{\underline{\beta}}{}_\beta \delta_{\underline{\alpha\beta}} , \qquad \delta_{\underline{\alpha\beta}} = h_{\alpha\beta} \varphi^\alpha{}_{\underline{\alpha}} \varphi^\beta{}_{\underline{\beta}}$$

In sect.8.3, using the above, we study the reduced Laplace operator (the "rough" laplacian, cf. sect.6.4) in two particular cases : scalar fields and vector fields. In the first case, we get:

$$\Delta_{eff} \Phi = \gamma^{\mu\nu} \mathbb{D}_\mu D_\nu \Phi + h^{\alpha\beta} T_\alpha T_\beta \Phi + v^\mu D_\mu \Phi$$

where the effective field $\Phi = (\Phi_I)$ gets an index I refering to a representation of G and is constrained by $T_{\hat{\alpha}} \Phi = 0$, $T_{\hat{\alpha}} \in$ Lie(H) . Also,

$$D_\mu = \partial_\mu + A_\mu{}^{\hat{a}} T_{\hat{a}} , \; T_{\hat{a}} \in \mathrm{Lie}(N(H)|H)$$

and

$$v_\mu = 1/2 \, \mathrm{Tr}(h^{-1} D_\mu h),$$

also, $\mathbb{D}_\mu$ denotes the covariant derivative with respect to both the Levi -Civita connection and the N(H)|H connection.

The case of spinors is handled in the same way by introducing adapted spin frames **APinE** and here we start with a structure group T which is locally **Pin**(m) × **Pin**(s), with s=dim G/H - to avoid complications with orientation and time orentation we have replaced **Spin₊** by **Pin** . This is studied in more details in sect.8.3 where we also repeat the argument of sect.11.2 and find that the effective structure group is $(\mathbf{Pin}(m)\times N(\overline{H})|\overline{H})/Z_2$ where $\overline{H}=\mathbf{diag}(\overline{\lambda}(H),H) \subset \mathbf{Pin}(s) \times G$ and $N(\overline{H})$ is the normalizer of $\overline{H}$ in **Pin**(s) × G.

Therefore **M** is endowed with a generalisation of a $\mathbf{Spin}^c$ structure.

The expression of the affine connection in an orthonormal frame being already studied in 8.1, we find easily in 8.3 what the expression of the spin connection is (the generators of the group Pin(E) are $\Sigma^{AB} = 1/4\,[\Gamma^A,\Gamma^B]$ where the Γ^A generate the Clifford algebra). In the same subsection we also find the structure of the dimensionally reduced Dirac operator acting on effective spinor fields $\Psi = (\Psi_l)$ which get an extra index l from the representation α of G and satisfy the constraint

$$(\alpha(T_{\hat\alpha}) + 1/2\, C_{\hat\alpha,\beta\gamma}\, \phi^{\beta}{}_{\underline{\beta}}\, \phi^{\gamma}{}_{\underline{\gamma}}\, \Sigma^{\underline{\beta}\underline{\gamma}})\, \Psi = 0$$

with $T_{\hat\alpha} \subset$ Lie H and where β,γ are S=G/H indices (this describes H invariance under G × Spin).

Notice that (Ψ_l), for a given l has $2^{[d/2]}$ components where d = dim E = m+s and [ ] denotes the integer part. The dimensionally reduced Dirac operator can itself be written as $D_{eff} = D_M + D_S + \Delta_1 + \Delta_2$ with the four terms described as follows :

$$D_M = \Gamma^{\mu} D_{\mu}$$

with $D_\mu = \partial_\mu + 1/2\ \omega_{\mu\nu\sigma}\ \Sigma^{\nu\sigma} - A_\mu{}^{\hat{a}}\ \alpha(T_{\hat{a}}) - 1/2\ B_{\mu\underline{\alpha}\underline{\beta}}\ \Sigma^{\underline{\alpha}\underline{\beta}}$

and

$$B_\mu{}^{\underline{\alpha}}{}_{\underline{\beta}} = \phi^{\underline{\alpha}}{}_\gamma\ e_\mu\ (\phi^\gamma{}_{\underline{\beta}}) + 1/2\ \phi^{\underline{\alpha}}{}_\gamma\ h^{\gamma\delta}\ (D_\mu h_{\delta\lambda})\ \phi^\lambda{}_{\underline{\beta}}$$

$$e_\mu(\phi^\gamma{}_{\underline{\beta}}) = \partial_\mu\ \phi^\gamma{}_{\underline{\beta}} + A_\mu{}^{\hat{a}}\ C^\gamma{}_{\hat{a}\delta}\ \phi^\delta{}_{\underline{\beta}}$$

$$D_S = \Gamma^{\underline{\alpha}}\ (\phi^\beta{}_{\underline{\alpha}}\ \alpha(T_\beta) + 1/2\ \omega_{\underline{\hat{\alpha}}\underline{\beta}\underline{\delta}}\ \Sigma^{\underline{\beta}\underline{\delta}})$$

$$\Delta_1 = 1/4\ \Gamma^{\underline{\alpha}}\ \phi^\beta{}_{\underline{\alpha}}\ h_{\beta\gamma}\ F_{\mu\nu}{}^\gamma\ \Sigma^{\mu\nu}$$

$$\Delta_2 = 1/2\ \Gamma^\mu v_\mu \qquad \text{with } v_\mu = 1/2\ \mathrm{Tr}\ (h^{-1}\ D_\mu\ h)$$

The study of the Laplace and Dirac operators on groups and homogeneous spaces for G-invariant metrics (not necessarily bi-invariant in the group case and not necessarily normal in the G/H case) fits in the previous framework as a particular case and examples of calculation of the spectrum of those operators are provided in sect.8.4 and 8.5. In sect.8.6, we give more general examples.

"Warning to the reader"

In this chapter as well as in the rest of the book, the same object (metric on the fibers) is sometimes written $g_{\alpha\beta}$ and sometimes $h_{\alpha\beta}$. If it is written as $h = (h_{\alpha\beta})$ it should not be confused with elements of the group H which are usually denoted by a letter "h".

8.1. The space of adapted orthonormal frames

Let E (being locally M × G/H) be equipped with a G-invariant metric g_E.

Adapted orthonormal frames

We denote by OE the bundle of orthonormal frames of E, its structure group will be denoted O(d), d=dim E; it consists of matrices Λ satisfying $\Lambda^T \eta_E \Lambda = \eta_E$ where η_E is the canonical diagonal form of g_E; if E is a "multidimensional universe" then it is natural to take $\eta_E = \text{diag}(-1,+1,+1,\ldots,+1)$. Since G acts on E by isometries the action of G lifts automatically to OE. But the bundle OE is unecessarily large and we shall use the extra information that we have to reduce its structure group. Recall (Ch.5) that because of G invariance, g_E generates a metric on M = E/G, a G-invariant metric $h_{\alpha\beta}(x)$ on each fiber S_x, $x \in M$, $S \simeq G/H$, and a connection $A_\mu{}^{\hat{a}}$ with gauge group N(H)|H. The Lie algebra of G can be decomposed into Lie(G) = Lie(H) + Lie(N(H)|H) + $\mathcal{L}$ and $\mathcal{S}$ = Lie(N(H)|H) + $\mathcal{L}$, can be identified with the tangent space to S = G/H at the origin; N(H) is the normalizer of H in G and H is assumed to be connected.

As in Ch.5 we call P the subbundle of E, fibers of which are made of points of E for which the isotropy subgroup is exactly H; P is a principal bundle with gauge group N(H)|H. To reduce the structure group O(E), we define for every $p \in P$ the subspaces K_p and L_p as the subspaces of the tangent space T_pE spanned by the Killing vectors from Lie(N(H)|H) and $\mathcal{L}$ respectively. For an arbitrary $y \in E$ there exist always $p \in P$ and $a \in G$ such that $y=pa$, we then define $K_y := K_p a$ and $L_y := \mathcal{L}_p a$. Because Ad(H) leaves invariant the subspace Lie(N(H)|H) and $\mathcal{L}$ of Lie(G) (with Lie(N(H)|H) being even pointwise invariant), our definitions are unambiguous. We know therefore how to split the tangent space $K_y E$ into three pieces : K_y, L_y and a third one orthogonal to K_y+L_y .

Definition : An orthogonal frame $e_{\underline{A}}(y)$ at $y \in E$ is called adapted if $e_{\underline{A}} = (e_{\underline{\mu}}, e_{\underline{\hat{a}}}, e_{\underline{a}})$ with $e_{\underline{\hat{a}}} \in K_y$, $e_{\underline{a}} \in L_y$ and $e_{\underline{\mu}}$ orthogonal to $K_y + L_y$.

Notice that for $p \in P \subset E$ the vectors $e_{\underline{\mu}}(p)$ span the horizontal subspace of the induced $N(H)|H$ connection A_μ.

The set of all adapted orthogonal frames (called AOE) is obviously a principal bundle with structure group $T = O(m) \times O(k) \times O(s-k)$, $m = \dim(M)$, $s = \dim(G/H)$, $k = \dim(N(H)|H)$, $s-k = \dim(G/N(H)) = \dim(\mathcal{L})$.

Notice that we do not loose anything by this reduction of the structure group from $O(d)$, $d = m+s$ to T, indeed associated bundles (and corresponding matter fields) will be the same.

Let $\{T_i\}$ be a basis in Lie(G) with $T_{\hat{\alpha}} \in$ Lie(H), $T_{\hat{a}} \in$ Lie(N(H)|H), $T_a \in \mathcal{L}$ and $T_\alpha = (T_{\hat{a}}, T_a)$; we denote (as in Ch.5) by $\epsilon_i(y)$ the fundamental vector fields on E (Killing vectors) generated by T_i. The vectors $\epsilon_\alpha(y)$ then span an s-dimensional subspace S_y, provided y is not too far from P. However one should stress that the $\epsilon_\alpha(y) = (\epsilon_{\hat{a}}(y), \epsilon_a(y))$ do not a priori coincide with the $e_{\underline{\alpha}}(y) = (e_{\underline{\hat{a}}}(y), e_{\underline{a}}(y))$; in a neighbourhood of P we can therefore write $(e_{\underline{a}}(y) = \epsilon_\beta(y)\, \varphi^\beta{}_{\underline{a}}(y)$. Actually the internal vielbein $\varphi^\beta{}_{\underline{\alpha}}$ is a function of $e_{\underline{\alpha}}$ since it describes the change of base between the (fixed) base $\{\epsilon_\alpha\}$ and an arbitrary orthonormal adapted frame $\{e_{\underline{\alpha}}\}$: we will sometimes write it as $\varphi(e;y)$. Here and below the underlined indices $\underline{\alpha}$, $\underline{\beta}$, $\underline{a}$, $\underline{b}$, $\underline{\hat{a}}$, etc... will correspond to an orthonormal internal frame; thus $\varphi^\beta{}_{\underline{\alpha}}$ has inverse $\varphi^{\underline{\beta}}{}_\alpha$ and we have $h_{\alpha\beta}\, \varphi^\alpha{}_{\underline{\gamma}} \varphi^\beta{}_{\underline{\delta}} = \eta_{\underline{\gamma\delta}}$, $\varphi^\alpha{}_{\underline{\gamma}} \varphi^{\underline{\gamma}}{}_\epsilon = \delta^\alpha{}_\epsilon$ etc...

If $\Lambda \in T$ is a change of adapted orthonormal frame, i.e. if $e_{\underline{\alpha}'} \to e_{\underline{\beta}} \Lambda^{\underline{\beta}}{}_{\underline{\alpha}}$ then $\varphi(e\Lambda) = \varphi(e)\Lambda$ etc... Notice also that $\varphi(e)$ has the block diagonal form $(\varphi^\alpha{}_{\underline{\beta}}) = (\varphi^a{}_{\underline{b}}, \varphi^{\hat{a}}{}_{\underline{\hat{a}}})$.

The effective gauge group.

We will now apply the dimensional reduction mechanism discussed in Ch.7, but first, we have to identify the homomorphism $\lambda : H \to T = O(m)\times O(k)\times O(s-k)$. Let $p \in P$ and $e(p) \in AOE$, then the action of any $h \in H$ leaves p invariant and thus rotates $e(p)$ by an orthogonal transformation. Thus we have $e_{\underline{\alpha}}\, h = e_{\beta}\, \lambda(e;h)^{\beta}{}_{\underline{\alpha}}$, $h \in H$. In fact it is only the $e_{\underline{a}} \in L_p$ which are rotated, ie $(\lambda^{\beta}{}_{\underline{\alpha}}) = (\delta^{\underline{b}}{}_{\underline{i}}, \lambda^{\underline{b}}{}_{\underline{a}})$; $\lambda : H \to O(s-k)$ is a group homomorphism. The transformation of φ under a change Λ of adapted frames implies that $\lambda(e\Lambda;h) = \Lambda^{-1}\, \lambda(e;h)\Lambda$; let then $(A^i{}_j)$ be the matrix of the adjoint representation of G in its Lie algebra, then, since $\epsilon_j h^{-1} = \epsilon_i\, A(h)^i{}_j$, we find that $\lambda(e;h) = \varphi^{-1}(e)A(h)\,\varphi(e)$ or $\lambda^{\underline{b}}{}_{\underline{a}} = A^b{}_a\, \varphi^{\underline{b}}{}_b \varphi^a{}_{\underline{a}}$: the matrix of the rotation $\lambda(h)$ in the orthonormal frame $(e_{\underline{\alpha}})$ differs only from the matrix of the adjoint representation by the change of frame φ (with $e_{\underline{\alpha}} = \epsilon_{\beta}\, \varphi^{\beta}{}_{\underline{\alpha}}$). Although it is by no means necessary, it is very convenient to choose a basis T_i in G in such a way that, for some $p_0 \in P$, the Killing vectors $\epsilon_{\alpha}(p_0)$ are orthonormal : the $(\epsilon_{\alpha}(p_0))$ are therefore "adapted" at p_0 and $\varphi^{\alpha}{}_{\beta}(p_0) = \delta^{\alpha}{}_{\beta}$; we then get $\lambda = A$ i.e. $\lambda(\epsilon(p_0);h) = A(h)$, $A(h)$ being the matrix of the adjoint representation of H in S (actually in L). We now apply the results of Ch.7 to the case U=AOE, $T=O(m)\times O(k)\times O(s-k)$ and $\lambda = A$ to get the following result :

The effective gauge group is $N(\overline{H})|\overline{H}$ where $\overline{H} = \mathrm{diag}(H, \lambda(H)) \subset G\times T$ and $N(\overline{H})$ the normalizer of $\overline{H}$ in $G\times T$. It is therefore locally isomorphic to the product $N(H)|H\times O(m)\times O(k)\times Z$, where Z is the centraliser of $Ad(H)$ in $O(L)$; we set $m=\dim(M)$, $s=\dim(G/H)$, $k=\dim(N(H)|H$, $L = \dim(G/N(H)) = s-k$. We will make precise the structure of $N(\overline{H})|\overline{H}$ below.

Although we will not need this fact, it is worthwhile to notice that one can easily get rid of the O(k) factor of the effective gauge group. Indeed, we can always reduce the bundle by demanding that $e_a(p)$ are fixed, e.g. by a standard orthonormalisation process of $\epsilon_a(p)$. That means that the effective internal gauge group is locally isomorphic with $N(H)|H \times Z$. In the particular, well known, case when E is a principal bundle, the internal spaces are group manifolds, H is trivial, L is trivial, Z is trivial and $N(H)|H = G$ as expected.

As discussed in sect.7.5, the total space U=AOE of adapted orthogonal frames can be seen as a non principal bundle over **M** and typical fiber $(G\times T)/\overline{H}$, this last bundle is itself associated with a principal bundle Q of base **M** and typical fiber $N(\overline{H})|\overline{H}$; we now describe Q in details. According to the general algorithm of dimensional reduction, Q is a subbundle consisting of those pairs $(p, e_A(p))$ for which $\lambda(e;h)=A(h)$. The following results gives an explicit description of the reduced frame bundle.

The frames $e_A(p) \in$ AOE are characterised by φ satisfying

(i) $\varphi\, A(h)=A(h)\, \varphi$, $h\in H$

(ii) $\varphi^T g\, \varphi = \eta$

where $g(p)=(g_{\alpha\beta})$ are the components of the metric in the ϵ basis, and $\eta = \pm 1$ depending on whether the metric in the orbit is positive or negative definite. Such frames exist for every $p \in P$ and any two such frames at the same $p \in P$ are connected by an orthogonal transformation $\Lambda_1 \times \Lambda_2$ with $\Lambda_1 \in O(T)$ and $\Lambda_2 \in Z$, Z being the centraliser of Ad(H) in O(L).

Proof. The non-evident part of the statement is the existence result. To prove the existence of a φ satisfying (i) and (ii) one starts with an arbitrary adapted frame (i.e. the result of a standard orthonormalisation of ϵ), with $e = \epsilon\, \varphi$ (we omit the non-interesting e_μ vectors and observe that φ satisfies (ii)) and $\lambda(e)=\varphi^{-1} A\, \varphi$ is an orthogonal matrix. It follows that $\varphi\, \varphi^T$ commutes with $A(h)$. Let φ =

$\varphi_0 \Lambda$ be the polar decomposition of φ, with $\varphi_0 > 0$ and Λ orthogonal. Then $\varphi_0 = (\varphi\varphi^T)^{1/2}$ commutes with A(h), and it also satisfies (ii) since $\varphi_0 = \varphi \Lambda^{-1}$ and Λ is an orthogonal transformation.

With the information given by the above theorem we can also explicitly describe the reduced structure group $N(\bar{H})|\bar{H}$. Indeed, let e(p) and e(p') be in Q, p' and p being in the same fibre of P. Then we first choose $n \in N(H)$ such that p'=pn, and then define $\Lambda \in O(L)$ by the relation

$$e(p)\, n = e(p')\ \Lambda^{-1}$$

(we again omit e_μ and $e_{\hat{a}}$, which are uninteresting here). Since both e and e' are in Q, Λ must satisfy the constraint that $A(n^{-1})\,\Lambda$ commutes with A(H) on L. But n is unique only modulo $h \in H$. Thus (n, Λ) should be identified with $(hn, A(h)\,\Lambda)$. Consequently we have the following result.

The effective structure group $N(\bar{H})|\bar{H}$ is the product $O(M)\times O(K)\times J$, where J consists of pairs (n, Λ) with $n \in N(H)$ and $\Lambda \in O(L)$ such that $A(n^{-1})\Lambda$ commutes with A(H); the pairs (n, Λ) and $(hn, A(h)\,\Lambda)$, $h \in H$, define the same element of J. (Notice that J is locally isomorphic with $N(H)|H \times Z$).

We can now interpret the φ as a cross section of a bundle associated to Q. Indeed, with $e \in Q$ and (n, Λ) as above, we find

$$\varphi\,(en\,\Lambda) = A(n)^{-1}\,\varphi\,(e)\,\Lambda$$

and the condition (i) ensures that the above transformation law effectively depends on the equivalence class of (n, Λ) only.

The affine connection in an adapted orthonormal frame

In order to calculate the dimensionally reduced Laplace-Beltrami operators and Dirac operators, we shall need explicit expressions for the affine connection $\omega_A{}^B{}_C(\mathbf{p})$, $\mathbf{p} \in \mathbf{P}$. The Christoffel symbols in the $(\mathbf{e}_\mu, \boldsymbol{\epsilon}_\alpha)$ basis have been given in Ch.5, here we transform them to an orthonormal basis $(\mathbf{e}_{\underline{\mu}}, \boldsymbol{\epsilon}_{\underline{\alpha}})$ (this calculation is not strictly necessary when we discuss for example the Laplace operator on scalar fields but it is necessary when we come to discuss the Dirac operator). The expressions will in general depend on a moving frame $\mathbf{e}_A$ which we will choose in $Q \subset U$ at the points of $P \subset E$. More specifically we proceed as follows

(i) Choose a local cross section $\sigma : \mathbf{x} \longrightarrow \sigma(\mathbf{x})$ of the bundle P.

(ii) At every point $\sigma(\mathbf{x}) \in P$ choose an adapted frame $\mathbf{e}_A(\sigma(\mathbf{x})) \in$ Q. Then $(\sigma(\mathbf{x}), \mathbf{e}_A(\sigma(\mathbf{x})))$ is a local cross section of the bundle Q.

(iii) Extend $\mathbf{e}_A(\sigma(\mathbf{x}))$ to an open neighbourhood of σ in E defining $\mathbf{e}_A(\sigma(\mathbf{x})\, e^{\xi}) = \mathbf{e}_A(\sigma(\mathbf{x}))\, e^{\xi}$ for ξ running through a neighbourhood of zero in S.

With the moving frame $\mathbf{e}_A$ introduced as above we write $\mathbf{e}_{\underline{\alpha}} = \boldsymbol{\epsilon}_\beta\, \varphi^\beta{}_{\underline{\alpha}}$ and from (iii) we easily find

$$\mathbf{e}_\alpha(\varphi^\beta{}_{\underline{\delta}}) = -\, C^\beta{}_{\alpha\gamma}\, \varphi^\gamma{}_{\underline{\delta}}$$

Observe that now $\mathbf{e}_\mu$ is a horizontal (i.e. orthogonal to the orbits of G in E) lift of a certain moving frame, denoted by ∂_μ, on **M**.

From the expression of the Christoffel symbols given in Ch.5 and from the change of basis : $\mathbf{e}_{\underline{A}} = \boldsymbol{\epsilon}_B\, \varphi^B{}_{\underline{A}}$,

$$\omega_{\underline{A'}}{}^{\underline{B'}}{}_{\underline{C'}} = \omega_A{}^B{}_C\, \varphi^A{}_{\underline{A'}}\, \varphi_B{}^{\underline{B'}}\, \varphi^C{}_{\underline{C'}} + \varphi^{\underline{B'}}{}_D\, \partial_{\underline{A'}}(\varphi^D{}_{\underline{C'}}) \quad ,$$

we get:

$\omega_{\mu\nu\rho}$ = the affine connection of **M** in the ∂_μ basis (in spite of the notation ∂_μ denotes here an anholomic orthonormal moving frame of **M** -from now on, we do not underline the μ-indices-)

$$\omega_{\mu\underline{\alpha}\nu} = -\omega_{\mu\nu\underline{\alpha}} = -1/2\ \eta_{\underline{\alpha}\beta}\ \varphi^{\beta}{}_{\underline{\gamma}}\ F_{\mu\nu}{}^{\gamma}$$
$$\omega_{\underline{\alpha}\underline{\beta}\mu} = -\omega_{\underline{\alpha}\mu\underline{\beta}} = 1/2\ \varphi^{\gamma}{}_{\underline{\alpha}}\ D_\mu\,(h^{\gamma\delta})\ \varphi^{\delta}$$
$$-\omega_{\mu\underline{\alpha}\underline{\beta}} = 1/2\ \varphi_{\underline{\alpha}}{}^{\beta}\,(D_\mu h_{\gamma\delta})\ \varphi^{\delta}{}_{\underline{\beta}} + \eta_{\underline{\alpha}\underline{\gamma}}\ \varphi^{\underline{\gamma}}{}_{\delta}\ e_\mu(\varphi^{\delta}{}_{\underline{\beta}})$$

where $F_{\mu\nu}{}^{\alpha} \neq 0$ for $\alpha = \hat{a}$ only and $F_{\mu\nu}{}^{\hat{a}}$ is the field strengh for the N(H)|H connection,

$$D_\mu h_{\alpha\beta} = \partial_\mu h_{\alpha\beta} + C_{\alpha\hat{c}}{}^{\delta}\ A_\mu{}^{\hat{c}}\ h_{\delta\beta} + C_{\beta\hat{c}}{}^{\delta}\ A_\mu{}^{\hat{c}}\ h_{\alpha\delta}\ ,$$
$$e_\mu(\varphi) = \partial_\mu\varphi - A_\mu{}^{\hat{a}}\ \epsilon_{\hat{a}}(\varphi)\ \ \text{i.e.}\ e_\mu(\varphi^{\delta}{}_{\underline{\beta}}) = \partial_\mu\varphi^{\delta}{}_{\beta} + C_{\hat{a}\gamma}{}^{\delta}\ A_\mu{}^{\hat{a}}\ \varphi^{\gamma}{}_{\beta}\ ,$$

we also used the "flat" metric η_{AB} of E to lower the upper index of $\omega_A{}^B{}_C$:

$$\omega_{ABC} = \eta_{BD}\ \omega_A{}^B{}_C\ .$$

The effective gauge fields

We already know that the effective gauge group is the structure group of the bundle Q i.e., the product O(m)×O(t)×J where J is locally isomorphic with N(H)|H×Z. In term of gauge fields, it is clear that we can identify the O(m) and N(H)|H parts with the fields $\omega_{\mu\nu\rho}$ and $A_\mu{}^{\hat{a}}$ respectively. It still remains to find the O(t)×Z part : with respect to the local moving frame chosen above we have $B_\mu{}^{\underline{\alpha}}{}_{\beta} = \omega_\mu{}^{\underline{\alpha}}{}_{\underline{\beta}}$, where we identify the Lie algebra of (O(t)×Z) with a subalgebra of $\mathcal{S}$ = Lie(N(H)|H)+ $\mathcal{L}$. Explicitly, one has

$$B_\mu{}^{\underline{\alpha}}{}_{\underline{\beta}} = -\varphi^{\underline{\alpha}}{}_{\gamma}\ e_\mu(\ \varphi^{\gamma}{}_{\underline{\beta}}) - 1/2\ \varphi^{\underline{\alpha}}{}_{\gamma}\ h^{\gamma\delta}\,(D_\mu\ h_{\delta\lambda})\ \varphi^{\lambda}{}_{\underline{\beta}}$$

with $e_\mu(\varphi^\gamma{}_{\underline{\beta}})$ and $D_\mu h_{\delta\lambda}$ given above.

8.2 Dimensionally reduced Laplace-Beltrami operators

Laplace operator on scalar fields.

A (real) scalar field on E is a section of the trivial bundle $\mathfrak{F} = E \times \mathbb{R}$ which is associated to the trivial representation of O(E) on $\mathbb{R}$. According to the discussion in sect.11, the effective fibre bundle $\overline{\mathfrak{F}}_\alpha$ over M has the fibre $\overline{F}_\alpha = F \otimes_H W^*_\alpha = \mathbb{R} \otimes_H W^*_\alpha = (W^*_\alpha)_0$, i.e. F_α consists of H singlets in W^*_α. The effective field on M therefore gets the index L (Φ becomes Φ_L) from the representation space W^*_α and is constrained by

$$\alpha(h)^L{}_M \Phi_L = \Phi_M \ , \quad h \in H$$

where $\alpha(h)^L{}_M$ is the matrix of the representation α. Knowing the connection coefficients $\omega_A{}^B{}_C$ one easily finds the effective Laplace operator acting on $\Phi = (\Phi_L)$:

$$\Delta eff\ \Phi = \gamma^{\mu\nu} \mathbb{D}_\mu D_\nu \Phi + h^{\alpha\beta} T_\alpha T_\beta \Phi + v^\mu D_\mu \Phi$$

where

$$D_\mu \Phi = \partial_\mu (\Phi) + A^{\hat{a}}{}_\mu T_{\hat{a}} \Phi$$

is the N(H)|H-covariant derivative. $\mathbb{D}_\mu$ contains also the gravitational part and v^μ is given by

$$v_\mu = 1/2 \ \mathrm{Tr}(h^{-1} D_\mu h) = 1/2 \ \mathrm{Tr}(h^{-1} e_\mu(h))$$

which measures rate of change of the volume of the internal space. In the derivation of the above formulae we used the fact that G, being compact, is unimodular; thus, in particular, $C^{\alpha}{}_{\alpha\beta} = 0$, $C^{i}{}_{jk}$ being the structure constants of G. In the particular case when $h_{\alpha\beta}$ comes from a fixed Killing metric h_{ij} on G the effective mass term in (5.5) is nothing but the Casimir operator of G acting on $(W^{*}{}_{\alpha})_{0}$. Indeed, owing to the constraint on Φ_{L} we can extend the summation of the indices α,β in $h^{\alpha\beta} T_{\alpha} T_{\beta}$ to $h^{ij} T_{i} T_{j}$ which is the Casimir operator. In general, however, $h_{\alpha\beta}$ is non-constant and the terms $D_{\mu} h_{\alpha\beta}$ contributes to the effective Laplacian.

Laplace operator on 1-forms

Here the vector bundle $\mathcal{F}$ is the cotangent bundle $T^{*}E$ with the typical fibre $F = \mathbb{R}^{N^{*}} = \mathbb{R}^{m^{*}} \oplus \mathbb{R}^{k^{*}} \oplus \mathbb{R}^{l^{*}}$. The effective fibre bundle $\mathcal{F}_{\alpha}$ over M has the fibre $F_{\alpha} = \mathbb{R}^{N^{*}} \otimes_{H} W^{*}{}_{\alpha} = [\, \mathbb{R}^{m^{*}} \otimes (W^{*}{}_{\alpha})_{0} \,] \oplus + [\, \mathbb{R}^{k^{*}} \otimes (W^{*}{}_{\alpha})_{0} \,] \oplus [\mathbb{R}^{l^{*}} \otimes_{H} W^{*}{}_{\alpha}]$. In other words a 1-form Φ_{A} on E gives rise to a multiplet $(\Phi_{\mu L}, \Phi_{aL}, \Phi_{aL})$ where each term has gotten an extra index from the representation space $W^{*}{}_{\alpha}$ constrained by $\alpha(h)^{L}{}_{M} \Phi_{.L} = \Phi_{.M}$ for the first two members of the multiplet and

$$\alpha(h)^{L}{}_{M} \Phi_{aL} = A(h)^{b}{}_{a} \Phi_{bM}$$

for the last one. Each member of the multiplet will, in general, split into submultiplets according to the colour charge of the gauge group N(H)|H.

8.3 Dimensional reduction of spinor fields

The group O (E) is not simply-connected and we denote by **Pin(E)** its two-fold simply-connected spin covering. To avoid complications with orientation and time orientation we have replaced **Spin+** (E) by **Pin(E)**. A spin structure on E consists of a principal bundle **PinE** over E with structure group **Pin(E)** and of a covering bundle homomorphism

PinE $\rightarrow$ OE commuting with the group homomorphism $\sigma : Pin(E) \rightarrow O(E)$. The elements of PinE are called spin frames. Thus over each orthonormal frame $e \in OE$ there sit two spin frames which differ by the transformation $(-I) \in Pin(E)$. When a spin frame s is rotated by a transformation $\Lambda \in Pin(E)$ then the corresponding orthonormal frame e is rotated by an orthogonal transformation $\sigma(\Lambda) \in O(E)$. The group homomorphism σ is given explicity by

$$\Lambda \Gamma_A \Lambda^{-1} = \Gamma_B \sigma(\Lambda)^B{}_A$$

where Γ_A are generators of the Clifford algebra of the metric η_E, satisfying $\{\Gamma_A, \Gamma_B\} = 2 \eta_{AB}$. As we have seen in sect.4 the G-space structure of E allows us to reduce the bundle OE to the subbundle AOE of adapted frames, the structure group being reduced from O(E) to AO(E). Here we will take $AO(E)=O(M)\times O(S)$ as there is no gain in a subtler reduction $O(S) \rightarrow O(K)\times O(L)$. We denote by APinE the counterimage of AOE under the covering homomorphism. The elements of APinE are called adapted spin frames. Thus over each adapted orthonormal frame there sit two adapted spin frames. The structure group of APinE is $(Pin(M)\times Pin(S)/Z_2$, where the quotient identifies the two "minus identities" of Pin(M) and Pin(S) respectively. Let $\partial_\mu(x)$ be a fixed orthonormal frame at $x \in M$ and consider the set of all spin frames at the points of the internal space S_x above x which project onto (e_μ, e_α) with e_μ being the horizontal lift of ∂_μ. This set of spin frames is a spin structure for S_x. Thus existence of a spin structure on E implies existence of a spin structure on the internal spaces S_x, $x \in M$, but not, in general, on M.

Let us assume that the action of G on AOE lifts to an action on APin(E); then, since H is assumed to be connected, the homomorphism $\lambda : H \rightarrow O(S)$ induces a unique homomorphism $\tilde{\lambda} : H \rightarrow Pin(S)$. Repeating now an argument used several times, we deduce that the resulting effective gauge group is now $(Pin(M)\times N(\bar{H})|\bar{H})/Z_2$ where $\bar{H} = diag(\lambda(H),H) \subset Pin(S)\times G$. Therefore M is endowed with a generalisation of a **Spin**^{c} structure (for a discussion of **Spin**^{c} structures see, e.g., [91]. To derive the dimensionally reduced Dirac operator we use a local moving spin frame corresponding to the orthonormal moving frame introduced at the beginning of this section. Let **F** be a representation space (real or complex) of the Clifford algebra C(**E**). The generators of the representation of **Pin**(**E**) on F are then

(1) $$\Sigma^{AB} = 1/4\, [\Gamma^A, \Gamma^B]$$

and the Dirac operator is

(2) $$D_E = \Gamma^A (e_A + 1/2\, \omega_{A,\,BC}\, \Sigma^{BC})$$

The typical fibre $\overline{F}_\alpha$ of the dimensionally reduced spinor of type α is then $F \otimes_H W^*_\alpha$. Thus the spinor Ψ gets an extra index ($\Psi = (\Psi_L)$) of the representation space W^*_α and satisfies the constraint

(3) $$(\alpha(T_{\hat{\alpha}}) + \tilde{\lambda}(T_{\hat{\alpha}}) = 0\ , \text{ i.e.:}$$

$$(\alpha(T_{\hat{\alpha}}) + 1/2\, C_{\hat{\alpha},\beta\gamma}\, \varphi^{\beta}{}_{\underline{\beta}}\, \varphi^{\gamma}{}_{\underline{\gamma}}\, \Sigma^{\underline{\beta}\,\underline{\gamma}})\, \varphi = 0$$

where $T_{\hat{\alpha}}$ ($=1,2,\ldots,\dim H$) are the generators of **H** . The dimensionally reduced Dirac operator can now be written down as

$$D_{eff} = D_M + D_S + \Delta_1 + \Delta_2$$

with the four terms given in the summary section (sect.8.0) .

Let us consider now the case of **M** four-dimensional of signature (-+++) and S the typical internal space of signature (++...+). The Clifford algebra C(**E**) of **E** is then isomorphic to a tensor product C(E) = C(**M**) $\otimes$ C(-S), the isomorphism being given by $\Gamma_\mu = \gamma_\mu \otimes 1$, $\Gamma_\alpha = \gamma_5 \otimes \gamma_\alpha$ with $\{\gamma_\mu, \gamma_\nu\} = 2\eta_{\mu\nu}$ and $\{\gamma_\alpha, \gamma_\beta\} = -2\, \eta_{\alpha\beta}$, where $\gamma_5 = \gamma_0\gamma_1\gamma_2\gamma_3$. Since

$C(M) \cong L(\mathbb{R}^4)$- the 4×4 real matrix algebra -it follows that the real spinor space F can be realised as the tensor product $F = \mathbb{R}^4 \otimes_{\mathbb{R}} F'$ where F' is a representation space of C(-S). The simplest example is the five-dimensional Kaluza-Klein theory. Here S is one-dimensional and the Clifford algebra C(-1) is isomorphic to the algebra $\mathbb{C}$ of complex numbers. Thus $F = \mathbb{R}^4 \otimes_R \mathbb{C} = \mathbb{C}^4$. The group G is, in this example, U(1) and all its irreducible representations α are realised on $\mathbb{C}$. Since H is now trivial the effective fibre is therefore $\overline{F}_\alpha = F \otimes \mathbb{C} = \mathbb{C}^4$. The effective structure group is then $Pin^c(3,1) := (Pin(3,1) \times U(1))/Z_2$.

8.4 The spectrum of Laplace operator for G invariant metrics on groups and homogeneous spaces.

Generalities

Let S be a homogeneous space endowed with an homogeneous metric and call G the isometry group, then $S = G/H$; H can be in particular the identity. Calling $s = \dim S$ we consider the O(s) bundle of orthonormal frames over G/H; we study first the scalar fields and therefore construct the associated bundle $\mathcal{F} = G/H \times \mathbb{R}$ and its sections $\Gamma\mathcal{F}$. According to the general discussion, the space $\Gamma\mathcal{F}$ can be analysed in term of the field harmonics (section of $\Gamma_\sigma\mathcal{F}$, $\sigma \in \hat{G}$); here the effective fiber bundle $\overline{\mathfrak{F}}_\alpha$ is a bundle over a point $M = (G\backslash G/H)$. Effective fields (sections of $\overline{\mathfrak{F}}_\alpha$) are just elements of the vector space $\mathbb{R} \otimes_H W^*_\sigma$ where W^*_σ is the dual of the representation space W_σ of σ. But $(W^*_\sigma)_H = \mathbb{R} \otimes_H W^*_\sigma$ is just the space of H singlets in W^*_α. Let $Lie(H) + \mathbf{S}$, a reductive decomposition of Lie(G), and let us split accordingly the generators $(T_i) = (T_{\hat{\alpha}}, T_\alpha)$. In a neigbourhood of the identity of G/H - the (inverse) metric reads $g^{\alpha\beta}\ e_\alpha \otimes e_\beta$ where e_α are the fundamental fields associated to the G-action, the Laplace operator is then $\Delta = g^{\alpha\beta}\, e_\alpha\, e_\beta$ where e_α are now considered as 1st order differential operators acting on the functions on G/H (on elements of $\Gamma\mathcal{F}$). The effective Laplace operator acting on an effective field (element of $(W^*{}_\sigma)_H$) is just :

$$\Delta_{eff}\, \Phi_\sigma{}^{eff} = g^{\alpha\beta}\, \sigma(T_\alpha)\, \sigma(T_\beta)\, \Phi_\sigma{}^{eff}$$
$$\text{with} \quad \Phi_\sigma{}^{eff} \in (W^*{}_\sigma)_H \;,\; T_\alpha \in S$$

and owing to the H invariance we can extend the summation to the whole of G :

$$\Delta_{eff}\, \Phi_\sigma{}^{eff} = g^{ij}\, \sigma(T_i)\, \sigma\, (T_i)\, \Phi_\sigma{}^{eff} \;.$$

When (g^{ij}) is a <u>bi</u>-invariant metric on G, the above Laplacian acting on $\Phi_\sigma{}^{eff}$ is (minus) the Casimir operator, in the representation σ, which is proportional to the unit matrix (cf 2-11) and we get

$$\Delta_{eff}\, \Phi_\sigma{}^{eff} = - C^G{}_\sigma\; \Phi_\sigma{}^{eff}$$

where the Casimir constant $C^G{}_\sigma$ depends upon our convention for the normalization of the volume. In the examples we will assume that the metric is normalised in such a way that $C^G{}_\sigma = \dim(G)/\dim(\sigma) \times i^G{}_\sigma$, $i^G{}_\sigma$ being the index defined in sect.2.11 ($i^G{}_\sigma$ can be found from tables [171] via the relation $i^G{}_\sigma = I_2(\sigma)/_{\text{rank}}\,(G)$).

Remark : Below we will introduce a global minus sign in the definition of the Laplacian in order to get positive eigenvalues (this amounts to considering the de-Rham Laplacian on 0-forms rather than the usual Laplacian on functions).

When (g^{ij}) is not bi-invariant but only $G\times H$ invariant, it is usually possible (as we shall see in the examples) to recast the problem in a way which can be handled easily (for example writing $g^{ij}\; e_i \otimes e_j = g_o{}^{ij}\; e_i \otimes e_j\; - k.\, g_o{}^{\underline{\alpha}\underline{\beta}}\; e_{\underline{\alpha}} \otimes e_{\underline{\beta}}$,$k \in \mathbb{R}$, where both $g_o{}^{ij}$ and $g_o{}^{\underline{\alpha}\underline{\beta}}$ describe <u>bi</u>-

invariant metrics on G and H). Notice that, the isomorphism $\Gamma_\sigma \mathcal{F} = \Gamma_{inv} \mathcal{E}_\sigma \otimes' \mathbf{W}_\sigma$, with proper case of $\otimes'$, just means in this context that the eigenstate Φ_σ^{eff} corresponds to an eigenstate of Δ which has degeneracy $\dim(\mathbf{W}_\sigma)$.

The Laplacian on a homogeneous 2 sphere.

Here $S=S^2$, $G=SU(2)$, $H=U(1)$. An irreducible representation of G is labelled by L which is integer or half integer and $\dim \mathbf{W}_L = 2L+1$; a basis of $\mathbf{W}_L$ is denoted by $|L,m\rangle$, $|m| \leq L$. Calling T_3 the generator of U(1) we have $T_3 \, |Lm\rangle = m \, |Lm\rangle$; when L is an integer the effective space $\overline{\mathbf{F}}_L = (\mathbf{W}_L)_H$ is 1 dimensional and is isomorphic with the space generated by $|L,0\rangle$ but, when L is half integer $\overline{\mathbf{F}}_L$ is zero. We get

$$\begin{aligned} \Delta_{eff} \, | \, L \, 0 \rangle &= (\sigma(T_1{}^2)+\sigma(T_2{}^2)) \, |L \, 0\rangle \\ &= (\sigma(T_1{}^2)+\sigma(T_2{}^2)+\sigma(T_3{}^2))|L \, 0\rangle = L(L+1) \, |L \, 0\rangle , \end{aligned}$$

with the physicist normalisation of (T_i). If we introduce a radius ($g_{\alpha\beta} \rightarrow \rho^2 g_{\alpha\beta}$) , the eigenvalue(s) are $L(L+1)/\rho^2$. ρ^2 would, of course, become a function of $x \in M$ in the case of an arbitrary SU(2) invariant metric on $M \times S^2$. By taking the tensor product of $\overline{\mathbf{F}}_L$ by $\mathbf{W}_L$, we recover of course $2L+1$ eigenstates of Δ - for integer L -, all degenerate with eigenvalue $L(L+1)$ -, these particular sections of $\mathcal{F} = G/H \times \mathbb{R}$ are of course the spherical harmonics $Y_L^m(\theta, \varphi)$.

The Laplacian on S^3 *with a* $SU(2)\times SU(2)$ *invariant metric.*

Here $S = S^3 = SU(2)$, $G=SU(2)_L \times SU(2)_R$, $H = SU(2)_{diag}$, $N|H = \{e\}$. Call $\mathbf{L}$ the generators of $SU(2)_L$, $\mathbf{R}$ those of $SU(2)_R$, $T_{\underline{i}} = (L_i + R_i)$ those of $H=SU(2)_{diag}$ and $T_i = 1/2 \, (L_i - R_i)$ the tangent vectors to $S = G/H$ at the origin. The adapted basis in Lie(G) is $T_{\underline{1}}, T_{\underline{2}}, T_{\underline{3}}, T_1, T_2, T_3$ and all generate isometries of S; in a neighbourhood of the origin a G-invariant metric on S is written as $g^{-1} = e_1^2 + e_2^2 + e_3^2$.

Let $|L, m ; L', m'>$ be a base in the representation $W_{L,L'}$ of G, of dimension $(2L+1)(2L'+1)$.

The singlets under H have to satisfy $T_i \mid L\, m \; ; L'\, m' > =0$, i.e. $\vec{L} = -\vec{R}$ in the effective vector space, then, $L = -L' = L$ and $m = -m' = m$, this is why we add a coefficient 1/2 in the definition of T_i. The effective space $\overline{F}_{L,L'}$ is therefore zero if $L \neq L'$ and is isomorphic with the $(2L+1)$ dimensional space $|L, m; L, -m>$. We get $\Delta_{eff} = T_1{}^1 + T_2{}^2 + T_3{}^2 = 1/4\ (L^2 + R^2 - 2L.R)$ and $\Delta_{eff}|L,m,L,-m> = +1/4\ (4L(L+1)) = L(L+1)$ [here we write T_i rather then $\sigma(T_i)$ to shorten the notation] . The degeneracy of this eigenvalue is given by $\dim W_{L,L} = (2L+1)^2$.

The Laplacian on S^3 with a $SU(2)_L \times U(1)_R$ invariant metric

- Here $S = S^3$, $G = SU(2)_L \times U(1)_R$, $H = U(1)_{diag}$. We introduce the following notation for the generators :

$SU(2)_L : \vec{L}$; $U(1)_R$: R_3 ; $U(1)_{diag} : \tilde{T}_3 = L_3 + R_3$; $S : T_1 = L_1, T_2 = L_2$, $T_3 = 1/2\ (L_3 - R_3)$.

Here we think of R_3 as the third generator of the group $SU(2)_R$; therefore R_3 will have integer or half integer eigenvalues. Notice that $\mathcal{S}$ = Lie(N|H)+$\mathcal{L}$ with N|H = U(1) generated by T_3 . The adapted base in Lie(G) is $(\tilde{T}_3, T_i)$ and all these vectors generate Killing vectors (e_3, e_i) on S^3 . A G-invariant metric in a neighbourhood of the identity must be of the form $g^{-1} = e_1^2 + e_2^2 + 1/\lambda^2\ e_3^2$, $\lambda \in \mathbb{R}$. Let $|L,m;s>$ be a base in the representation $W_{L,s}$ of G, L , m and s being integer or half integers; we have $L^2 \mid L,m;s> = L(L+1) \mid L,m;s>$, $L_3 \mid L,m;s> = m \mid L,m;s>$, $R_3 \mid L,m;s> = s \mid L,m;s>$.

The singlet under H have to satisfy $T_3 \mid L,m;s> = 0$ i.e. $m+s=0$, therefore, <u>in the effective space</u> $F_{L,s} = (W_{L,s})_H$ we have $T_3 = 1/2\ (L_3 - R_3) = L_3$ - this is why we introduced the coefficient 1/2 in the definition of T_3 -. The effective Laplacian in $\Gamma\mathcal{F}$ is $\Delta = T_1{}^2 + T_2{}^2 +$

$1/\lambda^2 T_3^2$ but it simplifies in $\overline{F}_{L,s}$ to $\Delta_{eff} = L_1^2 + L_2^2 + 1/\lambda^2 L_3^2 = L^2 + L_3^2(1/\lambda^2 - 1)$.

We get the eigenvalues $E_{L,m} = L(L+1)+m^2(1/\lambda^2-1)$ which should be $(2L+1)$ degenerate $(=\dim W_{L,s})$ but in fact the degeneracy is $2(2L+1)$ when $m = 0$ (because of the "accidental" degeneracy $m \to -m$) and $2L+1$ if $m \neq 0$. Notice that the smallest eigenvalues are $E_{1/2,1/2} = E_{1/2,-1/2} = 1/2 + 1/4\lambda^2$: there are no zero modes for the scalars, whatever λ is.

The Laplacian on S^3 with an SU(2) - invariant metric

Here $S = S^3$, $G = SU(2)$, $H = \{e\}$. We call L the generators of SU(2) and write in a neighbourhood of e, the (inverse) metric $g^{-1} = e_1{}^2/\lambda_1{}^2 + e_2{}^2/\lambda_2{}^2 + e_3{}^2/\lambda^2{}_3$ where e_1, e_2, e_3 are the Killing vector fields associated with $\mathbf{L}$. Here all vectors of W_L are invariant under $H = \{e\}$ therefore $\overline{F}_L = W_L$ where we choose a base $|L,m\rangle$. The effective Laplacian reads

$$\Delta eff = L_1{}^2/\lambda_1{}^2 + L_2{}^2/\lambda_2{}^2 + L_3{}^2/\lambda_3{}^2$$
$$= 1/4\,(1/\lambda_1{}^2 - 1/\lambda^2{}_2)\{L^2{}_+ + L^2{}_-\}$$
$$+\ 1/4(1/\lambda_1{}^2 + 1/\lambda_2{}^2)\{L_+ L_- + L_- L_+\} + L_3{}^2/\lambda_3{}^2$$ where we introduced

$$L_+ = 1/2\,(L_1 + iL_2), \quad L- = 1/2\,(L_1 - iL_2).$$

Using the well known properties $L_+ |Lm\rangle = \{(L+m)(L+m+1)\}^{1/2} |L,m+1\rangle$ or 0 when $m=+L$, it is then a straightforward exercise to compute the eigenvalues. Let us just give those corresponding to the lowest representations.

For $L=1/2$, the states $m=\pm\ 1/2$ are degenerate and $E_{1/2,\pm 1/2} = 1/4\ (1/\lambda_1{}^2 + 1/\lambda_2{}^2 + \lambda_3{}^2)$.

For $(L,m) = (1,0)$, $E_{1,0} = 1/\lambda^2{}_1 + 1/\lambda^2{}_2$; also $E_{1,-1} = E_{1,-1} = 1/2\,(1/\lambda_1{}^2 + \lambda_2{}^2) + 1/\lambda_3{}^2$.

For $L=3/2$, eigenstates will be either $\alpha\ |3/2\rangle + \beta\ |-1/2\rangle$ or $\alpha'\ |-3/2\rangle + \beta'\ |1/2\rangle$, the coefficient as well as the eigenvalues have to be determined. For example,

$$\Delta|3/2\rangle = 1/4\,(1/\lambda_1^2 - 1/\lambda_2^2)\,[0 + \sqrt{3}\,L_-|1/2\rangle]$$
$$+ 1/4\,(1/\lambda_1^2 + 1/\lambda_2^2)\,.3\,|3/2\rangle + 1/\lambda_3^2\,.\,9/4\,|3/2\rangle$$
$$= 1/2\,\sqrt{3}\,(1/\lambda_1^2 - 1/\lambda_2^2)\,|-1/2\rangle$$
$$+ [\,3/4(1/\lambda_1^2 + 1/\lambda_2^2) + 9/4\ 1/\lambda_3^2\,]\,|3/2\rangle$$

$$\Delta|-1/2\rangle = 1/4\,(1/\lambda_1^2 - 1/\lambda_2^2)\,[\,2L_+|1/2\rangle + 0\,]$$
$$+ 1/4\,(1/\lambda_1^2 + 1/\lambda_2^2)\,.7\,|-1/2\rangle + 1/\lambda_3^2\,.1/4\,|-1/2\rangle$$
$$= \sqrt{3}/2\,(1/\lambda_1^2 - \lambda_2^2)\,|3/2\rangle$$
$$+ [\,7/4\,(1/\lambda_1^2 + 1/\lambda_2^2) + 1/4.(1/\lambda_3^2)]\,|-1/2\rangle$$

Then, the eigenvalues E are given by :

$\Delta(\alpha|3/2\rangle + \beta|-1/2\rangle) = E\,(\alpha|3/2\rangle + \beta|-1/2\rangle)$ which implies

$$[\,3/4(1/\lambda_1^2 + 1/\lambda_2^2) + (9/4).(1/\lambda_3^2)\,]\alpha + [\sqrt{3}/2(1/\lambda_1^2 - 1/\lambda_2^2)]\beta = E\,\alpha$$
$$[1/2\,\sqrt{3}(1/\lambda_1^2 - 1/\lambda_2^2)]\,\alpha + [(7/4)(1/\lambda_1^2 + 1/2^2) + (1/4)(1/\lambda^2{}_3)]\beta = E\,\beta$$

therefore,

$$[3/4(1/\lambda_1^2 + 1/\lambda_2^2) + 9/4\ 1/\lambda_3^2] + [\sqrt{3}/2(1/\lambda_1^2 - 1/\lambda_2^2)]\,\beta/\alpha = E$$
$$[1/2\sqrt{3}(1/\lambda_1^2 - 1/\lambda_2^2)\,]\,\alpha/\beta + [7/4(1/\lambda_1^2 + 1/\lambda_2^2) + 1/4\lambda_3^2\,] = E$$

Then it is a straightforward exercise (that we leave to the reader) to do it again for $\Psi = \alpha|-3/2\rangle + \beta\,|\,1/2\rangle$, to solve the above coupled equations and to finds the two solutions (α_+, β_+, E_+) and (α_-, β_-, E_-) i.e. the two different eigenstates $\alpha_\pm\ |3/2\rangle + \beta_\pm|-1/2\rangle = \psi_\pm$ of respective eigenvalue E_+ and E_-. One also find two other eigenstates $\Psi'_\pm = \alpha'_+\ |-3/2\rangle + \beta'_+|1/2\rangle$ of respective eigenvalues E'_+ and E'_- by the same method.

For $L > 3/2$, one could of course find explicitly the eigenvalues by a similar technique.

Laplacian on G *with a bi -nvariant metric* .

Here, $S = (G\times G)/H$ with $H = G_{diag} \subset G\times G$.

Using a standard (physicist) normalisation of generators, for the representation ρ of the group G, the Laplacian has eigenvalue $C^G{}_\rho$ = (dim G / dim ρ) $\times$ i_ρ (this is the Casimir constant in the representation ρ, cf. sect 2.11), its degeneracy is $\nu^2 = (\dim \rho)^2$. For example, with G = SU(3), we get, for the first representations $\nu=3$, C = 4/3 ; ν = 6, C = 10/3 ; ν = 8, C = 3 ; ν =10 ; C = 6.

Laplacian on G *for a metric invariant under* $G\times H$

Here,we write $S = (G\times H)/H_{diag}$, S is diffeomorphic with G, and we get this family of metrics by squashing a bi-invariant metric in the H direction. Let $\mathbf{W}(\rho, \sigma)$ be an irreducible representation space for $G\times H$. In order to determine those representations of interest, i.e., those for which the effective vector space $\overline{\mathbf{F}}(\rho ; \sigma) = (\mathbf{W}(\rho; \sigma))_H$ is not zero, we decompose the representation (ρ; σ) with respect to H and keep only those which contain singlets for H. Let us take for example G = SU(3) and H = SU(2), denote by (L_i), i=1,2...8, (R_i), i=1,2,3 the generators of G and H respectively, also $T_{\underline{i}}=L_i+R_i$, i=1,2,3 those of H_{diag} and $\{T_i = 1/2\,(L_i - R_i)$, i=1,2,3 ; $T_j = L_j$, j=4...8} those of the supplementary subspace $\mathbf{S}$. The Laplacian associated with this "squashed" metric on S^3 is $\Delta = 1/\lambda^2\,(T_1{}^2+T_2{}^2+T_3{}^2) + (T_4{}^2+T_5{}^2+T_6{}^2+T_7{}^2+T_8{}^2)$ but in the effective space $\overline{\mathbf{F}}(\rho, \sigma)$ we have $\overline{T}_{\underline{i}} = 0$ i.e. $L_i = -R_i$, i =1,2,3. In this subspace, the Laplacian reads

$$\Delta = 1/\lambda^2\,(L_1{}^2+L_2{}^2+L_3{}^2) + (L_4{}^2+L_5{}^2+L_6{}^2+L_7{}^2+L_8{}^2)$$

$$= \Sigma_{i=1..8}\, L_i{}^2 + (1/\lambda^2 - 1)\;\Sigma_{i=1..3}\, L_i{}^2$$

The embedding of SU(2) in SU(3) being standard (Dynkin index is 1), we get the eigenvalue $E = C_\rho{}^G + (1/\lambda^2 - 1)C_\sigma{}^H$, whose degeneracy is $(\dim\rho)\times(\dim\sigma)$.

Laplacian on S^7 *with an* SO(8) *-invariant metric* .

Here, S = G/H with G = SO(8) and H = SO(7). G being of rank 4 has representation spaces $W(a_1,a_2,a_3,a_4)$ labelled by four integers. The interesting representations are those which contain a H-singlets in the branching rule G $\longrightarrow$ H.

In the present case they are of the type (n,o,o,o); the eigenvalues of Δ coincide here with the Casimir constants of the corresponding representation.

The following table can be computed from the table of indices [171] It is easy to see that E(n,o,o,o)= n(n+6)/2.

representation	$\dim(\rho)$	i_ρ	$C_\rho{}^G = \dim G \times i_\rho/\dim\rho = E$
1000	8	1	7/2
2000	35	10	8
3000	112	54	27/2
4000	294	210	20

S^7 *with a metric invariant under* U(2, ℍ)×SU(2)

Here, S^7 = G/H with G = U(2, ℍ)×SU(2) and H = U(1, ℍ)×SU(2); ℍ denoting, as usual the field of quaternions. Therefore, U(2, ℍ) is also the symplectic compact group Sp(2) or Spin(5) and U(1, ℍ) is also SU(2) . As usual, only those representations of G which contain H-singlets will contribute. In this particular case we can take advantage of the previous study since the interesting representations of G will themselves appear in the branching rule of the allowed representations of SO(8) for the inclusion G $\subset$ SO(8). For example the representation (2,0,0,0)$\longrightarrow$(2,0;2)+(0,1;0) i.e. 35$\longrightarrow$10×3+5×1; indeed the representation

(2,0;2) of G for example contains H-singlets (decompose (2,0) of U(2, ℍ) with respect to U(1, ℍ)).

The Laplacian associated with such a metric will be $\Sigma_{a=1..4} T_a^2$ + $1/\lambda^2 \ \Sigma_{\hat{a}=1..3} T_{\hat{a}}^2$ in terms of the adapted basis and, taking into account the constraint of H invariance in the effective space(s) $\overline{F}(a,b;c)$, it can be written

$$\Delta_{eff} = \Sigma_{a=1..4} L^2_a + 1/\lambda^2 \ \Sigma_{\hat{a}=1..3} L^2_{\hat{a}}$$
$$= \Sigma_{a=1..10} \ L_a^2 + (1/\lambda^2 - 1) \Sigma_{\hat{a}=1..3} L_{\hat{a}}^2$$

The eigenvalue(s) are

$$E(a,b;c) = 1/\rho^2 \ (C^{U(2,\mathbb{H})}(a,b) + \ (1/\lambda^2 \ -1) C^{SU(2)}(c) \ .$$

Notice that the embedding is standard and there is no need to introduce a Dynkin index in the above. We also introduced ρ^2 in order to adjust the "radius" to some normalised value. For example, choosing the representation (2,0;2)=10×3, we get

$$E = 1/\rho^2 \ (3 + (1/\lambda^2 - 1)2) \ ,$$

Also the metric is Einstein for $\lambda^2 = 2$ and $\lambda^2 = 2/5$ (see sect 3.6) therefore, for the usual SO(8)-invariant Einstein sphere ($\lambda^2 = 2$), we get $E = 2/\rho^2$ (2). In order to recover the eigenvalue E=8 (found in the previous example) associated with the representation (2000)≅35 of SO(8), we have to choose $\rho^2 = 1/4$. Then, eigenfunction associated to the representation (2,0;2) for a "squashing" deformation of the "standard" sphere have eigenvalue $E = 4(3+2(1/\lambda^2 - 1))$ and a degeneracy equal to dim(2,0;2)=10×3=30. A straightforward exercise in representation theory shows that only representations (n-2m,m; n-2m) of U(2, ℍ)×SU(2) will be associated with eigenfunctions; the corresponding eigenvalue being (choosing $\rho^2 = 1/4$),

$$E = 4 \{ [1/8 \ n(n+6) + 1/2 \ j(j+1)] + (1/\lambda^2 - 1) \ j \ (j+1) \}$$

with $j = n/2 - m$, and can be written

$$E = 1/2 \ n \ (n+6) + ((1-\mu^2)/\mu^2) \ \ 2j \ (j+1) \qquad \text{with } \mu^2 = \lambda^2/2$$

This results agrees with [60] where this particular case was studied .

85 The spectrum of the Dirac operator for G-invariant metrics on groups and homogeneous spaces.

The Dirac operator on the two-sphere

Here $S = S^2$, $G = SU(2)$, $H = U(1)$, $T=\mathbf{Spin}(2)$, the two fold covering of $U(1)$,

$F=\mathbb{R}^2 \cong \mathbb{C}$. We give to S^2 the metric obtained from the bi-invariant metric on $SU(2)$: $g^{-1} = e_1{}^2 + e_2{}^2 + e_3{}^2$, $[e_1, e_2] = e_3$ by going to the quotient. We call $\mathbf{L}$ the generators of $SU(2)$, with $[L_1, L_2] = L_3$, L_3 generating $H= U(1)$; $\mathbf{L}^2$ acts on the basis vectors $|L\ m\rangle$ of the representation space $\mathbf{W}(L)$ of $SU(2)$ by multiplication by $-L(L+1)$. The Lie algebra of $\mathbf{Spin}(2)$ is a subalgebra of the Clifford algebra $\mathrm{Cliff}(0,2)$ generated by γ^1 and γ^2 with $\gamma^1\gamma^2 + \gamma^2\gamma^1 = 0$ and $(\gamma^1)^2 = (\gamma^2)^2 = -1$; the generator of $\mathbf{Spin}(2)$ is $S = \Sigma^{12} = 1/4\ [\gamma^1, \gamma^2] = 1/2\ \gamma^1\gamma^2$ and therefore $S^2 = -1/4$. We have then a basis $|L,m;s\rangle$ in the representation space $\mathbf{W}(L,s) = \mathbf{W}(L)\otimes\mathbb{C}$ of $G = G\times T = SU(2)\times\mathbf{Spin}(2)$, L is integer or half integer and $s=+1/2$. We have also a map $\overline{\lambda} : H= U(1) \mapsto T=\mathbf{Spin}(2)$ with $\overline{\lambda}(L_3) = 1/2 C_{3ab}\ \varphi^a{}_a \varphi^b{}_b \Sigma^{ab}$ with the notations of 8.3, i.e. $\lambda\ (L_3) = C_{312}\ \Sigma^{12} = S$. The effective spinor space $F(L,s) = (\mathbf{W}(L,s))_H$ contains only $\overline{H}$ singlets with $\overline{H} = \mathrm{diag}(\overline{\lambda}(H),H) \subset T\times G$ generated by $L_3 + \lambda(L_3) = L_3 + S$ - i.e. they satisfy the constraint (8.3.3). In this subspace, $(L_3 + S)\ |\ L,m;s\rangle = 0$,i.e. $m=-s$; this implies that $F(L,s)$ is one dimensional (this last property should not be surprising in vue of the fact that G/H is symmetric irreducible - cf Ch.3 -). The effective Dirac operator on $F(L,s)$ is $D = \gamma^1 L_1 + \gamma^2 L_2 + 1/4\ \omega_{\alpha\beta\gamma}\gamma^\alpha\gamma^\beta\gamma^\gamma$ where $\omega_{\alpha\beta\gamma}$ is the spin connection in the base $\{e_1, e_2\}$ of fundamental fields generated on S^2 by L^1, L^2 in a neighbourhood of the north pole; the range of the α indices being $\{1,2\}$, we get $\omega_{\alpha\beta\gamma} = 0$ (notice that at the origin of S^2, we have $e_3 = 0$ and the curvature of S^2 in this base enters only via the derivatives $\partial_\alpha f_{\beta\gamma}{}^\delta = -C_{\beta\gamma}{}^3\ C_{\alpha 3}{}^\delta$, for instance $\partial_2 f_{12}{}^1 = -C_{12}{}^3\ C_{23}{}^1$). Therefore the Dirac equation in the effective space $F(L,s)$ reads

$$\gamma^1 L_1 + \gamma^2 L_2\ |L,-s;s\rangle = E(L,s)\ |\ L,-s\ ;\ s\ \rangle.$$

Calling $P = \gamma^1 L_1 + \gamma^2 L_2$, we get $P^2 = -L_1{}^2 - L_2{}^2 + 2SL_3 = -\mathbf{L}^2 + L_3{}^2 + 2SL_3$; in the effective space, we have $L_3 + S = 0$ hence $P^2 = -L^2 - S^2 = L(L+1) + 1/4$. We therefore get eigenvalues $E(L,s) = [L(L+1)+1/4]^{1/2}$.

The Dirac operator on S^3 *for a bi-invariant metric under* SU(2).

Here $G = SU(2)_L \times SU(2)_R$, $H = SU(2)_{diag}$, $T = Spin(3)$, $N|H = \{e\}$, $F = \mathbb{C}^2$.

The Clifford algebra Cliff(0,3) is generated by $\gamma^1, \gamma^2, \gamma^3$ with $(\gamma^i)^2 = -1$, $\gamma^i\gamma^j + \gamma^j\gamma^i = 0$ if $i \neq j$; the Lie algebra of Spin(3) is generated by $S_k = \Sigma^{ij} = 1/2\ \gamma^i\gamma^j$ then $[S_i, S_j] = \epsilon_{ijk}\ S_k$. $SU(2)_L$ and $SU(2)_R$ are generated by $\mathbf{L}$, $\mathbf{R}$ with $[L_i, L_j] = \epsilon_{ijk}\ L_k$, $[R_i, R_j] = \epsilon_{ijk}\ R_k$, $[L_i, R_j] = 0$. $H = SU(2)_{diag}$ is generated by $T_{\underline{i}} = L_i + R_i$ and the complementary subspace $\mathfrak{P}$ by $T_i = L_i - R_i$; then $[T_{\underline{i}}, T_{\underline{j}}] = \epsilon_{ijk}\ T_{\underline{k}}$ and $[T_i, T_j] = \epsilon_{ij\underline{k}}\ T_{\underline{k}}$. We choose the following (inverse) metric at the origin of G :

$$g^{-1} = 8(\mathbf{L}^2 + \mathbf{R}^2) = 4(T_1{}^2 + T_2{}^2 + T_3{}^2 + T_{\underline{1}}{}^2 + T_{\underline{2}}{}^2 + T_{\underline{3}}{}^2)$$

Writing $Lie(G) = Lie(H) + \mathfrak{P}$ with adapted basis $\{T_{\underline{i}}, T_i\}$, we get in a neighbourhood of the origin of the quotient G/H : $g^{-1} = 4(e_1{}^2 + e_2{}^2 + e_3{}^2)$, where the e_i are fundamental vector fields associated to the T_i ; notice that this base is not orthonormal and that an adapted orthonormal moving frame differs from the e_i basis by a "vielbein" φ with $\varphi^i{}_{\underline{i}} = 2$. We call $\overline{G} = G \times T$ and $\overline{H} = diag(H, \overline{\lambda}(H))$ where $\overline{\lambda}: H \to T$ is the homomorphism which traduces the fact that H acts by spin rotations at the origin, the adapted basis for the decomposition $Lie(\overline{G}) = Lie(\overline{H}) + \mathfrak{P}$ is $(T_{\underline{i}} + \overline{\lambda}(T_{\underline{i}}), T_i)$.

We know that $\overline{\lambda}(T_{\underline{i}}) = 1/2\ C_{ijk}\ \varphi^j{}_{\underline{i}}\ \varphi^k{}_{\underline{k}}\ \Sigma^{\underline{ik}}$. For example, using $C_{31}{}^2 = 1$ (then $C_{31\underline{2}} = 1/4$) and $\varphi_1{}^1 = \varphi_2{}^2 = 2$ we get

$$\overline{\lambda}(T_3) = 1/2(C_{312}\varphi^1{}_1\varphi^2{}_2\,\Sigma^{12} + C_{321}\,\varphi^2{}_2\,\varphi^1{}_1\,\Sigma^{21}) = S_3$$

We have therefore to restrict the representation space $W(L,m;\ r,m;\ \epsilon=\pm 1/2)$ of $\overline{G}$ spanned by the basis vectors $|L,m_L\ ;\ r,m_r\ ;\ \epsilon=\pm 1/2\rangle$ to the effective subspaces F_+ and F_- spanned by the basis vectors

$$|L,\ m=m_L\ ,\ r=L+1/2,\ m_r=-m-1/2,\ 1/2\ \rangle$$
$$|L,\ m=m_L\ ,\ r=L-1/2;\ m_r=-m+1/2,\ -1/2\rangle$$

and which satisfy the constraint $T_{\underline{i}} + \overline{\lambda}(T_{\underline{i}}) = 0$, i.e. $\mathcal{L}+\mathcal{R}+\mathcal{S} = 0$. The effective Dirac operator in an orthonormal basis T_i is $P=\gamma^i T_i + 1/4\ \omega_{ijk}\ \gamma^i\gamma^j\gamma^k$; for example $\omega_{123} = 1/2 C_{123} = 1/2 C_{12}{}^3 = 1$ since $T_i = 2T_i$, we get $P = 2(\gamma^1 T_1 + \gamma^2 T_2 + \gamma^3 T_3\ ) + 3/2\ \gamma^1\gamma^2\gamma^3$. It is convenient to introduce the operator $Q = -\gamma^1\gamma^2\gamma^3 P$, (multiplying P by $\Pi\gamma^i$ is a trick which is standard in odd dimensions); then $Q=4\ \{T_1 S_1 + T_2 S_2 + T_3 S_3\ \} + 3/2$ which can be written (using $T_i = L_i - R_i$ and the constraint $S_i = -L_i - R_i$), $Q = -4\ (L^2 - R^2) + 3/2$. We get two kinds of eigeinstates for the operator Q : those Ψ_- for which $r=L-1/2$ and those Ψ_+ for which $r=L+1/2$; using $\mathcal{L}^2 = L(L+1)$ and $\mathcal{R}^2 = r(r+1)$ we obtain immediately the eigenvalues $E_-=-4r-3/2=-4L+3/2$ and $E_+=+4r+1/2=4L+5/2$. It may be useful to introduce $j=L+r$ then $E_-=-2j-1/2$ and $E_+ = 2j+3/2$. The degeneracy corresponds to the dimension of the isometry group.

The Dirac operator on S^3 *with a metric invariant under* $SU(2)\times U(1)$.

Here $G = SU(2)\times U(1)$ is generated by L_1, L_2, L_3, R_3 and $H = U(1)_{diag}$ by $L_3 + R_3 = T_3$.

The adapted basis in Lie(G) is $\{T_{\underline{3}}, T_1 = L_1, T_2 = L_2, T_3 = L_3 - R_3\ \}$; the group T = Spin(3) is generated as before. We choose the following (inverse) metric at the origin of G :

$$g^{-1} = a\ \mathcal{L}^2 + b\ R_3{}^2$$
$$= a(T_1{}^2 + T_2{}^2 + 1/4(T_3{}^2 + T_{\underline{3}}{}^2 + 2T_3\ T_{\underline{3}}\ )) + b\ 1/4\ (T_3{}^2 + T_{\underline{3}}{}^2 - 2T_3\ T_{\underline{3}}).$$

In a neighbourhood of the origin of $S^3 = G/H$, we get, by going to the quotient :

$$g^{-1} = a(e_1{}^2 + e_2{}^2 + 1/4\, e_3{}^2) + b/4\, e_3{}^2 = a(e_1{}^2 + e_2{}^2) + (a+b)/4\, e_3{}^2$$

where e_i are the fundamental fields associated to the T_i - those corresponding to T_3 vanish in this neighbourhood. We set $a=4$, $b= 1/\lambda^2 -1$, then $g^{-1} = 4(e_1{}^2 + e_2{}^2 + e_3{}^2/\lambda^2)$ and coincides with the previously studied case for $\lambda = 1$. The study is very similar to the previous case but now $\phi_1{}^1 = \phi_2{}^2 = 2$, $\phi_3{}^3 = 2/\lambda$; also $\overline{G} = G{\times}T$, $H = \mathbf{diag}(U(1), \overline{\lambda}(U(1))$ and the constraint $T_3 + \overline{\lambda}(T_3) = 0$ in the effective spinor space; since $\overline{\lambda}(T_3) = S_3$ we get $L_3 + R_3 + S_3 = 0$. This constraint is weaker than in the previous case.

The Dirac operator now reads

$$P = 2\,(\gamma^1 T_1 + \gamma^2 T_2 + \gamma^3/\lambda\, T_3) + (\lambda^2 + 2)\ /2\lambda\ \gamma^1\gamma^2\gamma^3$$

and

$$Q = -\gamma^1\gamma^2\gamma^3\ P = 4\,(T_1 S_1 + T_2 S_2 + T_3 S_3/\lambda) + (\lambda^2 + 2)/2\lambda$$

Using the definition of T_i and the constraint $L_3 + R_3 + S_3 = 0$, we get

$$Q = 4(L_1S_1 + L_2S_2 - (L_3-R_3)(L_3+R_3)/\lambda) + (\lambda^2+2)/2\lambda$$
$$= 2(L_+S_- + L_-S_+) - 4/\lambda\,(L_3{}^2 - R_3{}^2) + (\lambda^2+2)/2\lambda$$

where we introduced $L_\pm = 1/2(L_1 \pm iL_2)$ and $S_\pm = 1/2(S_1 \pm iS_2)$.

Eigeinstates of Q are a priori of the form

$$\Psi = a\,|L, m_L = n-1/2, r = L+1/2, m_r = -n, +1/2\rangle$$
$$+ b\,|L, m_L = n+1/2, r = L-1/2, m_r = -n, -1/2\rangle$$

then, using the well known property

$$L_\pm |L,m\rangle = [L(L+1) - m(m\pm 1)]^{1/2}\,|L, m\pm 1\rangle\ ,\ \text{if } m \neq \pm L$$
$$= 0 \qquad \text{if } m = \pm L,$$

the eigenvalue equation $Q\Psi = E\Psi$ leads to the system :

$$\begin{cases} 2\alpha a - 4/\lambda(n+1/4)b + (\lambda^2+2)\, b/2\lambda = Eb \\ 2\alpha b - 4/\lambda(-n+1/4)a + (\lambda^2+2)\, a/2\lambda = Ea \end{cases} \Rightarrow \begin{cases} 2\alpha a/b - 4n/\lambda + \lambda/2 = E \\ 2\alpha b/a + 4n/\lambda + \lambda/2 = E \end{cases}$$

where we set $\alpha = \{ L(L+1) - (n+1/2)(n-1/2)\}^{1/2}$
$= \{(L+n+1/2)(L-n+1/2)\}^{1/2}$.

The above implies in particular $2\alpha(b/a-a/b) + 8n/\lambda = 0$; we find two solutions Ψ_+ and Ψ_- with respective eigenvalues E_+ and E_- :

$$\begin{cases} a_+ = 2n/\lambda + \{n^2/\lambda^2 + \alpha^2\}^{1/2} \\ b_+ = \alpha \\ E_+ = \lambda/2 + 2\,\{n^2/\lambda^2 + \alpha^2\}^{1/2} \end{cases} \quad \text{and} \quad \begin{cases} a_- = 2n/\lambda - \{n^2/\lambda^2+\alpha^2\}^{1/2} \\ b_- = \alpha \\ E_- = \lambda/2 - 2\{n^2/\lambda^2 + \alpha^2\}^{1/2} \end{cases}$$

The degeneracy is $(2L+1)$ in general but there exist a particular case where we get two eigenfunctions

$$\Psi_1 = | L, m_L = +L;\ r = L+1/2,\ m_r = -L-1/2;\ +1/2 >,$$
$$\Psi_2 = | L, m_L = -L;\ r = L-1/2,\ m_r = L+1/2;\ -1/2 >,$$

corresponding to the same eigenvalue $E = (2L+1)/\lambda + \lambda/2$ which is therefore degenerated $2(2L+1)$ times.

The above results agree with [91] (provided we change our notations and set $2n=p-q$ and $2L+1=p+q$).

It is interesting to notice cf. [91] that for some values of L, n and of the squashing parameter λ, the eigenvalue E_- may vanish (we recall that the scalar curvature corresponding to this metric is $R=2(4-\lambda^2)$), indeed:

if $\lambda^2 < 4$ (in particular $\lambda^2 = 1$), $R > 0$ and there are no harmonic spinors (solutions of $Q\ \Psi = 0$); this is in agreement with the Lichnerowicz theorem.

if $4 < \lambda^2 < 16$, $R < 0$ (the Lichnerowicz theorem does not prevent occurence of harmonic spinors) but there are no solutions.

if $\lambda^2 \geq 16$, $R \leq -24$, one can find several solutions, let us mention:

$\lambda^2 = 16$ unique solution $l = 1/2$, n=0, degeneracy $\nu = e$, $\tau = -24$

$\lambda = 4\mu$ with μ, a prime number such that $\mu \equiv 3 \bmod 4$, then, the only solution is $l = \mu - 1/2$, n=0, degeneracy $\nu = 2\mu$; this happens for example, for $\mu = 3, 7, 11, 19 \ldots$

$\lambda = 4\mu$ where μ is an integer (not in the above class) then the equation has a solution $l = \mu - 1/2$, n=0, but is not the only one, for example if $\mu = 65$, $\lambda = 260$, then $l = 535/2$, n=260 is also an harmonic spinor.

$S = S^3$ *with full isometry group* SU(2)

Here, $S = G/H$ with $G = SU(2)$ and $H = \{e\}$.

This is the "completely squashed" case (but still homogeneous !), the (inverse) metric on S^3 is $g^{-1} = 4\{e_1{}^2/\lambda_1{}^2 + e_2{}^2/\lambda_2{}^2 + e_3{}^2/\lambda_3{}^2\}$ - It can be discussed along the same lines and we leave it as an exercise for the reader.

8.6 Examples of dimensional reduction of spinor fields

Call $E \cong M \times S$, $S = G/H$, m = dim(M) and Lie $\bar{K}$ the Lie algebra of the effective gauge group (remember that $\bar{K}$ is locally isomorphic with $N|H \times Spin(m) \times Spin(dim(N|H)) \times Z(\lambda(H) \subset Spin(dim(G/N))$ where Z stands for "centralizer")

G	H	S	K is <u>locally</u> isomorphic with
G	{e}	G	$G \times SO(m) \times SO(dim(G))$
SU(2)	{e}	S^3	$SU(2) \times SO(m) \times SO(3)$
SU(4)	{e}	SU(4)	$SU(4) \times SO(m) \times SO(15)$
SU(2)	U(1)	S^2	$1 \times SO(m) \times 1 \times U(1)$
SU(3)	SU(2)	S^5	$U(1) \times SO(m) \times 1 \times SU(2)$
U(2,IH)	U(1,IH)	S^7	$SU(2) \times SO(m) \times SO(3) \times SU(2)$
SU(4)	SU(3)	S^7	$U(1) \times SO(m) \times 1 \times SU(2)$

SO(8)	SO(7)	S^7	1×SO(m)×1×1
SU(2)×U(1)	diag(U(1))	S^3	U(1)×SO(m)×1×U(1)
SU(2)×SU(2)	diag(SU(2))	S^3	1×SO(m)×1×1

8.7 Pointers to the literature

$Spin^c$ - structures: cf. References in chapter 6.
Harmonic analysis on vector bundles: cf. References in chapter 7.
Spectrum of differential operators in Riemannian manifolds:
12, 22, 29, 55, 91, 170, 175
Dimensional reduction and spinor fields:
33, 44, 123, 140, 187, 205, 232, 233

IX

G-INVARIANCE OF EINSTEIN-YANG-MILLS SYSTEMS

9.0. Summary section

9.0.1 Introduction

Symmetry properties of gravity (metric structure) and Yang-Mills fields (connections) have been often studied separately, both by physicists and mathematicians. These two kinds of geometrical structures are however deeply inter-related and several techniques of "dimensional reduction" allow us to cast a new light on the subject. Let us suppose that we live in an extended universe U endowed with a metric g(U) invariant under a group $\overline{G}$ (description 1), then, in many cases, we can also describe the same situation by saying that we live in an universe **E** (dim E < dim U) endowed with a metric **g**(E) and a Yang-Mills field **A**(E), both invariant under a subgroup of $\overline{G}$ (description 2). We can finally describe the same situation by saying that we live in a universe **M** (space-time, dim **M** < dim E < dim U) endowed with a metric **g**(M), a new Yang-Mills field **A**(M), a few scalar fields and no symmetries left (description 3). The study of the link between descriptions 2 and 3 is the aim of this chapter. The method that we shall use is the following : the reduction theorem of chapter 5 allows us to obtain the link between descriptions 1 and 2 as well as the link between descriptions 1 and 3, we will therefore obtain the desired results by comparing the above two relations. Particular examples of the general situation have been studied in [33], [139], providing interesting phenomenological models; the interpretation of Higgs fields as Yang-Mills fields has been emphasized in [32], [65], [99], [141], where some properties of symmetric Yang-Mills fields are also studied. The construction given here is a natural application of the methods developed in Ch.5 and may be thought of as an alternative to the construction of [108], where a direct analysis of the link between the descriptions 2 and 3 is made (see also [95]).

Symmetries can be studied globally (group actions) or locally (vector fields); here we study symmetric configurations of coupled Yang-Mills and Einstein fields in the most general case and we perform this analysis both from the local and global point of view. Also, the discussion can be carried out at the bundle level or at the base level via the choice of (local) gauge.

9.0.2. Content of the chapter

In sect.9.1., we define and study symmetries of bundles and connections. In sect.9.2., we analyze the geometrical structure for a space E endowed with an G-invariant metric and a G-invariant, Lie(T)-valued Yang-Mills field; we obtain a generalized reduction theorem and a "dimensional reduced" Einstein-Yang-Mills action. In sect.9.3., we discuss examples and model building recipes. The reader who only wants to get the main ideas of the present paper may read only the next sub-section (as well as Tables 1,2).

9.0.3. Summary of the results

Space-time is, in this paper, identified with the manifold **M** of orbits of a compact group G (global symmetry group) acting on a manifold E (extended space-time, multidimensional universe). Thus each space-time point $\mathbf{x} \in \mathbf{M}$ has internal stucture of a homogeneous space G/H. H is the isotropy group characterizing the orbit of G over $\mathbf{x}$.

Let (**g**(E), **A**(E)) be an Einstein-Yang-Mills system in **E**, consisting of a (pseudo) Riemannian metric **g**(E) and a Yang-Mills field **A**(E) on E with gauge group **T**. We show how such a system can be interpreted in terms of fields on **M**, when a constraint of G-symmetry is imposed. It is proved that:

There is a one-to-one correspondence between G-invariant Einstein-Yang-Mills systems (**g**(E),**A**(E)) and quadruples (**g**(**M**),**A**(**M**),$\mathbf{\Phi}$,**h**), where **g**(**M**) is a metric on **M**, **A**(**M**) is a Yang-Mills field on **M** with the effective gauge group $N(\overline{H})/\overline{H}$ described below, while $\mathbf{\Phi}$ and **h** are scalar fields. **g**(**M**) and **h** originate from **g**(E), $\mathbf{\Phi}$ originates from **A**(E), while **A**(**M**) takes its origin from both **g**(E) and **A**(E).

The effective gauge group is the quotient $N(\overline{H})|\overline{H}$, where $N(\overline{H})$ is the normalizer of $\overline{H}$ in $\overline{G}=G\times T$, and $\overline{H} \subset \overline{G}$ is defined as $\overline{H}=\mathbf{diag}(H\times\lambda(H))$, where $\lambda : H \rightarrow T$ is the group homomorphism determined by the action of symmetry group G on the gauge field (see below). Locally $N(\overline{H})|\overline{H}$ is isomorphic to the product $(N(H)|H)\times Z$, where N(H) is the normalizer of H in G, and Z is the centralizer of $\lambda(H)$ in T.

The homomorphism $\lambda : H \rightarrow T$, where H is the stability subgroup of G at $y \in E$, is defined by the action of G on the principal T-bundle U over E on which the initial gauge field A(E) lives. Let $u \in U$ be such that $\pi(u)=y$. Then for each $s \in H$, $s.u=u.\lambda(s)$.

In the particular case where M is restricted to a point, we are in the situation of the Wang theorem [127].

The derivative $\lambda': \mathcal{H} \to \mathcal{T}$ of λ is a homomorphism of Lie algebras. λ' and Φ are parts of the map $\Lambda: \mathcal{G} \to \mathcal{T}$ defined as follows : every $\xi \in G$ is an infinitesimal symmetry of A(E). Thus the Lie derivative L_ξ (A(E)) is an infinitesimal gauge transformation - there exists a function $\Lambda_y: \xi \to \Lambda_y(\xi)$ such that $L_\xi(A(E)) = D\Lambda(\xi)$. For $\xi \in H$, $\Lambda(\xi) = \lambda'(\xi)$. On the other hand write $\mathcal{G} = \mathcal{H} + \mathcal{S}$ with $\mathcal{S}$ such that $Ad(H)\mathcal{S} \subset \mathcal{S}$. Then $\Phi(\xi;x) = \Lambda_y(\xi)$ for $x = \pi(y)$. Thus $\Phi(x)$ is a linear map from $\mathcal{S}$ to $\mathcal{T}$. It is constrained to satisfy

(1) $\Phi \circ Ad(s) = Ad(\lambda(s)) \circ \Phi, \quad s \in H.$

The field h, originating from g(E), describes a G-invariant metric on the homogeneous space G/H at x; algebraically $h=(h_{\alpha\beta})$ is an Ad(H)-invariant scalar product on $\mathcal{S}$.

The Einstein-Yang-Mills lagrangian EYM(E) for (g_B, A_B)[1] on E when represented in terms of (g_M, A_M, Φ, h) is given in Eq. 2.

(2) $EYM(E) = R(\mathfrak{T}) + YM(A_\mu{}^i, A_\mu{}^{\hat{\alpha}}) + KE(h_{\alpha\beta}) + KE(\Phi_\alpha{}^i) - V(h) - V(\Phi)$

The first term $R(\mathfrak{T})$ is the scalar curvature of M; the other terms are explicitly written in eqs: 37,40,45-48,50,51,59. The field Φ splits into two parts $\Phi_{\mathcal{K}}$ and $\Phi_{\mathcal{L}}$, where $\mathcal{K}$ is the space of H-singlets in $\mathcal{S}$ and $\mathcal{L}$ is a complement of $\mathcal{K}$ in $\mathcal{S}$. The fields h and $\Phi_{\mathcal{L}}$ interact with A(M) by the minimal interaction while $\Phi_{\mathcal{K}}$ interacts also directly with the Yang-Mills strength. The potential energy term for $\Phi_{\mathcal{K}}$ and $\Phi_{\mathcal{L}}$ is quartic. KE are kinetic energy terms.

---

[1] Often we will write g_B, or A_B, rather than g(E), A(E).

The results improve those of refs [69], [80], [81]: the new ingredient is the N(H)|H of the gauge group (with Lie algebra isomorphic to $\mathcal{K}$) which may be present if G/H is not isotropy irreducible.

9.0.4. The Space of Higgs Fields

Let us first remember that we have a linear map Λ from $\mathcal{G} = \mathcal{H} + \mathcal{K} + \mathcal{L} = \mathcal{H} + \mathcal{S}$ into $\mathcal{T}$ =Lie(T) and that Λ has to coincide on the subalgebra $\mathcal{H}$ of $\mathcal{G}$ with the algebra homomorphism λ' which characterizes the G action. The Higgs field Φ was defined as the restriction of Λ to the complement $\mathcal{S}$ of $\mathcal{H}$ in $\mathcal{G}$; it is also convenient to split Φ in $\Phi = \Phi_{\mathcal{K}} + \Phi_{\mathcal{L}}$, where $\Phi_{\mathcal{K}}$ maps $\mathcal{K}$ into $\mathcal{Z} \subset \mathcal{T}$ and $\Phi_{\mathcal{L}}$ maps $\mathcal{L}$ into a complement $\mathcal{W}$ of $\mathcal{Z}$ in $\mathcal{T}$ (see also sect. 9.3.4). Schematically :

$$
(3) \qquad
\begin{array}{ccccccc}
\mathcal{G} & = & \mathcal{K} & + & \mathcal{H} & + & \mathcal{L} \\
 & & | & & | & & | \\
 & & \Phi_{\mathcal{K}} & & \lambda' & & \Phi_{\mathcal{L}} \\
 & & | & & | & & | \\
 & & \mathcal{Z} & & \lambda'(\mathcal{H}) & & \mathcal{W}
\end{array}
$$

and,

$$
\mathcal{T} = \overbrace{\mathcal{N}(\lambda'(\mathcal{H}))/\lambda'(\mathcal{H}) + \mathcal{C}(\lambda'(\mathcal{H}))}^{\mathcal{Z}} + \overbrace{\underbrace{\lambda'(\mathcal{H})/\mathcal{C}(\lambda'(\mathcal{H}))}_{\lambda'(\mathcal{H})} + \mathcal{W}'}^{\mathcal{W}}
$$

where $(\mathcal{N}'(\mathcal{H}))$ is the normalizer of $\lambda'(\mathcal{H})$ in $\mathcal{T}$, $\mathcal{C}(\lambda'(\mathcal{H}))$ the center of $\lambda'(\mathcal{H})$, and $\mathcal{W}'$ a supplementary subspace.

It is easy to show that, under the action of the group N($\overline{\text{H}}$), the map Λ transforms as follows : let $(\sigma,\rho) \in N(\overline{H}) \subset G{\times}T$ and $s \in G$. Then

(4) $\qquad \Lambda(s) \in \mathcal{T} \rightarrow \rho^{-1}\Lambda(\sigma s \sigma^{-1})\rho \in \mathcal{T}$

Moreover (see (1)), Λ satisfies the constraint

(5) $\quad \Lambda(hsh^{-1}) = \lambda(h)\Lambda(s)\lambda(h)^{-1}$ for $h \in H$

From this we can deduce the transformation properties of $\Phi_{\mathcal{K}}$ and $\Phi_{\mathcal{L}}$ under the gauge group $N(\overline{H})/\overline{H}$ whose Lie algebra is $\mathcal{Z} + \mathcal{K}$; in particular :

- K acts on $\Phi_{\mathcal{K}}$ via the coadjoint representation, and Z via the adjoint - this can also be seen from the fact that $\Phi_{\mathcal{K}}$ has lower indices in $\mathcal{K}$ and upper indices in $\mathcal{Z}$. There are no constraints on $\Phi_{\mathcal{K}}$ coming from (5).

- K and Z act as above on the space of $\Phi_{\mathcal{L}}$ fields, however, $\Phi_{\mathcal{L}}$ has to satisfy the constraint $\Phi_{\mathcal{L}}(hsh^{-1}) = \lambda(h)\, \Phi_{\mathcal{L}}(s)\, \lambda(h)^{-1}$ for $h \in H$ and $s \in G$.

The representation of the gauge group on the space of Φ fields is usually reducible; in order to find which irreducible representations appear, it is convenient to decompose $\mathcal{L}$ into irreducible representations of the product H.K by looking at the branching rule for **ad** G into H.K and to decompose $\mathcal{W}'$ into irreducible representations of $\lambda(H)$.Z by looking at the branching rule for **ad** T into $\lambda(H)$.Z.

9.0.5 Extrema of Potentials

a) The potential $V(\Phi)$ for the Higgs field can be written as the norm of the Yang-Mills strength $F_{\alpha\beta}{}^{i}$ for a connection $\Phi_{\alpha}{}^{i}$ on $\overline{G}/\overline{H}$ considered as a T-principal bundle over G/H (see Sect. 9.4.5.); $V(\Phi)$ is therefore automatically invariant under T and in particular under the subgroup Z.

From the expression of $F_{\alpha\beta}{}^{i}$, -cf. eq.55,58- we will see that the zeros of $V(\Phi)$ are those maps Φ which extend λ' to a Lie algebra homomorphism Λ (in general, for an arbitrary Φ, $\Lambda = \lambda' + \Phi$ is a linear map but only the λ' piece preserves the Lie algebra structure and Λ is not a Lie algebra homomorphism); by a positivity argument, we see that these zeros are also absolute minima.

Then it is not too difficult to prove using (4) that the "unbroken" gauge group (the stability group of a minimum Φ of V) has Lie algebra isomorphic to $Z_\Lambda + \mathcal{K}$, where Z_Λ is the commutant (centralizer) of $\Lambda(G)$ in T, and $\mathcal{K}$ = Lie(N(H)|H).

b) The potential $V(h)=-R^{G/H}$ for the scalar fields $h_{\alpha\beta}(x)$ has been already analysed in sect.5.9.1. Remember that it is not necessarily of a fixed sign (even if G/H is compact and h is positive definite), also the saddle points of the functional $h \rightarrow \int R^{G/H}\, dvol(h)$ when h varies in the space of all metrics with fixed volume element, coincide with Einstein metrics on G/H, these saddle points are usually neither minima nor maxima. The potential V(h) is invariant under the group N(H)|H×G. All the metrics we are considering (all the fields h) are, by assumption, G-invariant; their full isometry group can of course be bigger: if h is a saddle point of V(h) and if the isometry group of h is included in N(H)|H×G it will then be of the kind F×G, with $F \subset$ N(H)|H we can say that the N(H)|H piece of the gauge group is "broken" to F.

9.1 Symmetries of a Principal Bundle

In this section we will introduce the principal concepts and notation used throughout the rest of the chapter.

9.1.1. Symmetries of Yang-Mills Fields (global description)

Let (U,π,E,T) be a principal bundle with base E, projection π, and structure group T acting on U from the right. Let G be a Lie group acting on U from the left by bundle automorphisms, i.e.

(6) $\quad g(ut) = (gu)t \;, \quad \forall\, g \in G \;, t \in T$ (see Fig. 1)

We shall assume that the action of G on U is effective, i.e. $su=u$, $u\in U$ implies $s=e$. The action of G on U induces an action of G on E : $s\,\pi(u)=\pi(su)$. This induced action on E need not be effective (thus we allow for pure gauge transformations also called vertical automorphisms (cf. sect.4.11)).

We will also use the right action of G on U and on E defined by

(7) $us := s^{-1}u \qquad u\in U\,,\ s\in G$
$\quad\ \ ys := s^{-1}y \qquad y\in E\,,\ s\in G$

When a possibility of confusion can arise, we shall write L and R for the left and right action respectively.

Fig. 1

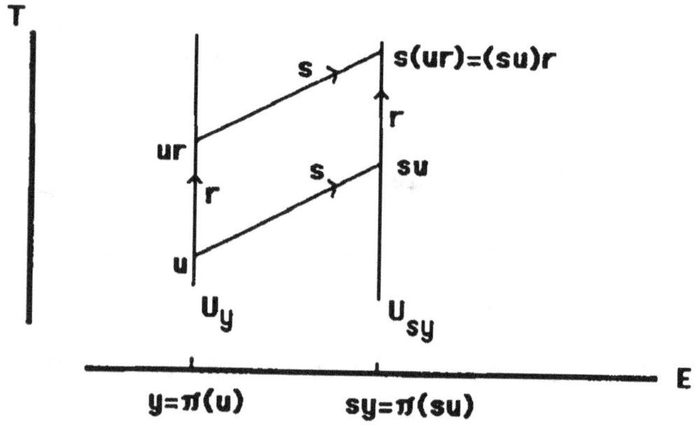

Denote by $\mathcal{T}$ the Lie algebra of T, and let, for each $v\in\mathcal{T}$, Z_v denote the fundamental vector field on U generated by v:

(8) $Z_v = d/dt\,[u\,\exp(tv)]_{t=0}$

Recall that a 1-form ω, defined on U and with values in $\mathcal{T}$, is a 1-form of a principal connection if $\omega(Z_v)=v$, and $R_v^*(\omega) = Ad(v)^{-1}\omega$, $v \in T$. The group G, introduced above, is said to be a symmetry group of ω if, for all $s \in G$,

(9) $\quad R_s^*(\omega) = \omega$

The content of the above equation was discussed in many papers and we refer the reader to the existing literature [13,69,80,81,99,141] . The brief discussion we give below has the purpose of introducing notation and concepts we will use later on.

Global description of a gauge field involves a principal connection 1-form ω(cf ch.6). Locally a gauge field is described by a $\mathcal{T}$ -valued 1-form A^σ on E rather than on U. Let $\sigma: E \rightarrow U$ be a local cross-section (gauge), then A^σ - the Yang-Mills potential in the gauge σ - is defined by $A^\sigma = \sigma^*\omega$.

Let G be a symmetry group of ω, and let us see what can be said about the local representative A^σ. Of course the cross-section σ will not be, in general, invariant under G. Its noninvariance is described by a compensating function $r = r^\sigma(s;y)$ taking values in the gauge group T, defined by (see Fig.2):

(10) $\quad \sigma(ys) = [\sigma(y)s]\, r$

Fig. 2

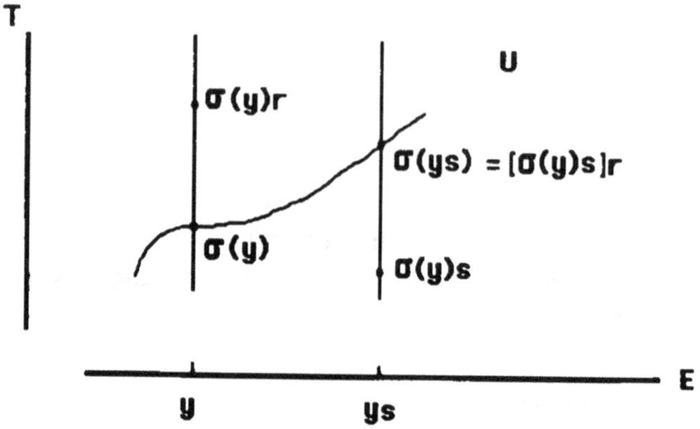

Because of the non-invariance of the gauge σ, also the Yang-Mills potential A^σ will not be invariant; indeed, using (9) we find

$$(11) \quad R_s^* A^\sigma = (R_s^{-1} \circ \sigma \circ R_s)^* \omega = (\sigma r)^* \omega = A^{\sigma r} = r^{-1} A^\sigma r + r^{-1} dr$$

therefore $R_s^* A^\sigma$ differs from A^σ by a gauge transformation.

Let $\mathfrak{g}$ be the Lie algebra of the symmetry group G. With $\xi \in \mathfrak{g}$ let $t \mapsto \exp(t\xi)$ be the 1-parameter subgroup generated by ξ. Denote by Z_ξ the fundamental vector field generated by ξ:

$$(12) \qquad Z_\xi(y) := d/dt \, [\, y \exp(t\xi) \,]_{t=0}$$

With $s=s(t)$ differentiate (11) with respect to t at t=0. From the very definitions of Lie and covariant derivatives, we obtain $L_{Z_\xi} A^\sigma = D\Lambda^\sigma(\xi)$, where

$$(13) \qquad \Lambda^\sigma(\xi;y) := d/dt \, [\, r^\sigma (\exp(t\xi);y) \,]_{t=0}$$

For a fixed $y \in E$, $\Lambda^\sigma(.;y)$ is a linear map from $\mathcal{G}$ to $\mathcal{T}$. It will play an important role later on. Let us analyze it a little bit closer. Choose $y \in E$ and denote by H the stability group of y, $H=\{s \in G : sy=y\}$. Choose $u \in \pi^{-1}(y)$, and let, for every $s \in H$, $\lambda_u(s)$ be the unique element of T satisfying

(14) $$su = u\lambda_u(s)$$

(see Fig. 3). Then $\lambda_u : H \rightarrow T$ is a homomorphism of Lie groups. Comparing (10) and (14) we find

(15) $$r^\sigma(s;y) = \lambda_{\sigma(y)}(s) \quad s \in G$$

and thus $s \rightarrow r^\sigma(s;y)$, when restricted to $s \in H$, is a group homomorphism. Let $\mathcal{H}$ =Lie(H) be the Lie algebra of H. It follows then that Λ^σ restricted to $\mathcal{H}$ is a Lie algebra homomorphism, which coincides with the derivative $\lambda'_{\sigma(y)}$ of $\lambda_{\sigma(y)}$. We thus see that Λ^σ restricted to $\mathcal{H}$ depends only on the action of G on U, and not on the connection. What does depend on the connection ω is the restriction of Λ^σ to a complement, say $\mathcal{P}$, of $\mathcal{H}$ in $\mathcal{G}$. Write $\mathcal{G} = \mathcal{H} + \mathcal{P}$ with $Ad(H)\mathcal{P} \subset \mathcal{P}$ (reductive decomposition), and define $\Phi^\sigma(.;y)$ to be the restriction of Λ^σ to $\mathcal{P}$. The field Φ defined in this way will be later on interpreted as the Higgs field resulting from dimensional reduction of the Yang-Mills field.

Fig. 3

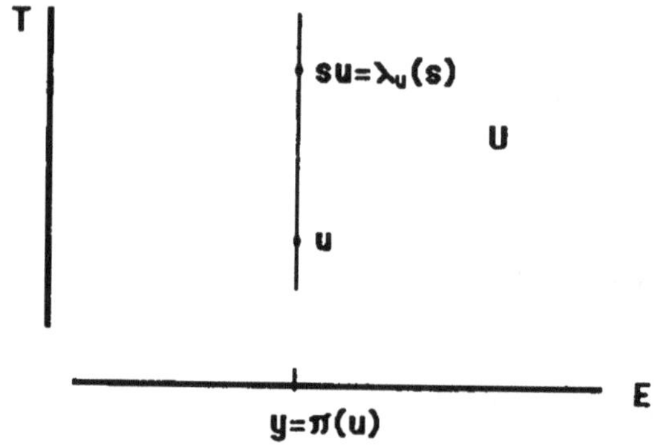

9.1.2 Symmetries of Yang-Mills fields (local description) :

Infinitesimal automorphisms of a principal bundle U are described by invariant vector fields on U. A vector field X on U is called invariant if Xt = X , for all $t \in T$, or infinitesimally, if

(15') $[X, Z_\xi] = 0, \quad \xi \in Lie(T)$,

Z_ξ being the fundamental vector field associated with ξ.

An invariant vector field X is a (local) symmetry of a connection ω if

(15") $\quad L_X\omega = 0$

i.e. if

(15"') $X\omega(Y) - Y\omega(X) - \omega([X, Y]) = 0$

If G is a symmetry group of ω then the fundamental vector fields X = Z_v, $v \in Lie(G)$ satisfy the equation (15"').

Any invariant vector field has a unique projection πX onto E and determines an infinitesimal diffeomorphism of E.

9.1.3 Several Bundle Structures of the Principle Bundle U

As before, consider a principal bundle (U, E, π, T) with base E and structure Lie group T, and let G be a Lie group acting on U by bundle automorphisms. We want to discuss now in more detail the structure arising from such an action. We will assume that both T and G are compact although, with proper case, our discussion could be carried through for non-compact groups admitting biinvariant, nondegenerate metric, with essentially the same results.

We recall from the previous subparagraph that there is an induced action of G on the base E. For every $y \in E$ denote by H_y the stability group of y: $H_y = \{ s \in G: sy = y \}$. There are then two natural equivalence relations in E. One is : $y \sim y'$, if H_y and $H_{y'}$ are conjugate; the other, stronger one, is : $y \cong y'$ iff $H_y = H_{y'}$. The equivalence classes $[y]$ for the first relation are called strata. In general E will be an union of several strata and, since G is compact, one of them will be open and dense (cf. ch.5). Let us restrict our further discussion to one of these strata. Or, better, let us assume that E consists of a single stratum only E is then a simple G space and the reader should now refer to sect 5.1 for a discussion of simple G spaces (for the moment, we consider a left G space whereas in sect.5.1 we studied a right G space).The equivalence class $[[y]]$ of $y \in E$ for the second relation is conveniently called the substratum of y. Using therefore the results of sect.5.1 (and replacing "right" by "left"), we choose, once and for all, one of these substrata, call it P, and denote by H the stability group common to all the points of P. P is a submanifold of E, and E can be thought of as a collection of orbits of type G/H, the collection being parametrized by a manifold M (the manifold of orbits). In other words E is a fiber bundle of base M and fiber G/H - see Fig.4. It is important to realize that P is a subbundle of E, it is a principal bundle with the same base M and structure group $K = N(H)|H$, N(H) being the normalizer of H in G (cf ch.5).

Fig. 4

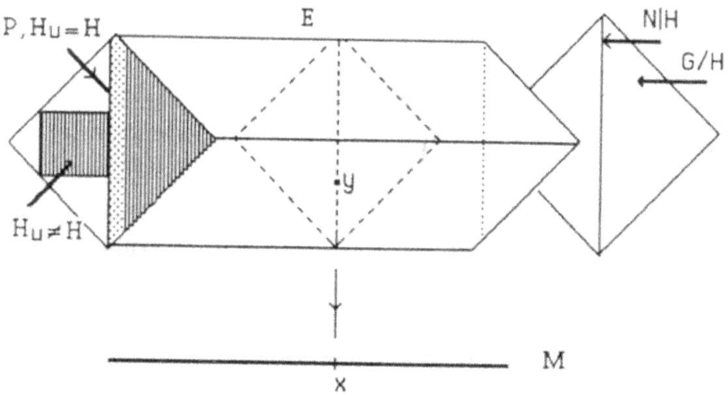

For our further discussion it is important to notice that the direct product group $\overline{G}=G\times T$ acts on U via the following <u>right</u> action :

(16) $(s,r) \in \overline{G}=G\times T : u \in U \to s^{-1}ur \in U$

With the terminology introduced above we will assume that this action is simple, i.e. that U consists of a single stratum. To fix a substratum Q of U let us choose $u \in U$ such that $\pi(u) \in P$ [see Fig.4] and let $\overline{H}=\overline{H}_u$ be the stability group of this u. We take for the substratum $Q \subset U$ the substratum characterized by $\overline{H}$:

(17) $Q = \{u \in U;\ s^{-1}ur = u \text{ if } (s,r) \in \overline{H}\ \}$

Since $Q = [[u]]$ and $\pi(u) \in P$, it follows that $\pi(Q) \subset P$. As before U is a fiber bundle with (the same as B) base M and fiber $\overline{G}/\overline{H}$, and $(Q,M,N(\overline{H})|\overline{H})$ is a principal bundle. The space U can therefore be fibrated in several ways[2] and Fig. 5 below summarizes the results.

[2] Other fibrations are possible but we will not need them here.

Fig. 5

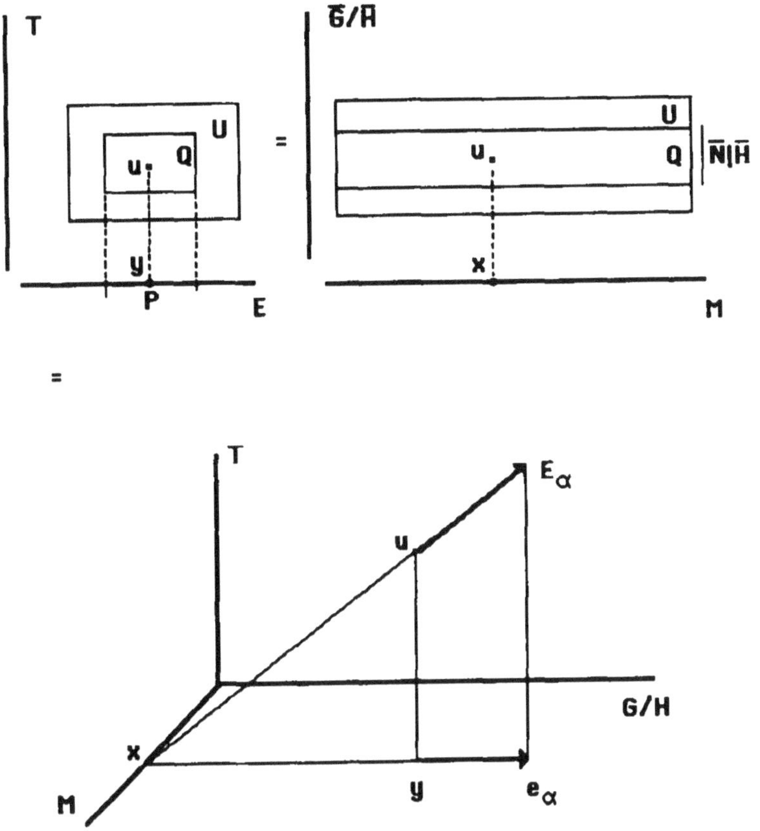

In sect. (9.1.1) we already introduced the group homomorphisms λ_u : $H_{\pi(u)} \to T$. Let us observe that λ_u remains constant while u varies in Q. Indeed, by the very definitions we have $\overline{H}_u = \{(s, \lambda_u(s)) : s \in G\}$. When u runs through Q, then $\overline{H}_u = \overline{H}$ and $H_{\pi(u)} = H$ are both constant, and thus $\lambda = \lambda_u$ is constant too. Thus we have

$$(18) \qquad \overline{H} = \{ (h, \lambda(h)) : h \in H \} = \mathrm{diag}(H, \lambda(H)) \subset \overline{G}$$

The stability group $\overline{H}$ is isomorphic with H, but it is not equal to $H \times \{e\}$. Because of this fact the normalizer of $\overline{H}$ in $\overline{G} = G \times T$ is not isomorphic with $N(H) \times T$. Let us see what is the relation between $N(\overline{H})$ and N(H). By using the definitions we find

$$(s,r) \in N(\overline{H}) \iff \{ s \in N(H) \text{ and } r\lambda(h)r^{-1} = \lambda(shs^{-1}) \ \forall h \in H \ \}$$

Consider the centralizer Z of the image $\lambda(H)$ of H in T :

$$(19) \qquad Z = \{ r \in T : r\lambda(h)r^{-1} = \lambda(h), \ \forall h \in H \}.$$

By the embedding $Z \rightarrow Z \times \{e\}$, Z can be considered as a subgroup of $N(\overline{H})|\overline{H}$, and one easily gets the following important result [108]:

Z is an invariant subgroup of $N(\overline{H})|\overline{H}$, and $N(\overline{H})|\overline{H}$ is locally isomorphic with $(N(H)|H) \times Z$, N(H) being the normalizer of H in G.

9.1.4 Lie Algebra Decomposition and the Vielbein

We introduce the Lie algebra $\mathcal{G}$, $\mathcal{H}$, $\mathcal{N}(H)$, $\mathcal{K}$ of G, H, N(H), $K = N(H)|H$, and decompose the Lie algebra $\mathcal{G}$ of G as follows :

$$\mathcal{G} = \mathcal{H} + \mathcal{K} + \mathcal{L}, \qquad \mathcal{S} \cong \mathcal{G}/\mathcal{H}, \qquad \mathcal{K} \cong \mathcal{N}(H)/\mathcal{H}$$
$$\mathcal{N}(H) = \mathcal{H} + \mathcal{K},$$
$$\mathcal{S} = \mathcal{K} + \mathcal{L},$$
$$\mathcal{K} = \{ \xi \in \mathcal{P} : [\xi, \mathcal{H}] = 0 \}, \qquad [\mathcal{H}, \mathcal{K}] = 0,$$
$$[\mathcal{K}, \mathcal{K}] \subset \mathcal{K}, \quad [\mathcal{N}(H), \mathcal{L}] \subset \mathcal{L}, \quad [\mathcal{H}, \mathcal{S}] \subset \mathcal{S},$$

and let us introduce the Lie algebra bases $T_{\underline{\alpha}}, T_{\hat{\alpha}}, T_{\alpha}, T_{\hat{i}}, T_{i}$ in $\mathcal{H}$, $\mathcal{K}$, $\mathcal{S}$, $\mathcal{Z}$ = Lie (Z), and $\mathcal{T}$ respectively -see Fig. 6.

The homomorphism λ: $H \rightarrow T$ introduced in sect.9.1.2 induces the homomorphism λ': $\mathcal{H} \rightarrow \mathcal{T}$ of Lie algebras, and we define the matrix elements $\lambda_{\underline{\alpha}}{}^{i}$ of λ' by

(20) $\lambda'(T_{\underline{\alpha}}) = \lambda_{\underline{\alpha}}{}^{i}\, T_i$.

Fig. 6

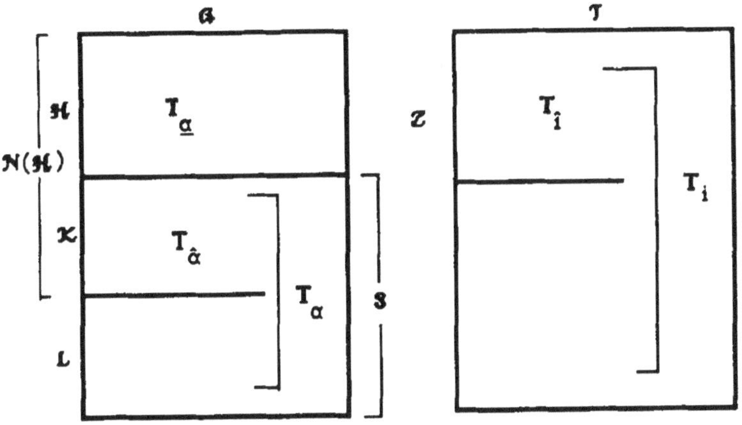

The Lie algebra of the direct product $\overline{G} = G \times T$ can be decomposed in two ways:

$$\mathrm{Lie}(\overline{G}) = \underbrace{\mathcal{H} + \mathcal{K} + \mathcal{L}}_{\mathcal{G}} + \mathcal{T} = \mathrm{Lie}(\overline{H}) + \underbrace{\mathcal{K} + \mathcal{L}}_{\mathcal{S}} + \mathcal{T}$$

where

(21) $\quad \mathrm{Lie}(\overline{H}) = \{\xi + \lambda'(\xi) : \xi \in \mathcal{H}\} = \mathrm{diag}(\mathcal{H}, \lambda'(\mathcal{H})) \subset \mathcal{S} + \mathcal{T}$

is the Lie algebra of the stability group $\overline{H}$. It is the second decomposition which is used in the reduction theorem for $\overline{G}/\overline{H}$. To apply this theorem we have to introduce a new basis in $\mathrm{Lie}(\overline{G})$, with tilda, adapted to the second, reductive, decomposition :

(22)

$$\left.\begin{array}{ll} \tilde{T}_{\underline{\alpha}} = T_{\underline{\alpha}} + \lambda'(T_{\underline{\alpha}}) & \text{in } \mathrm{Lie}(\overline{H}) \\ \tilde{T}_{\alpha} = T_{\alpha} & \text{in } \mathcal{S} \end{array}\right\}, \; \tilde{T}_A = T_A \text{ in } \mathrm{Lie}(\overline{G}) \bmod \mathrm{Lie}(\overline{H})$$
$$\tilde{T}_i = T_i \quad \text{in } \mathcal{T} \qquad , \; \tilde{T}_{\underline{A}} = T_{\underline{A}} \text{ in } \mathrm{Lie}(\overline{H}).$$

It is important to notice that the Lie algebra of $N(\overline{H})|\overline{H}$ is composed of $\mathcal{K} = \mathrm{Lie}(N|H)$ and $\mathcal{Z} = \mathrm{Lie}(Z)$ (see Proposition at the end of sect. 9.1.4).

(23) $\quad \mathrm{Lie}(N(\overline{H})|\overline{H}) = \mathcal{K} + \mathcal{Z}$

We shall use the index $\hat{A} = (\hat{\alpha}, \hat{\imath})$ to label generators of this algebra. Thus

(24) $\quad \tilde{T}_{\hat{A}} = T_{\hat{A}} = (T_{\hat{\alpha}}, T_{\hat{\imath}}).$

The structure constants of the adapted (non-product) basis in $\mathrm{Lie}(\overline{G}) = \mathcal{G} + \mathcal{T}$, which differ from those of the product (non-tilda) basis, are the following ones

(25)
$$\tilde{C}_{\alpha\beta}{}^{i} = - C_{\alpha\beta}{}^{\underline{\gamma}}\, \lambda_{\underline{\gamma}}{}^{i}$$
$$\tilde{C}_{\underline{\alpha}\underline{\beta}}{}^{i} = C_{jk}{}^{i}\, \lambda_{\underline{\alpha}}{}^{j}\, \lambda_{\underline{\beta}}{}^{k} - C_{\underline{\alpha}\underline{\beta}}{}^{\underline{\gamma}}\, \lambda_{\underline{\gamma}}{}^{i}$$
$$\tilde{C}_{\underline{\alpha}i}{}^{j} = \lambda_{\underline{\alpha}}{}^{k}\, C_{ki}{}^{j}$$

Observe that

(26) $\quad \tilde{C}_{\alpha\hat{\beta}}{}^{i} = C_{\alpha\hat{\beta}}{}^{i} = 0$

Let now g_U be a $\overline{G}$-invariant metric on U. In particular, being G-invariant, g_U induces a metric g_E on E, and g_E is G-invariant. Thus g_E induces a metric g_M on M. We recall that $M=E/G=U/\overline{G}$, therefore g_M can be also induced directly from g_U. For our purpose it will be enough to assume that g_U, restricted to the fibers of the principal bundle $U \rightarrow E$, induces a fixed bi-invariant metric k on T. We will call it the Killing metric.

Let u be a point in Q, y-its projection in P, and let x be the projection of y on M (see Fig.5). We introduce the following vector fields :

∂_μ - a holomorphic moving frame (vielbein) around $x \in M$,
e_μ - the horizontal, (i.e. orthogonal to the fibers of $E \rightarrow M$), lift of ∂_μ,
e_α - fundamental fields $e_\alpha = Z_{T_\alpha}$ in E,
$e_M = (e_\mu, e_\alpha)$ - the vielbein around y,
E_α - fundamental fields $E_\alpha = Z_{T_\alpha}$ in U,
E_i - fundamental fields $E_i = Z_{T_i}$ in U,
$E_A = (E_\alpha, E_i)$ - vertical vielbein around u in U,
$E_{\hat{A}} = (E_{\hat{\alpha}}, E_i)$ - vertical vielbein around u in Q,
$\hat{E}_\alpha$ - the horizontal, i.e. orthogonal to the fibers of $U \rightarrow E$, lift of e_α .

The components of the fields appearing in the following discussion will refer to the vielbeins introduced above. In particular we will use the following notation (some of the formulas will be explained later):

$g_{AB} := g_U(E_A, E_B) = (g_{\alpha\beta}, g_{\alpha i}, g_{ij})$, $\quad g^{AB} :=$ (the inverse of g_{AB}),
$h_{MN} := g_E(E_M, E_N) = (h_{\mu\nu}, h_{\alpha\beta})$, $\quad h^{MN} :=$ (the inverse of h_{MN}),
$h_{\alpha\beta} := g_E(e_\alpha, e_\beta) = g_U(\hat{E}_\alpha, \hat{E}_\beta)$, $\quad h^{\alpha\beta} :=$ (the inverse of $h_{\alpha\beta}$),
$\gamma_{\mu\nu} := g_M(\partial_\mu, \partial_\nu) = g_E(e_\mu, e_\nu)$, $\quad \gamma^{\mu\nu} :=$ (the inverse of $\gamma_{\mu\nu}$),

Let ω be the principal connection form on U induced by g_U (considered as T-invariant).

We define the fields $\Phi_\alpha{}^i$ by

$$(27) \qquad \omega(E_\alpha(u)) = - \Phi_\alpha{}^i(u)\, T_i$$

Then one easily finds the following relations

$$(28) \qquad E_\alpha = \hat{E}_\alpha - \Phi_\alpha{}^i E_i$$

$$\begin{aligned} g_{\alpha\beta} &= h_{\alpha\beta} + \Phi_\alpha{}^i \Phi_\beta{}^j k_{ij}, & g^{\alpha\beta} &= h^{\alpha\beta} \\ g_{\alpha j} &= - \Phi_\alpha{}^i k_{ij}, & g^{\alpha i} &= h^{\alpha\beta} \Phi_\beta{}^i \\ g_{ij} &= k_{ij}, & g^{ij} &= k^{ij} + \Phi_\alpha{}^i \Phi_\beta{}^j h^{\alpha\beta} \end{aligned}$$

where k_{ij} are the components of the Killing metric of T. The $\overline{G}=G\times T$-invariance of the metric g_U implies that $\Phi = (\Phi_\alpha{}^i)$, considered as a linear map from $\mathcal{S}$ to $\mathcal{T}$, satisfies the constraint of $Ad(\overline{H})$-invariance

$$(29) \qquad \Phi \circ Ad(i) = Ad(\lambda(i)), \qquad i \in H$$

or, infinitesimally,

$$(30) \qquad C_{\underline{\alpha}\beta}{}^{\gamma} \Phi_\gamma{}^k = \lambda_{\underline{\alpha}}{}^i \; \Phi_\beta{}^j \, C_{ij}{}^k$$

In particular $\Phi_{\hat{\alpha}}{}^i = 0$ if $i \neq \hat{\imath}$ and $\Phi_\alpha{}^{\hat{\imath}} = 0$ if $\alpha \neq \hat{\alpha}$, which means that $\Phi(\mathcal{K}) \subset \mathcal{Z}$ and that $\Phi(\mathcal{L}) \cap \mathcal{Z} = \{0\}$.

We also have

$$(31) \qquad C_{\underline{\alpha}\beta}{}^{\gamma} h_{\gamma\delta} + C_{\underline{\alpha}\delta}{}^{\gamma} h_{\beta\gamma} = 0,$$

which expresses the H-invariance of $h_{\alpha\beta}$.

9.2. Reduction of the Einstein Yang-Mills Action

9.2.1 Outline of the Method

As was already explained in the introduction, our aim is to investigate dimensional reduction of G-invariant Einstein-Yang-Mills fields in a multidimensional universe. Thus we start with an G-invariant metric g_E on E, and a G-invariant principal connection ω on U. We also fix a bi-invariant metric k_{ij} on the initial gauge group T. The logic followed in this section is summarized by the Tables 1 and 2.

Table 1 stresses the equivalence between three possible descriptions of the same geometrical structure (we called them Description 1,2, and 3 in sect. 9.0.1). The links between any two boxes of this chart are provided by the use of the Reduction Theorem (of ch.5). The main goal of our work is to connect Description 2 and 3 (Link No1), and this will be done via Description 1.

Table 2 summarizes the formulae associated to the links L1,L2,L3,L4,L5 of Table 1 via the use of the Reduction Theorem; all these relations are of course obtained and analyzed in the subsequent paragraphs. We introduce the following notations: **YM**("base","group") denotes the Lagrangian for a Yang-Mills field defined on the space "base", valued in the Lie-algebra of "group"; **KE**("base","fiber") denotes the kinetic energy term containing "base derivatives" of the metric on the space "fiber"; finally, the scalar curvature of a space **F** is denoted by R_F or **R(F)**.

Table 1

Description 1

→ $\overline{G} = S \times T$ invariant metric on U.

L2

Description 2

→ G invariant metric h_{MN} on E.
→ G invariant Yang Mills field $A_N{}^i$ on E valued in Lie(T).
→ Killing metric k_{ij} on T.

L3

L1

→ G invariant metric $h_{\alpha\beta}$ on G/H parametrized by $x \in M$.
→ Yang Mills field $A_\mu{}^{\hat{a}}$ on M valued in Lie(N(H)|H).
→ Metric $\delta_{\mu\nu}$ on M.

L4

Description 3

→ Metric $\delta_{\mu\nu}$ on M
→ Yang Mills field $(A_\mu{}^{\hat{a}}, A_\mu{}^i)$ on M valued in the Lie algebra of $N(H)|H \times Z(\lambda(H))$.
→ G invariant metric g_{AB} on G/H parametrized by $x \in M$.

L5

→ T invariant metric k_{ij} on T.
→ G invariant metric $h_{\alpha\beta}$ on G/H (parametrized by $x \in M$)
→ G invariant Yang Mills field $\Phi_\alpha{}^i$ on G/H, valued in Lie(T) and parametrized by $x \in M$: the Higgs field.

---

where

U =

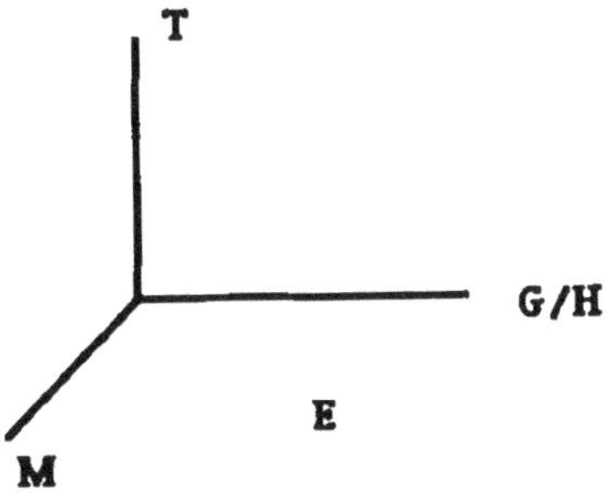

and

H = typical stabilizer of the G action on E,

N(H) = normalizer of H in G,

λ(H) = homomorphic image of H in T characterizing the G action on U,

Z(λ(H)) = centralizer of λ(H) in $\overline{G}$ = G×T.

9.2.2 The Link No.2

We start with Description 2 : a multidimensional universe E furnished with a G-invariant metric $g(E)=(h_{MN})$ and a G-invariant Yang-Mills field $A_N{}^i$ with values in the Lie algebra $\mathcal{T}$ of T; the G-invariance was discussed in sect. 9.3.1. The Einstein-Yang-Mills Lagrangian is given by the expression

$$(32) \qquad EYM(E) = R(E) - 1/4\, k_{ij} h^{MO} h^{NP} F_{MN}{}^i F_{OP}{}^j$$

where R(E) is the scalar curvature of E for the metric h_{MN}, and $F_{MN}{}^i$ is the Yang-Mills field strength associated to $A_N{}^i$. Provided that we add to this expression a constant R(T) with value equal to the scalar curvature of the group T (endowed with the Killing metric) we can use the Reduction Theorem of Ch.5 to construct, out of these three pieces, a T-invariant metric **g**(U) on U considered as an T-principal bundle over E. This metric **g**(U) will be actually G×T-invariant because we started with the ingredients which were themselves G-invariant (here one also exploits biinvariance of the Killing metric **k**). The scalar curvature of U, associated to this metric **g**(U) is

$$\text{(33)} \qquad R(U)_u = R(E)_y + R(T) + YM(E,M)_y$$

where

$$\text{(34)} \qquad YM(E,R)_y = -1/4\; \mathbf{k}_{ij} \mathbf{h}^{MO} \mathbf{h}^{NP} F_{MN}{}^{i}(y)\, F_{OP}{}^{j}(y)\,,$$

u being any point in the T-fiber of U over y. This geometrical structure, described in terms of the space U, constitues what we called Description 1 in our introduction.

Table 2

L2 :	R_U =	R_E	+ R_T		+ YM(E,T),	
L3 :	R_E =	R_M	+	$R_{G/H}$	+ YM(M,N(H)|H)	+ KE(M,G/H),
L4 :	R_U =	R_M	+	$R_{\bar{G}/\bar{H}}$	+ $YM(M,N(\bar{H})/\bar{H})$	+ $KE(M,\bar{G}/\bar{H})$,
L5 :	$R_{\bar{G}/\bar{H}}$ =	$R_{G/H}$	+ R_T		+ YM(G/H,T).	

Also

$YM(M,N(\bar{H})|\bar{H}) = YM(M,N(H)|H) + YM(M,Z(\lambda(H))) + \Delta$,

cf. Eqs.(45)-(48),

$YM(G/H,T) = - V(\lambda\ ,\ \Phi)$

cf. Eqs.(58) and (59),

$KE(M,\bar{G}/\bar{H}) = KE(M,G/H) + KE(\Phi)$

cf. Eqs.(40) and (49).

9.2.3 The Link No. 3

This link is a standard application of the Reduction Theorem of Ch.5 to a G-invariant metric $\mathbf{g}(E)$ on E. For the scalar curvature of E, endowed with this metric, we get:

$$(35) \qquad R(E) = R(M) + R(G/H) + YM(M,N(H)|H) + KE(M,G/H),$$

where,

$$(36) \quad R(M) = R(\gamma_{\mu\nu}),$$

$$(37) \quad R(G/H) = - h^{\alpha\alpha'}(1/2\ C_{\alpha\beta}{}^{\gamma}\, C_{\alpha'\gamma}{}^{\beta} + 1/4\ h^{\beta\beta'}\, h_{\gamma\gamma'}\, C_{\alpha\beta}{}^{\gamma}\, C_{\alpha'\beta'}{}^{\gamma'} + C_{\alpha\beta}{}^{\underline{\gamma}}\, C_{\alpha'\bar{\underline{\gamma}}}{}^{\beta}\ ),$$

$$(38)\quad YM(M,N(H)|H) = -1/4\, h_{\hat\alpha\hat\alpha'}\, \gamma^{\mu\mu'}\gamma^{\nu\nu'} F_{\mu\nu}{}^{\hat\alpha} F_{\mu'\nu'}{}^{\alpha'} ,$$

$$(39)\quad KE(M,G/H) = -1/4\, h^{\alpha\beta} h^{\gamma\delta} (D_\mu h_{\alpha\gamma} D^\mu h_{\beta\delta} + D_\mu h_{\alpha\beta} D^\mu h_{\gamma\delta}) - \nabla_\mu(h^{\alpha\beta} D_\mu h_{\alpha\beta}),$$

9.2.4 The Link No. 4

As explained in a sect. 9.3.3, U is also a $\overline{G}//\overline{H}$ bundle over M (recall that, since U=E×T, the manifold M of $\overline{G}$=G×T-orbits in U is the same as the manifold of G-orbits in E); here again we can use the Reduction Theorem. The $\overline{G}$-invariant metric g(U) on U can be expressed as being built out of the following three pieces : a metric $g(M)=(\gamma_{\mu\nu})$ on M (usually interpreted as space-time metric), a Yang-Mills field $A_\mu{}^{\mathring{A}}$ valued in Lie $(N(\overline{H})|\overline{H}) = \mathcal{K} + \mathcal{Z}$, and a scalar field $h_{AB}(x)$ which can be interpreted as a $\overline{G}$-invariant metric in the internal space $U_x \cong \overline{G}/\overline{H}$ above $x \in M$. The scalar curvature R(U) of U can now be written entrirely in terms of M-based quantities :

$$(40)\quad R_u(U) = R_x(M) + R(\overline{G}/\overline{H}) + YM(M,N(\overline{H})|\overline{H}) + KE(M,\overline{G}/\overline{H}),$$

where

$$(41)\quad R(\overline{G}/\overline{H}) = -g^{AA'}(1/2\, \tilde{C}_{AB}{}^{C}\tilde{C}_{A'C}{}^{B} + 1/4\, g^{BB'} g_{CC'} \tilde{C}_{AB}{}^{C}\tilde{C}_{A'B'}{}^{C'} + \tilde{C}_{AB}{}^{\underline{C}}\tilde{C}_{A'\underline{C}}{}^{B}),$$

$$(42)\quad YM(M,N(H)|H) = -1/4\, g_{AA'}\, \gamma^{\mu\mu'}\gamma^{\nu\nu'} F_{\mu\nu}{}^{A} F_{\mu'\nu'}{}^{A'} ,$$

$$(43)\quad KE(M,G/H) = -1/4\, g^{AB} g^{CD} (\tilde{D}_\mu g_{AC}\, \tilde{D}^\mu g_{BD} + \tilde{D}_\mu g_{AB}\, \tilde{D}^\mu g_{CD})$$

and

(44) $\tilde{D}_\mu g_{AB} = \partial_\mu g_{AB} + \tilde{C}_{A\hat{C}}{}^D A_\mu{}^{\hat{C}} g_{BD} + \tilde{C}_{B\hat{C}}{}^D A_\mu{}^{\hat{C}} g_{AD}$

These formulae may be understood as referring to a certain local cross-section

$$\sigma: x \in M \rightarrow \sigma(x) \in Q \subset U.$$

The tildas refer to the reductive basis in $\mathrm{Lie}(\overline{G})$ (cf. eq. 22). To reduce further the above expressions we apply the relations (25) and (28) with the result

(45) $YM(M,N(\overline{H})/\overline{H}) = YM(M,K) + YM(M,Z) + \Delta,$

where

(46) $YM(M,K) = -1/4\, \gamma^{\mu\mu'} \gamma^{\nu\nu'}\, h_{\hat{\alpha}\hat{\beta}}\, F_{\mu\nu}{}^{\hat{\alpha}}\, F_{\mu'\nu'}{}^{\hat{\beta}},$

(47) $YM(M,Z) = -1/4\, \gamma^{\mu\mu'} \gamma^{\nu\nu'}\, k_{\hat{i}\hat{j}}\, F_{\mu\nu}{}^{i}\, F_{\mu'\nu'}{}^{\hat{j}}$

and

(48) $\Delta = -1/4\, \gamma^{\mu\mu'} \gamma^{\nu\nu'}\, k_{\hat{i}\hat{j}}\, \Phi_{\hat{\alpha}}{}^{\hat{i}}\, F_{\mu\nu}{}^{\hat{\alpha}}\, (\Phi_{\hat{\beta}}{}^{\hat{j}}\, F_{\mu'\nu'}{}^{\hat{\beta}} - 2\, F_{\mu'\nu'}{}^{\hat{j}})$

We also get

(49) $KE(M,\overline{G}/\overline{H}) = KE(M,G/H) + KE(\Phi)$, where

(50) $KE(\Phi) = -1/2\, h^{\alpha\beta}\, \gamma^{\mu\nu}\, k_{ij}\, D_\mu\Phi_\alpha{}^i\, D_\nu\Phi_\beta{}^j$, with

(51) $D_\mu\Phi_\alpha{}^i = \partial_\mu\Phi_\alpha{}^i + C_{\alpha\hat{\gamma}}{}^{\delta}\, A_\mu{}^{\hat{\gamma}}\, \Phi_\delta{}^i + C_{\hat{j}k}{}^{i}\, A_\mu{}^{\hat{j}}\, \Phi_\alpha{}^k$

Notice that when $K = N(H)/H$ (respectively $Z = Z(\lambda(H))$) is discrete, then the 2nd (respectively 3rd) term of (51) vanishes as well as (48). We omit the result one gets for $R(\overline{G}/\overline{H})$ since it will be derived in a different way in the discussion of the Link No. 5 below.

9.2.5 The Link No 5

The term $R_x(\overline{G}/\overline{H})$ in (40) is the scalar curvature of the fiber U_x of the bundle $(U,M,\overline{G}/\overline{H})$. It is easy to see that the projection $\pi : U \rightarrow E$ makes $U_x \cong \overline{G}/\overline{H}$ into a principal bundle with base $E_x \cong G/H$ and structure group T (Fig. 7) :

Fig. 7

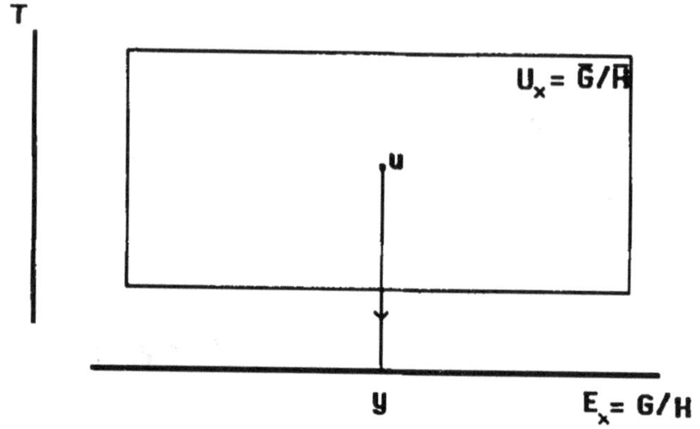

The metric g_{AB} (defined in 9.3.4) is G×T-invariant on U_x and therefore, a fortiori, T-invariant. Therefore we can apply the Reduction Theorem in its principal bundle version (Sect. 4.4.1) : g_{AB} can be expressed entirely in terms of an (G-invariant) metric $h_{\alpha\beta}$ on E_x , an (G-invariant) Yang-Mills field $\Phi_\alpha{}^i$ on E_x , valued in T, and the metric k_{ij} on T (the Killing metric). The explicit expression of g_{AB} in terms of its building blocks has been already given in (28). Applying the Reduction Theorem of Sect. 4.4.1, we also get

$$(52)\qquad R(\overline{G}/\overline{H}) = R(G/H) + R(T) + YM(G/H,T),$$

where $R(T)$ and $R(G/H)$ already appeared in (33) and (35) respectively. Let us discuss now the Yang-Mills term. Since G acts transitively on the $E_x \cong G/H$ of our bundle, we are in the situation of the Wang Theorem (see e.g. [127]), and we could simply refer to the literature. However, it is both instructive and convenient to obtain the desired expression directly. Denoting by $\omega = (\omega_\alpha{}^i)$ and $F=(F_{\alpha\beta}{}^i)$ the connection and curvature form of the induced Yang-Mills field, from the very definition of the curvature, we get:

$$(53)\qquad F_{\alpha\beta} = D\omega(E_\alpha,E_\beta) = E_\alpha\omega(E_\beta) - E_\beta\omega(E_\alpha) - \omega([E_\alpha,E_\beta]) + [\omega(E_\alpha),\omega(E_\beta)]$$

while the G-invariance of ω implies

$$(54)\qquad 0 = (L_{E_\alpha}\omega)(E_\beta) = E_\alpha\omega(E_\beta) - \omega([E_\alpha,E_\beta])$$

Combining the two formulae we get

$$(55)\qquad F_{\alpha\beta} = \omega([E_\alpha,E_\beta]) + [\omega(E_\alpha),\omega(E_\beta)]$$

Now

$$(56)\qquad \omega([E_\alpha,E_\beta]) = \omega(C_{\alpha\beta}{}^{\gamma} E_\gamma + C_{\alpha\beta}{}^{\underline{\gamma}} E_{\underline{\gamma}}) = C_{\alpha\beta}{}^{\gamma}\,\omega(E_\gamma) + C_{\alpha\beta}{}^{\underline{\gamma}}\,\omega(E_{\underline{\gamma}})$$

and we can use the formula (27) as well as the fact that $E_{\underline{\gamma}}+\lambda_{\underline{\gamma}}{}^i E_i$ vanishes on Q see (22), to get

$$(57)\qquad \omega([\acute{E}_\alpha,E_\beta]) = -(C_{\alpha\beta}{}^{\gamma}\,\Phi_\gamma{}^i + C_{\alpha\beta}{}^{\underline{\gamma}}\,\lambda_{\underline{\gamma}}{}^i)\,T_i$$

and consequently

$$(58)\qquad F_{\alpha\beta}{}^i = -(C_{\alpha\beta}{}^{\gamma}\,\Phi_\gamma{}^i + C_{\alpha\beta}{}^{\underline{\gamma}}\,\lambda_{\underline{\gamma}}{}^i - C_{jk}{}^i\,\Phi_\alpha{}^j\,\Phi_\beta{}^k)$$

The Yang-Mills term of (52) gives therefore the potential energy for the Higgs field Φ :

(59) $\quad -V(\Phi) = YM(G/H) = -1/4\ h^{\alpha\alpha'}\ h^{\beta\beta'}\ k_{ij}\ F_{\alpha\beta}{}^{i}\ F_{\alpha'\beta'}{}^{j}$

with $F_{\alpha\beta}{}^{i}$ given by (58).

9.2.6 The Link No. 1

By simply collecting the results obtained so far we find the following theorem.

Einstein-Yang-Mills Reduction Theorem.

There is one-to-one correspondence between pairs (g(E),A(E)) of G-symmetric Einstein-Yang-Mills systems on E and the quadruples (g(M),A(M),Φ,h) of fields on M. The Einstein-Yang-Mills Lagrangian EYM(E) of E, when expressed in terms of the component fields on M reads

(60) $EYM(E) = R(\gamma) + YM(A_{\mu}{}^{i},A_{\mu}{}^{\hat{\alpha}}) + KE(h_{\alpha\beta}) + KE(\Phi_{\alpha}{}^{i}) - V(h) - V(\Phi)$

where

$R(\gamma)$ is the scalar curvature of the metric $g(M)=(\gamma_{\mu\nu})$,
$YM(A_{\mu}{}^{i},A_{\mu}{}^{\hat{\alpha}})=YM(M,N(\overline{H})|\overline{H})$ is given by (45)-(48),
$KE(h_{\alpha\beta})=KE(M,G/H)$ is given by (40),
$KE(\Phi_{\alpha}{}^{i})=KE(\Phi)$ is given by (50), (51),
$V(h)= R(G/H)$ is given by (37),
$V(\Phi)$ is given by (59).

As a by-product we also get the reduction formulae for the Yang-Mills term alone :

$$(61) \qquad YM(E,T)=YM(M,Z(\lambda(H)) + KE(\Phi) + \Delta - V(\Phi),$$

where, however, the terms $KE(\Phi)$ and Δ depend on the connection $A_\mu{}^{\hat{\alpha}}$ in the (P,M,K) bundle too.

One has to remember that the scalar fields $\Phi_\alpha{}^i$ and $h_{\alpha\beta}$ satisfy the algebraic constraints (29,30,31).

Remark: The expressions (60) and (61) can be still multiplied by $|\det h_{\alpha\beta}|^{1/2}$.

9.3 Examples and Comments

Sign Conventions. If M is interpreted as a four dimensional space-time with signature -+++, then, the Einstein lagrangian with cosmological term is $1/(16\pi K) \times (R(M)-2\Lambda)$. For positivity reasons (to get positive kinetic energy), the signature of the "internal" metric on G/H has to be spacelike, i.e., of positive sign with the above convention. Notice that the cosmological "constant" is in our case a function of $x \in M$; however, if we expand the internal metric $h_{\alpha\beta}$ around some background $h_{\alpha\beta}{}^0$ we obtain indeed a constant (coming from the R(G/H) term) to be identified with the cosmological term; however, if $R^0(G/H) := R(G/H;h^0)$ is a positive scalar curvature, then Λ will be negative and vice versa. Notice that we do not get any contribution to the cosmological constant from the Higgs mechanism (when it exists) for $\Phi_\alpha{}^i$ since the potential is zero at the minimum.

If we choose the signature +--- on M, the "physical" conclusions are of course the same, but one has to remember that in order to be spacelike, the signature of the "internal" metric $h_{\alpha\beta}$ has to be taken negative; the scalar curvature of a standard 2-sphere S^2, for example, would therefore also be negative. In what follows, we assume that our choice for M is -+++, therefore $h_{\alpha\beta}$ (and k_{ij}) are positive definite.

Model building. One can distinguish two classes of models: those where the group N(H)|H is discrete and those where this is not the case; all models studied so far in the literature belong to the first category [33,65,139]. One has just to choose an extended spacetime E which can be written locally as M×G/H; when the pair (G,H) (symmetric or not) is an isotropy irreducible space, the group N(H)|H is indeed discrete and we are in the first situation; notice that Tables 1 and 2 of Ch.5 can be helpful to provide examples where N(H)|H is not discrete. In any case, this choice being made, one has to choose a group T containing a subgroup isomorphic with H (or H divided by a normal subgroup of H), this allows one to define the group homomorphism λ: $H \rightarrow \lambda(H) \subset T$. As already stressed, this map does not characterize the "geometry" but rather the action of G on the local product E×T. If we now choose a Lie(T) valued, G-invariant Yang-Mills field on E and a G-invariant metric on E, the dimensionally reduced Einstein-Yang-Mills field lagrangian will in particular contain a Lie(Z(λ(H)) + Lie(N(H)|H) valued Yang-Mills field on M. In general one chooses the group T big enough, in such a way that the centralizer Z(λ(H)) of λ(H) is not discrete, but one could of course find an extreme situation where Z(λ(H)) is discrete and where N(H)|H is the only piece left! (see example below). The Higgs field will now belong to some representation of the final gauge group Z(λ(H))×N(H)|H, and this representation can be found through the technique explained in sect. 9.0.4. If one is now interested in possible "symmetry breaking", one should use the comments of sect. 9.0.5; in particular, if one wants to be in a situation such that the potential V(Φ)=0, then, rather than choosing λ, one can directly construct a global homomorphism Λ from G into T by choosing a group T containing a subgroup isomorphic with G (or divided by a normal subgroup). The final work is of course to restore the dimensionfull constants and to analyze the physical spectrum of the model. Let us now analyze several examples :

1) G=SO(3) (respectively SU(2)) , H=SO(2)=U(1), T=SU(3). These models have already been studied in [139] . G/H is the two sphere S^2. The homomorphism λ maps H=SO(2)=U(1) onto a U(1) subgroup of SU(3).

From the one hand, the centralizer of the image is $Z=SU(2)\times U(1)/Z_2$, from the other hand, the normalizer $N(H)$ of SO(2) in SO(3) (respectively SU(2)) is $SO(2)\times Z_2$ (respectively SO(2)), therefore $N(H)|H=Z_2$ (respectively [e]) is discrete. The Lie algebra of the final gauge group emerging from the reduction of the Einstein-Yang-Mills system is Lie(SU(2)×U(1)); in this case the Einstein part of the Lagrangian on $M\times S^2$ does not bring anything new (but for a factor Z_2); also the $h_{\alpha\beta}$ field is quite trivial since S^2 admits only one - up to scale - SO(3) (respectively SU(2)) invariant metric : $h_{\alpha\beta}(x)$ is therefore a real valued function $h(x)$ and is even a constant if we keep fixed the volume of G/H; then $KE(h)=0$ and $V(h)$ is just a (cosmological) constant. The branching rules for the adjoint representation of SO(3) or SU(2) into U(1) and of the adjoint of SU(3) into SU(2) are $3\rightarrow 1^0 + [1^{+1} + 1^{-1}]$ and $8 \rightarrow 3^0 + 1^0 + [2^{+1} + 2^{-1}]$, where the upper subscript refers to the U(1) eigenvalues; these eigenvalues can be obtained by writing $3\times 3=8+1$ and by specifying the eigenvalues of the U(1) generator (hypercharge) in the fundamental representation (3) of SU(3); a conventional choice is diag (2/3;-1/3,-1/3). The most general $\Phi_\alpha{}^i$ field would map the $[1^{+1} + 1^{-1}]$ subspace of Lie(SU(2)) into the $[2^{+1} + 2^{-1}]$ subspace of Lie(SU(3)); however we can further specify the model by considering only those Φ fields which map 1 into 2 and whose restriction to 1 is just zero. This last choice allows us to make contact with the phenomenology of the Weinberg Salam model where Φ is a doublet of SU(2) with hypercharge +1. Notice that there is no direct coupling of Φ to the field strength Eq. (48), $\Phi_{\hat{\alpha}}{}^i$ is zero since $N(H)|H$ is discrete. Absolute minima $V(\Phi)=0$ for the Higgs potential are associated with the existence of algebra homomorphism Λ from Lie(G) into Lie(T) - see sect.9.05 and eq.55,59. The group SU(3) has two maximal subgroups (defined up to conjugacy) $(SU(2)\times U(1))/Z_2$ and SO(3); correspondingly, the Lie algebra of SU(3) has two maximal simple subalgebras that we call Lie(SU(2)) and Lie(SO(3)) although they are isomorphic. The homomorphism λ from $H=U(1)\subset G=SU(2)$ into Lie(SU(3)) can be extended in two possible ways: either we set Λ(Lie(G))=Lie(SU(2)) or we set Λ(Lie(G))=Lie(SO(3)). Only in the first case the stabilizer of the minimum (the "unbroken" gauge group), i.e. the centralizer of Λ(Lie(SU(2))) in Lie(SU(3)), is not zero, we get Lie(U(1)); indeed the centralizer of SU(2) in SU(3) is U(1) whereas the centralizer of SO(3) in SU(3) is discrete (SU(3)/SO(3) is an irreducible symmetric space). Notice that there is a difference between the two cases because an algebra homomorphism cannot necessarily be lifted to a group homomorphism: there are homomorphisms SO(3)→SO(3), SU(2)→SU(2), and SU(2)→SO(3) but no homomorphism SO(3)→SU(2). One can now compute phenomenological quantities in this model like the mixing angles [139].

2) G=U(2,ℍ)=Sp(2) -2 by 2 unitary matrices over the quaternions ℍ. H=U(1,ℍ)=Sp(1)=SU(2); T=SU(5). The "internal" space G/H is the seven sphere S^7. The maximal subgroups of SU(5) are SU(4)×U(1), SU(3)×SU(2)×U(1) and Sp(2) (in this section we no longer mention the discrete Z factors). We choose λ as a homomorphism from SU(2) onto an SU(2) subgroup of SU(5). From the one hand, the centralizer of the image is SU(3)×U(1), from the other hand, the normalizer N(H) of H=Sp(1) in Sp(2) is Sp(1)×Sp(1)=SU(2)×SU(2), as is clear by representing Sp(2) by 2×2 matrices over the quaternions; therefore we get in this case a non-discrete Lie group N(H)|H=SU(2). The full gauge group emerging from the reduction of the Einstein-Yang-Mills system is therefore SU(3)×SU(2)×U(1). Besides the SU(2), the Einstein part of the lagrangian on M×S^7 brings us also a non-trivial $h_{\alpha\beta}$ field; indeed, the space of Sp(2) invariant metrics on S^7 has dimension d=7 (one decomposes the tangent space at the origin of S^7 into Ad(Sp(1)) real-irreducible representations : 7=(1+1+1)+4 and construct an Ad(Sp(1)) invariant bilinear symmetric form, therefore d=(3×4)/2+1=7, cf. also Ch.3).

Let us now consider the Higgs field Φ; the branching rule for the adjoint representation of Sp(2) into Sp(1)=SU(2) and of SU(5) into SU(3)×SU(2) are [10] → [3] + [1+1+1] + [2+$\overline{2}$] and [24]→ $(8,1)^0 + (1,3)^0 + (3,\overline{2})^{5/3} + (\overline{3},2)^{-5/3} + (1,1)^0$ where the upper subscript refers to the eigenvalue of the U(1) generator Y. These eigenvalues are obtained by specifying (arbitrarily) Y=(2/3,2/3,2/3,-1,-1) in the fundamental representation [5] of SU(5) - as in the conventional SU(5) model of Georgi - Glashow - and writing $[5]\times[5]=[24]+[1]$. The most general Higgs field $\Phi_\alpha{}^i$ would map the $[1+1+1]$ + $[2+\overline{2}]$ subspace of Lie(Sp(2)) into the $(3,\overline{2})+(\overline{3},2)+(1,1)$ subspace of Lie(SU(5)); however we can further specify the model by imposing that the restriction of Φ to $1+1+1+\overline{2}$ vanishes; in such a way we obtain an SU(3) triplet of Higgs with hypercharge 5/3. In this theory we obtain a Weinberg angle equal to the one found in the conventional SU(5) model but the analogy stops there : indeed, we can find a homomorphism Λ of $G=Sp(2)$ onto $\Lambda(G)=Sp(2) \subset SU(4) \subset SU(5)$, the centralizer of $\Lambda(G)$ in SU(5) is then U(1), the SU(3) group is "broken" and the $N(H)|H=SU(2)$ piece of the gauge group stays unbroken; this example is therefore for illustration only but cannot be used in phenomenology (where, according to the standard model. SU(2) is broken and SU(3) is not). Notice finally that, with the above assignment for the Higgs field, there is no direct coupling between the Higgs field and $F_{\mu\nu}{}^{\hat{\alpha}}$ or $F_{\mu\nu}{}^i$; this would not be true if we suppose that $\Phi_{\hat{\alpha}}{}^i$ is not zero.

3) Let us conclude this section by just giving an example where $N(H)|H$ is not discrete but where $Z(\lambda(H))$ is discrete: $G=SO(5)$, $H=SO(3)$, and $T=SO(4)$. $G/H = V_{5,2}$ is the Stiefel manifold[3] of 2-planes in $\mathbb{R}^5$ and $N(H)|H=SO(2)=U(1)$, but if $\lambda(H)=SO(3) \subset SO(4)$, then $Z(\lambda(H))$ is discrete.

9.4 Pointers to the literature

Phenomenological models: 33, 139
Higgs fields as Yang-Mills fields: 43, 32, 65, 99, 141
Einstein-Yang-Mills systems: 43, 95, 108
Symmetries of connections: 13, 43, 69, 80, 81, 99, 141
Extrema of potentials of the Higgs type: 39,40, 74, 75, 87, 144
Geometrical analysis of symmetry breaking: 17

[3] Notice that $V_{5,2} = SO(5)/SO(3)$ is not homeomorphic with $S^7 = Sp(2)/Sp(1)$ although SO(5) - respectively SO(3) - is isomorphic with $Sp(2)/Z_2$ - respectively $Sp(1)/Z_2$.

X

ACTION OF A BUNDLE OF GROUPS

10.1 Motivations

So far, we have mostly studied how to get dimensional reduction from G-invariant metrics on fiber bundles (and from "matter fields" - for example spinor fields- associated with such structures). However, in sect. 4.10 and 5.11 we already discussed metrics which are not G-invariant but are, in some sense "dimensionally reducible". A dimensionally reducible geometrical object is an object defined on a fiber bundle E which can also be interpreted as a finite component field on the base M. As a rule, all objects on E which are G-invariant (G-singlets) are dimensionally reducible but also objects whose values transform under a finite representation of G are dimensionally reducible. Sometimes it may be, however convenient to consider objects transforming under an infinite representation of G as dimensionally reducible too. From the point of view of the "history" of Kaluza-Klein- like theories, it should be stressed that the "popular ansatz" (for instance [231]) is not Kaluza-Klein in spirit since it is not G-invariant (remember that, for instance, the popular ansatz on $M\times S^7$ leads to a gauge group SO(8) -actually $SO(8)\times Z_2$ - whereas the SO(8) invariant metric on $M\times S^7$ -which is Kaluza-Klein in spirit- leads to the gauge group Z_2 !).

We want to show that interesting phenomena may happen when we deal with dimensionally reducible -but not necessarily G-invariant- metrics; the examples that follow somehow generalise situations that have been proposed and used in the physical litterature. Let us warn the reader that the present chapter should not be considered at all as a definitive point of view on this subject but as an invitation to further study.

10.2 Examples of dimensionally reducible metrics (but not G-invariant)

10.2.1 Remark

Before discussing the next two examples, first remember that there exist in general two very different kinds of bundles with fibers H\G over a manifold M.

In the first situation we start with a *left* [1] principal bundle P with base M and structure group G and build the associated bundle E = $H \backslash G \times_G P$ (notice that E = H\P). Then G no longer acts on E since we have divided through the action and in this case it is meaningless to look for a G-invariant metric on E (unless P is trivial, in which case there would be another action on the right of P -and of E-). As an example, take $P=S^7$ as a G=SU(2) bundle over S^4 and E as a S^2 bundle over S^4 (with $S^2=U(1)\backslash SU(2)$).

In the second case we have a bundle E, also above M with typical fiber H\G, but this time there exists a G-action, and we saw in chapter 5 that E is associated with a *right* principal bundle P with structure group K = N(H)|H and E = $P \times_K H \backslash G$. Many examples have been given in chapter 5.

To simplify the discussion we assume G compact and semi-simple in the next two examples (sect. 10.2.2 and 10.2.3).

[1]Here and below, we could exchange everywhere the words right and left and at the same time the cosets H\G and G/H.

10.2.2 The case where G acts globally on $E=E(M,H\backslash G)$.

We are now in the second case of sect. 10.2.1 (the principal bundle is $P = P(M,N(H)|H)$) and we have seen in sect. 5.11.2 how to get dimensionally reducible metrics which are not G-invariant. The effective gauge group was $G\times N(H)|H$. This construction gives the geometrical meaning of the "popular Kaluza-Klein ansatz" in the case where E is a non trivial bundle endowed with a G-action.

10.2.3 The case where G does not act globally on $E=E(M,H\backslash G)$.

We are now in the first case of sect. 10.2.1 (the principal bundle is $P = P(M,G)$) and the constrution is the following:

1) Build $\overline{E} = P\times H\backslash G$. Notice that $G\times G$ acts on $\overline{E}$, the little group associated with this action is $\{e\}\times H$.

2) According to the general technique of chapter 5, we can build $G\times G$ invariant metrics on $\overline{E}$ and the effective Yang-Mills field emerging from this invariant ansatz will be $\overline{N}|\overline{H}$ where $\overline{N}$ is the normalizer of $\overline{H} = \{e\}\times H$ into $G\times G$, i.e. $\overline{N}|\overline{H} \cong N|H \times G$. Equations obtained from direct dimensional reduction from $\overline{E}$ to M will, of course be consistent with equations in $\overline{E}$. The above metrics on $\overline{E}$ are also G^{diag}-invariant and therefore go to the quotient $\overline{E}/G^{diag} \cong E$. The obtained metrics have usually no invariance left, and this describes a non-invariant ansatz on E.

3) One can now perform calculations of the scalar curvature associated with this more general, non-G-invariant ansatz. The scalar curvature is not constant along the fibers of E. For this reason, if we want to build a Lagrangian on M, we have to integrate this scalar curvature on the fibers of E, but then, it can be seen that the equations obtained from dimensional reduction from E to M will not, in general, be consistent with equations in E. These problems of consistency have already been discussed in sect. 4.6, 5.7 and are also studied in the article [44].

This construction gives the geometrical meaning of the "popular Kaluza-Klein ansatz" in the case where E is a non trivial bundle without global G-action.

Another method to get such non-invariant metrics on E would be to start with a G-invariant metric on the principal bundle P= P(M,G); it would be a fortiori H-invariant and we would get a metric on the quotient E = H\P.

Notice that in both cases (sect. 10.2.2 and 10.2.3) the metrics obtained on E have usually no invariances left and that in both situation we get an effective gauge group G×N(H)|H.

10.2.4 Double cosets.

Consider a homogeneous space H\G endowed with a G-invariant metric (cf. sect. 3.4). Let now K be another closed Lie subgroup of G, we can form the double coset H\G/K. The right action of K on H\G will have in general more than one orbit type: H\G decomposes into strata $H\backslash G = \cup E_i$ and each stratum E_i is a bundle with fibers of the type K/K_i, K_i beeing a typical stabiliser of the K action. In such a case we can restrict ourselves to an open dense submanifold of H\G which constitutes the principal stratum of the K action on H\G (cf. sect. 5.12.1); with this understanding, H\G/K becomes a manifold.

Every G-invariant metric on H\G is now a fortiori K-invariant and therefore determines its Kaluza-Klein projection on H\G/K. In this way, starting with a finite parameter family of G-invariant metrics on H\G, we obtain a finite parameter family of metrics on H\G/K which, in general, have no isometries at all. Observe that the group which survives the double quotient and still acts on H\G/K is N(H)|H × N(K)|K but the action of this group on H\G/K is in general not effective.

As an example we may take $G = U(2,\mathbb{H})$, and for H and K the following two subgroups each isomorphic with $U(1,\mathbb{H}) \cong SU(2)$:

$$H = \{ \begin{pmatrix} a & 0 \\ 0 & a \end{pmatrix} \quad a \in \mathbb{H},\ a \neq 0 \};\ K = \{ \begin{pmatrix} b & 0 \\ 0 & 1 \end{pmatrix} \quad b \in \mathbb{H},\ b \neq 0 \}.$$

then $G/K = S^7$, $H\backslash G/K = S^4$ and the projection $G/K \to H\backslash G/K$ is nothing but the Hopf fibration of S^7 (in this case there is only one stratum). Observe that the residual group which still acts on S^4 is $N(H)|H \times N(K)|K = O(2) \times SU(2)$.

The groups H and K being naturally isomorphic with $U(1,\mathbb{H})$, we could also take the quotient of G by the diagonal $U(1,\mathbb{H})$ acting on both sides of G; the resulting double coset is an exotic seven sphere Σ^7 (on which $O(2) \times SU(2)$ still acts).

10.2.5 The Klein-bottle

Both the two-torus and the Klein-bottle are S^1 fibrations over S^1; both carry a flat riemannian metric but the internal (i.e. vertical) $U(1)$ acts globally on the two-torus but does not act globally on the Klein-bottle.

10.3 An extended Kaluza-Klein scheme

In order to deal with situations where no global G action can be defined on E, an extended Kaluza-Klein scheme has been proposed and studied in [105]. We will just here sketch the construction and give the results.

The Klein-bottle is an archetype for this new structure which can be defined by the following axioms:

1) There are two fibrations $\mathfrak{G}$ and E over M.

2) The fibers G_x ($x \in M$) of $\mathfrak{G}$ are groups which act transitively (from the right) on the fibers E_x of E (which are all diffeomorphic with a homogeneous space $H\backslash G$).

3) There exist local trivialisations $(U_\alpha, \varphi_\alpha)$ of E and (U_α, ψ_α) of $\mathcal{G}$ such that $\varphi_\alpha(ya) = \varphi_\alpha(y)\psi_\alpha(a)$ for all $y \in E_x$, $a \in G_x$, $x \in U_\alpha$.

In the case of the Klein bottle, the group G is U(1) and the bundle of groups $\mathcal{G}$ coincides with E i.e. with the Klein-bottle itself. The model of a global G-action considered in most of the previous chapters of the book is a particular case corresponding to $\mathcal{G}$ beeing the global product $\mathcal{G} = M \times G$.

In such a situation, one can construct "dimensionally reducible" metrics; we will only quote the results taken from [105]: the effective gauge group is $G_{eff} = N(\overline{H})|\overline{H}$ with $\overline{G} = G \circledS \mathrm{Aut}_H G$, $\overline{H} = H \circledS 1$, where $\circledS$ denotes the semi-direct product[2], $\mathrm{Aut}_H G$ is the subgroup of AutG consisting of those automorphisms φ of G for which $\varphi(H)$ is conjugated to H (notice that all inner automorphisms of G belong automatically to $\mathrm{Aut}_H G$) and $N(\overline{H})$ is the normalizer of $\overline{H}$ in $\overline{G}$.

$G_{eff} = N(\overline{H})|\overline{H}$ is an extension of $\mathrm{Aut}_H G$ by $N(H)|H$ and in the compact case we get $G_{eff} \simeq_{loc} N(H)|H \times \mathrm{Aut}_H G \simeq_{loc} N(H)|H \times G/Z(G)$, where Z(G) is the center of G.

Notice that cases discussed previously fit into this framework.

The geometrical construction leading to the previous results can be suggested by the following picture:

2 The group $\overline{G}$ consists therefore of pairs $(a, \varphi) \in G \times \mathrm{Aut}_H G$, with the multiplication law: $(a,\varphi)(a',\varphi') = (a\varphi(a'), \varphi\varphi')$.

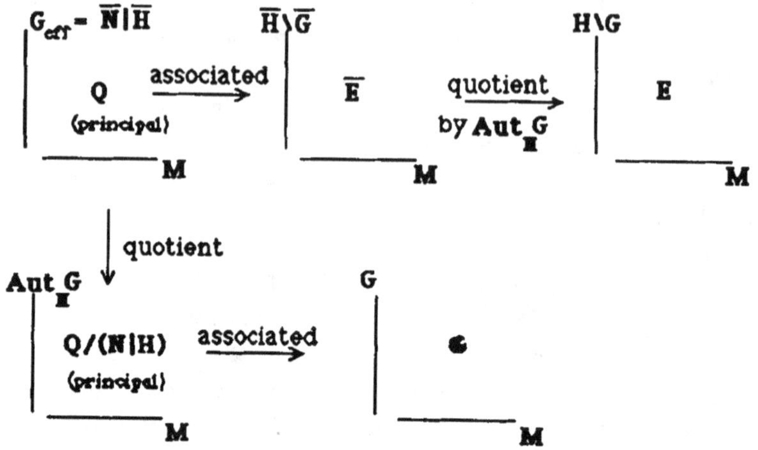

The class of metrics on E that interest us is now defined as the Kaluza-Klein projections of $\overline{G}$ -invariant metrics on $\overline{E}$. E is a quotient of $\overline{E}$, and $\overline{E}$ istself is a bundle endowed with a $\overline{G}$-action associated with the G_{eff}-principal bundle Q.

BIBLIOGRAPHY AND REFERENCES

1 Atiyah M.,Hirzebruch F.
Spin-manifolds and group actions
In: Essays on topology and related topics. Ed. A. Haefliger and R. Narasimhan, Springer
1970

2 Atiyah M.F.,Hitchin N and Singer I.
Self duality in four dimensionnal geometry
Proc. Roy. Soc. London A362,425
1978

3 Atiyah M.F.,Singer I.M.
Dirac operator coupled to vector potentials
Proc.Natl.Acad.Sci.USA,81,2597
1984

4 Avis S.J. and Isham C.J.
Generalised Spin structures on four dimensional Space Times
Comm. Math. Phys. 72, 103 118
1980

5 Awada M.A.
Kaluza-Klein theory over coset spaces
Phys. Lett. 127B, 415-418
1983

6 Awada M.A., Duff M.J. and Pope C.N.

Phys. Rev. Lett. 50,294
1983

7 Bais F.A.,Nicolai H.,van Nieuwenhuizen P.

CERN preprint TH 3577
1983

8 Barut A.O.,Raczka R.
Theory of group representations and applications
Warszawa:PWN Polish Sci. Publ.
1980

9 Berard Bergery L.
sur la courbure des metriques riemanniennes invariantes des groupes de Lie et des espaces homogenes
Ann. Sci. Ecole Norm. Sup. 11,543
1978

10 Berger M.
Les varietes riemanniennes homogenes normales simplement conexes a courbure strictement positive
Ann. Sc. Norm. Pisa 15,179
1961

11 Berger M.
Rapport sur les varietes d'Einstein
In:Analyse sur les varietes.Metz1979.Soc. Math. de France,Asterisque 80
1979

12 Berger M. Gauduchon P. and Mazet E.
Le spectre d'une variete riemannienne
Lecture Notes in Math. 194, Springer
1971

12' Besse A.
Einstein Manifolds
Springer Verlag
1986

13 Bergmann P.,Flaherty E.
Symmetrys in gauge theories
J. Math. Phys. 19,212
1978

14 Bichteler K.
Global existence of spin structures for gravitational fields
J. Math. Phys. 9,813
1968

15 Bleecker D.

Int. J. Theor. Phys. 22, 557-74
1983

16 Bleecker D.
Harmonic analysis on the space-time continuum
Int. J. Theor. Phys. 22,557
1983

17 Bleecker D.
Symmetry Breakdown in Principal Bundles and Physical Applications
Hawai Math Prep.
1983

18 Bleecker D.

Int. J. Theor. Phys. 23, 735-50
1984

19 Bleecker D.
Physics from the G-bundle viewpoint
Int. J. Theor. Phys. 23,735
1984

20 Bore A.,Bredon G.,Floyd E.,Montgomery D.,Palais R.
Seminar on transformation groups
Ann. Math. Stud. Vol 46.Princeton.Princeton Univ. Press
1960

21 Borowiec A. and Jadczyk A.
Particle motion in a multidimensional Universe and nonlinear Higgs charges
Preprint N598, ITP Wroclaw Univ
1984

22 Bott R.
The index theorem for homogeneous differential operators
In: Differential and combinatorial topology. S. Cairns Ed., Princeton University Press
1965

23 Bourguignon J.P.
Une stratification de l'espace des structures riemanniennes
Compositio Math. 30
1975

24 Bourguignon J.P.
Deformations de metriques d'Einstein
In: Analyse sur les varietes.Metz 1979.Soc. Math. de France,Asterisque 80 (1980)
1979

25 Bourguignon J.P. and Karcher H.
Curvature operators: pinching estimates and geometric examples
Ann. Sci. Ecole Norm. Sup. 11, 71
1978

26 Bradfield T. and Kantowski R.
Jordan-Kaluza-Klein type unified theories of gauge and gravity fields
J. Math. Phys. 23,128-131
1980

27 Bredon G.
Seminar on transformation groups
Ann. Math. Stud. Vol 46.Princeton.Princeton Univ. Press
1960

28 Bredon G.E.
Introduction to compact transformation groups
Acad. Press, New York London
1972

29 Brunning J.,Heintze E.
Representation of compact Lie groups and elliptic operators
Inv. Math. 169
1979

30 Castellani L. and Warner N.P.

Phys. Lett. B130, 47
1983

31 Castellani L.,Romans L.J.,Warner N.P.

Nucl. Phys. B241 ,429
1984

32 Chapline G.,Manton N.S.
The geometrical signifiance of certain Higgs potentials
Nucl. Phys. B184,391
1981

33 Chapline G.,Slanski R.
Dimensional reduction and flavour chirality
Nucl. Phys. B209 461
1982

34 Chichlinsky C.
Group actions on manifolds
Trans.Am.Math.Soc. 172,307
1972

35 Cho Y.M.
Higher-dimensionnal unifications of gravitation and gauge theories
J. Math. Phys. 16, 2029
1975

36 Cho Y.M.
Gauge theory on homogeneous fiber bundle
Cern-TH 3414
1982

37 Cho Y.M.,Freund P.G.O.
Non abelian gauge fields as Nambu-Goldstone fields
Phys. Rev. D12,1711
1975

38 Cho Y.M.,Jang P.S.
Unified geometry of internal space with space-time
Phys. Rev. D12,3789
1975

39 Combe Ph., Sciarrino A., Sorba P.
On the directions of spontaneous symmetry braking in SU(n) gauge theories
Nucl. Phys. B158,452-468
1979

40 Combe Ph.,Sciarrino A.,Sorba P.
On the directions of spontaneous symmetry breakingon SU(N) gauge theories.
Nucl. Phys. B158,452
1979

41 Coquereaux R, Jadczyk A.
Geometry of multidimensional universes
Commun. Math. Phys. 90,79-100
1983

42 Coquereaux R.
Dimensional reduction,Kaluza-Klein,Einstein spaces and symmetry breaking (Szczyrk Lectures 1983)
Acta Phys. Pol. Vol. B15,821,1984
1983

43 Coquereaux R. and Jadczyk A.
Symmetries of Einstein-Yang-Mills fields and dimensional reduction
Commun. Math. Phys. 98, 79-104
1985

44 Coquereaux R. and Jadczyk A.
Consistency of the G-invariant Kaluza-Klein ansatz
Nucl. Phys B
1986

45 Coquereaux R. and Jadczyk A.
Harmonic expansion and dimensional reduction in G/H Kaluza-Klein theories
Class. Quantum Grav. 3, 29-42
1986

45' Coquereaux C.
Spinors and Clifford algebras
In: Spinors in Physics and Geometry. Ed G. Furlan, A. Trautman. World Scientific
1987

46 Cremmer E. and Julia B.

Nucl. Phys. B135, 149
1978

47 Cremmer E. and Julia B.
The SO(8) supergravity
Nucl. Phys. B159, 141
1979

48 Cremmer E. and Scherk J.
Spontaneous compactification of extra space dimensions
Nucl.Phys.B118,61-75
1975

49 D'Atri J.E.,Ziller W.
Naturally reductive metrics and Einstein metrics on compact Lie groups
Memoirs of rhe Amer. Math. Soc. Vol 18,no 215
1979

50 Dabrowski L.
Introducing spinors, isospinors, etc. in globally non trivial Space Times
Geometrodynamics Proceedings pp139-148
1985

51 De Witt B.S.
In "Relativity, Groups and Topology"
ed. B.S. De Witt and C. De Witt, Gordon and Breach
1964

52 Dieudonne J.
Elements d'analyse Vol.4
Gauthiers-Villars ed. Paris
1971

53 Dieudonne J.
Treatise on analysis,Vol III
Academic Press
1972

54 Dieudonne J.
Elements d'Analyse Vol.5
Gauthier-Villars,Bordas
1975

55 Diubeck H., Friedrich T.H.
Spektraleigenschaften des Dirac operator
J. Diff. Geom. 15, 1-26
1980

56 Domokos G.,Kovesi-Domokos S.
Gauge field on coset spaces
Nuovo Cimento 44 A,318
1978

57 Duff M.J.
Invited Lecture at the third M.Grossman meeting,Shanghai
Cern-TH 3451
1982

58 Duff M.J.
See: Awada M.A.

1983

59 Duff M.J. and Pope C.N.
Consistent truncations in Kaluza-Klein theories
Nucl. Phys. B255, 355-364
1985

60 Duff M.J., Nilsson B.E.W.,Pope C.N.

Phys. Rev. Lett. 50,2043
1983

61 Duff M.J., Nilsson B.E.W.,Pope C.N. and Warner N.P.
On the consistency of the Kaluza-Klein ansatz
Phys. Rev. Lett.149B, 90-94
1984

62 Duval C.

In: Proc. Conf. Aix en Provence and Salamanca 1979, LNM 836, Springer-Verlag
1980

63 Ellis G.F.R.
See: Hawking S.W.

1973

64 Englert F., Rooman M. and Spindel P.

Phys. Lett. 130B, 50
1983

65 Fairlie D.B.
The interpretation of Higgs fields as Yang-Mills fields
In:Lecture Notes in Physics,Vol 129,p45.Springer
1980

66 Flaherty E.
See: Bergmann P.

1978

67 Floyd E.
See: Bore A.

1960

68 Forgacs P.
See: Manton N.S.

1980

69 Forgacs P.,Manton N.S.
Space-time symmetries in gauge theories
Commun. Math. Phys. 72,15
1980

70 Freund P.G.O.
See: Cho Y.M.

1975

71 Freund P.G.O.
Higher dimensional unification
Preprint, Enrico Fermi Institute, EFI 84/7
1984

72 Gauduchon P.
See: Berger M.

1971

73 Geroch R.
Spinor structures of space -times in general relativity I
J. Math. Phys. 9,1739
1968

74 Girardi G., Sciarrino A., Sorba P.
Some relevant properties of SO(n) representations for grand unified theories
Nucl. Phys. B182, 477-504
1981

75 Girardi G.,Sciarrino A.,Sorba P.
Some relevant properties of SO(n) representations for grand unified theories
Nucl. Phys> B182,477
1981

76 Greub W. Halperin S. and Vanstone R.
Connections, Curvature, and Cohomology, Vol II
Academic Press
1972

77 Gromoll D. and Meyer W.
An exotic sphere with nonnegative sectional curvature
Ann. of Math. 100, 401-406
1974

78 Gromov M. and Lawson H.B.
Positive scalar curvature and the Dirac operator on complete riemannian manifolds
IHES Math. Pub., N58, 83-196
1983

79 Halperin S.
See: Greub W.

1972

80 Harnad J.,Schnider S.,Tafel J.
Group action on principal bundles and dimensionnal reduction
Lett. Math. Phys. 4,(107)
1980

81 Harnad J.,Schnider S.,Vinet L.
Group actions on principal bundles and invariance conditions for gauge fields
J. Math. Phys. 21,(12)
1980

82 Hattori A.
Spin-c structures and S1 actions
Inv.Math.48, 7
1978

83 Hawking S.W. and Ellis G.F.R.
The Large -Scale Structure of Space -Time
Cambridge University Press
1973

84 Hawking S.W.,Pope C.H.
Generalized spin structures in quantum gravity
Phys.Lett.73B,42
1978

85 Heintze E.
See: Brunning J.

1979

86 Helgason S.
Differential geometry and symmetric spaces
Academic press
1962

87 Henneaux M.
Remarks on spacetime symmetries and nonabelian gauge fields
J. Math. Phys. 23, 830-833
1982

88 Hilbert D.
Die grundlagen der Physik
Nackr. Akad. Wiss Gottingen Math. Phys. KL II, 395
1915

89 Hirzebruch F.
See Atiyah M.F,

1970

90 Hitchin N.
See Atiyah M.F,

1970

91 Hitchin N.
Harmonic spinors
Adv. in Math.14,1
1974

92 Hitchin N.
Compact four dimensional Eistein manifolds
J. Diff. Geometry 9, 435
1974

93 Hsiang W.Y.
On the bound of the dimensions of the isometry groups of all possible metrics on an exotic sphere
Ann. of Math. 85,352-358
1967

94 Hsiang W.Y.
In: Conference on transformation groups, New Orleans 1967
Ed. Mostert. Springer 1968
1968

95 Hudson L.B.,Kantowski R.
Higgs fields from symmetric connections,the bundle picture.
Oklahoma prep
1983

96 Husemoller D.
Fiber bundles
Springer
1975

97 Isham C. J.
See: Avis S.J.

1980

98 Isham C.J.
Spinor fields in four dimensional space-time
Proc. R. Soc. London A 364, 591-99
1978

99 Jackiw R.
Schladming Lectures
Acta Physica Austriaca Suppl. XXII,383.Springer
1980

100 Jadczyk A.
Conservation laws and string-like matter distributions
Ann. Inst. H. Poincare A38, 99-111
1983

101 Jadczyk A.
See: Coquereaux R. and Jadczyk A.

1983

102 Jadczyk A.
Fiber bundles and Kaluza-Klein theory
In: Proceedings of the XIX Karpacz winter school,Milewski B., World Scientific
1983

103 Jadczyk A.
Colour and Higgs charges in G/H Kaluza-Klein Theory
Classical and Quantum gravity, 1, 517-530
1984

104 Jadczyk A.
Symmetry of Einstein-Yang_mills systems and dimensional reduction.Banach Center Lectures 1983.
J. Geom. Phys.,Vol.1, n.2, 97-126
1984

105 Jadczyk A.
On the effective gauge group from G/H spontaneous compactification
Lectures at Torino Meeting. CERN-TH.4332/85
1985

106 Jadczyk A.
See: Coquereaux R. and Jadczyk A.

1985

107 Jadczyk A.
See: Coquereaux R. and Jadczyk A.

1986

108 Jadczyk A., Pilch K.
Geometry of gauge fields in a multidimensional universe
Lett. Math. Phys. 8,97
1984

109 Jang P.S.
See: Cho Y.M.

1975

110 Jensen G.R.
The scalar curvature of left-invariant Riemannian metrics
Ind. Univ. Math. J. 20,1125
1971

111 Jensen G.R.
Einstein metrics on principal bundles
J. Diff. Geom. 8,599
1973

112 Julia B.
See: Cremmer E.

1978

113 Julia B.
See: Cremmer E.

1978

114 Julia B.
See Cremmer E.

1979

114 Just K.
The motion of Mercury according to the theory of Thiry and Lichnerowicz
Z. Nat., 14, 1959, p. 751
1959

115 K Warner F.
See: Kazdan J.

1975

116 Kaluza T.
Zum Unitatesproblem der Physik
Sitzber. Preuss. Akad. Wiss, 966-972
1921

117 Kantowski R.
See: Bradfield T.

1980

118 Kantowski R.
See: Hudson L.B.
1983

119 Karcher H.
See: Bourguignon J.P.
1978

120 Kastler D.
Lectures on differential geometry
Marseille (unpublished)
1981

121 Kazdan J. and Warner F.
Existence and conformal deformations of metrics with prescribed Gaussian and scalar curvature
Ann. of Math. 101, 317
1975

122 Kerner R.
Generalization of the Kaluza-Klein theory for an arbitrary non abelian gauge group
Ann. Inst. Poincare 9,143
1968

123 Kerner R.
Spinors on fibre bundles and their use in invariant models
In: Methods in Mathematical Physics. Lecture Notes in Maths 836, Springer
1980

124 Kerner R.
Geometrical background for the Einstein-Cartan theory over a principal bundle
Ann.Inst. Henri Poincare,Vol XXXIV,437
1981

125 Kirillov A.A.
Elements of the theory of representations
Springer Verlag
1976

126 Klein O.
Quantentheorie und funfdimensionaler Relativitatstheorie
Z. Phys. 37 , 895-906
1926

127 Kobayashi S.,Nomiz K.
Foundations of differential geometry,Vol I/II
Interscience Publ
1963

128 Koiso N.
Non deformability of Einstein metrics
Osaka Math. J.15,419
1979

129 Koiso N.
Rigidity and stability of Einstein metrics
Osaka Math. J.17, 51
1980

130 Kolar I.

In:Global differential geometry and global analysis. Lecture Notes in Math.,Springer
1981

131 Kosmann Y.

These, Paris.
1969

132 Kovesi-Domokos S.
See: Domokos G.

1978

133 Lashof R.K.
Equivariant bundles
Ill. J. Math. 26,257
1982

134 Lawson H.B.,Michelson M.L.
Clifford bundles,immersions of manifolds and the vector field problem
J.Diff.Geom.15,237
1980

134' Leutwyler L.
Sur une modification des théories pentadimensionnelles..
Comptes Rendus à l'Acad. des Sciences, Paris. T251 P2292
1960

135 Lichnerowicz A.
Spineurs harmoniques
C.R. Acad. Sci. Paris Ser A-B 257, 7-9
1963

136 Lichnerowicz A.
Geometry of groups of transformations
Leyden,Noordhoff Intern. Publ.
1977

136' Lichnerowicz L.
Les théories relativistes de la gravitation et de l'electromagnétisme
Masson. Paris.
1954

137 Luciani J.F.
Space-time geometry and symmetry breaking
Nucl. Phys. B135, 11-130
1978

138 Manton N.S.
See: Chapline 1981

139 Manton N.S.
A new six dimensional approach to the Weinberg -Salam model
Nucl. Phys. B158,141
1979

140 Manton N.S.
Fermions and parity violation in dimensional reduction schemes
Nucl.Phys.B193,502-16
1981

141 Mayer M.E.
Geometrical aspects of symmetry breaking in gauge theories
In:Lecture Notes in Physics,Vol 116,p291.Springer
1980

142 Mazet E.
See: Berger M.

1971

143 Meyer W.
See: Gromoll D.

1974

144 Michel L.
Symmetry defects and broken symmetry,configurations,hidden symmetry
Rev. Mod. Phys. 52,617
1980

145 Michelson M.L.
See: Lawson H.B.

1980

146 Milnor J.
On manifolds homeomorphic with the 7-sphere
Ann. of Math 64, 399-405
1956

147 Milnor J.
Spin structures on manifolds
L'enseignement mathematique 9,128
1963

148 Milnor J.
Curvatures of left invariant metrics on Lie groups
Adv. Math.21,21
1976

149 Misner C.W., Thorne K.S. and Wheeler J.A.
Gravitation
Freeman (San Francisco)
1970

150 Montgomery D.,
See: Borel A.

1960

151 Montgomery D.,Yang C.T.
The existence of a slice
Ann. Math. 65,108
1956

152 Montgomery D.,Yang C.T.
Orbits of highest dimension
Trans. Am. Soc. 87,284
1958

153 Nagano T.
A problem of existence of an Eintein metric
J. Math. Soc. Jpn.19,30
1967

154 Nagano T.
A problem on the existence of an Einstein metric
J. Math. Soc. Jpn., 19, 30
1967

155 Nilsson B.E.
See: Duff M.J.

1983

156 Nilsson B.E.
See: Duff M.J.

1984

157 Nomizu K.
See: Kobayashi S.

1963

158 O'Neill
Semi Riemannian geometry
Academic Press,1983
1983

159 Oniscik A.L.
Transitive compact transformation groups
Mat. Sb. 60(102),447-485, (Translated in AMS Transl. vol 50)
1962

160 Orzalesi C.A.
Multidimensional Unified theories
Fortschr. Phys. 29,413
1981

161 Orzalesi C.A. and Pauri H.
Geodesic motion in multidimensional unified gauge theories
Nuovo Cimento B68, 193-202
1982

162 Orzalesi C.A. and Pauri M.
Spontaneous compactification, gauge symmetry and the vanishing of the cosmological constant
Phys. Lett. 107B, 186-190
1981

163 P. van Nieuwenhuizen
Proceedings of the Les Houches Summer School 1983
North Holland
1983

164 Palais R.
See: Bore A.

1960

165 Palais S.R.
The classification of G-spaces
Memoirs of the AMS No.36.Providence, RI, AMS
1960

166 Palais S.R.
On the existence of slices for action of non compact Lie groups
Ann. Math. 73,295
1961

167 Palais S.R.
Differential operators on vector bundles
In: Seminar on the Atiyah-Singer theorem,ed, R.Palais,Princeton University Press
1965

168 Palla L.
On dimensional reduction of gauge theories;symmetric fields and harmonic expansions
Z. Phys. C24,195
1984

169 Parthasaraty K.
Dirac operator and the discrete series
Ann.Math.96,1
1972

170 Patera J.,Sankoff D.
Tables of branching rules and representations of simple Lie algebras
Montreal:les Presses de l'Universite de Montreal
1974

171 Pauri M.
See: Orzalesi C.A.

1981

172 Percacci R. and Randjbar-Daemi S.
Kaluza-Klein theories on bundles with homogeneous fibres
J. Math. Phys.,24, 807-814
1983

173 Pilch K.
See: Jadczyk A. 1984

174 Pilch K. and Schellekens A.N.
Formulae for the eigenvalues of the Laplacian on tensor harmonics on symmetric coset spaces
Stony Brook Preprint ITP-SB-84-20
1984

175 Pilch K., van Nieuwenhuizen P. and Towsend P.K.

Preprint ITP-SB-83-51
1984

176 Poncet J.
Groupes de Lie compacts de transformations
Commun. Math. Helv. 33,109
1959

177 Pope C.H.
See: Hawking S.W.

1978

178 Pope C.N.
See: Awada M.A.

1983

179 Pope C.N.
See: Duff M.J.

1983

180 Pope C.N.
See: Duff M.J.

1984

181 Pope C.N.
See: Duff M.J.

1985

182 Price J. F.
Lie groups and compact groups
London Math. Soc. Lecture Notes Series 25. Cambridge University Press
1977

183 Raczka R.
See: Barut A.O.

1980

184 Rayski J.
Unified field theory and modern physics
Acta Phys. Polon. 27,89-97
1965

185 Romans L.
See: Castellani L.

1984

186 Romer E.
Dimensional reduction,spinor fields and characteristic classes
Freiburg preprint THEP 82/8
1982

187 Rooman M.
See: Englert F.

1983

188 Salam A.,Strathdee J.
On Kaluza-Klein theory
Ann. Phys 141,316
1982

189 Sally J.P., Jr
Harmonic analysis and group representations

190 Sankoff D.
See: Patera J.

1974

191 Scherk J.
See: Cremmer E.

1975

192 Scherk J.,Schwartz J.H.
How to get masses from extra dimensions
Nucl. Phys. B153,61
1979

193 Schnider S.
See: Harnad 1980

194 Schwartz J.H.
See: Scherk J.

1979

195 Sciarrino A.
See: Combe Ph.

1979

196 Sciarrino A.
See: Girardi G.

1981

197 Segal G.
Equivariant K-theory
IHES Publ. Math. No 34,129-151
1968

198 Segal G.
The representation ring of a compact group
IHES Publ. Math. No34, 113-128
1968

199 Singer I.
See Atiyah M.F,

1978

200 Singer I.
See Atiyah M.F,

1984
201 Slanski R.
See: Chapline 1982

202 Sorba P.
See: Combe Ph.

1979
203 Sorba P.
See: Girardi G.

1981
204 Sorokin D.P.
See: Volkov D.V.

1983
205 Souriau J.M.
Five-Dimensional Relativity
Nuovo Cimento Vol. XXX N.2
1963
206 Souriau J.M.
Relativite multidimensionnelle non stationnaire
Colloque de Royaumont In: Les theories relativistes de la gravitation.Paris CNRS 1962
1959
207 Souriau J.M.
in,"Les Theories relativistes de la gravitation"
Paris,CNRS
1962
208 Spindel P.
See: Englert F.

1983
209 Steenrod N.
The topology of fiber bundles
Princeton Univ. Press,New Jersey
1951
210 Stora R.
Yang-Mills instantons,geometrical aspects
Erice lectures
1977
211 Strathdee J.
See: Salam A.

1982

212 Strathdee J.
On Kaluza-Klein Theories
in: Unification of the fundamental particle interactions II, eds. J. Ellis and S. Ferrara, Plenum Press
1983

213 Tafel J.
See: Harnad 1980

214 Thorne K.S.
See Misner C.W.

1970

215 Tkach V.I.
See: Volkov D.V.

1983

215' Tonnelat A.
Les théories unitaires de l'electromagnétisme et de la gravitation
Dunod. Paris.
1960

216 Towsend P.
See: Pilch K.

1984

217 Trautman A.
Fibre bundles associated with space-time
Rep. Math. Phys. 1, 29-62
1970

218 Trautman A.
On groups of gauge transformations.
In: Lecture Notes in Physics,Vol. 129.Springer
1980

219 van Nieuwenhuizen P.
See: Pilch K.

1984

220 Vanstone R.
See: Greub W.

1972

221 Vinet L.
See: Harnad 1980

222 Volkov D.V.,Sorokin D.P.,Tkach V.I.

Theor. and Math. Phys. 56,171
1983

223 Wallach N.R.
Harmonic analysis on homogeneous spaces
Marcel Dekker,New York
1973

224 Wang M.,Ziller W.
On the Isotropy representation of a symmetric space

1984

225 Warner N.P.
See: Castellani L.

1983

226 Warner N.P.
See: Castellani L.

1984

227 Warner N.P.
See: Duff M.J.

1984

228 Weiss G.
Harmonic analysis on compact groups

229 Wetterich C.
Realistic KJaluza-Klein Theories
in: Perspectives in Particles and Fields, Eds. M. Levy et al. , Plenum Press
1983

230 Wheeler J.A.
See Misner C.W.

1970

231 Witten E.
Search for a realistic Kaluza Klein theory
Nucl.Phys. B186,412
1981

232 Witten E.
Fermion quantum numbers in Kaluza-Klein theory
Princeton preprint
1983

233 Wolf A.J.
Correction to Ref: 235
Acta. Math. 152,141
1984

234 Wolf J.A.
Spaces of constant curvature
McGraw-Hill
1967

235 Wolf J.A.
The geometry and structure of isotopy irreducible homogeneous spaces
Acta Math. 120,59
1968

236 Wong S.K.

Nuovo Cimento A65, 689
1970

237 Yang C.T.
See: Montgomery D.

1956

238 Yang C.T.
See: Montgomery D.

1958

239 Ziller W.
See: D' Atri

1979

240 Ziller W.
Homogeneous Einstein metrics
In:
1984

www.ingramcontent.com/pod-product-compliance
Lightning Source LLC
Chambersburg PA
CBHW061621220326
41598CB00026BA/3836
9789971504274